PRACTICAL SUPPORT
FOR LEAN SIX SIGMA
SOFTWARE
PROCESS DEFINITION
Using IEEE Software Engineering Standards

IEEE Computer Society Publications
The world-renowned IEEE Computer Society publishes, promotes, and distributes a wide variety of authoritative computer science and engineering texts. These books are available from most retail outlets. Visit the CS Store at *http://computer.org/cspress* for a list of products.

IEEE Computer Society / Wiley Partnership
The IEEE Computer Society and Wiley partnership allows the CS Press authored book program to produce a number of exciting new titles in areas of computer science and engineering with a special focus on software engineering. IEEE Computer Society members continue to receive a 15% discount on these titles when purchased through Wiley or at wiley.com/ieeecs.

To submit questions about the program or send proposals e-mail j.wilson@computer.org. Telephone +1-714-821-8380. **Additional information regarding the Computer Society authored book program can also be accessed from our web site at** *http://computer.org/cspress*.

PRACTICAL SUPPORT FOR LEAN SIX SIGMA SOFTWARE PROCESS DEFINITION
Using IEEE Software Engineering Standards

SUSAN K. LAND
DOUGLAS B. SMITH
JOHN W. WALZ

IEEE computer society

WILEY

A JOHN WILEY & SONS, INC., PUBLICATION

Published by John Wiley & Sons, Inc., Hoboken, New Jersey
Published simultaneously in Canada.

For general information on our other products and services please contact our Customer Care Department within the U.S. at 877-762-2974, outside the U.S. at 317-572-3993 or fax 317-572-4002.

Wiley also publishes its books in a variety of electronic formats. Some content that appears in print, however, may not be available in electronic format.

Library of Congress Cataloging-in-Publication Data is available.

ISBN 978-0470-17080-9

Printed in the United States of America.

10 9 8 7 6 5 4 3 2 1

CONTENTS

PREFACE

Lean Six Sigma emphasizes that speed is directly related to process excellence. Many books have been written in support of streamlining core software processes or making core software processes and practices "lean." Additionally, numerous books and articles have been written in support of Six Sigma statistical measurement. This book begins by providing software developers with the fundamentals of software engineering. It then goes on to supply software project managers and development staff with the materials needed so that they can define their core software development processes and practices, and then proceeds to help support the refinement of these software and systems engineering processes using Lean Six Sigma methods.

Many times, organizations head down the road toward the implementation of the Lean Six Sigma methodologies without a common understanding of the firm underpinnings of software and systems engineering. Before organizations can have processes that rapidly (using lean production) produce products of high quality (high sigma), they must first define their basic software engineering processes and practices.

This book provides support for the foundation of a Lean Six Sigma program by providing detailed process and practice documentation guidance in the form of:

- An overview of Lean Six Sigma methodologies
- Detailed organizational policy examples
- An Integrated set of over 40 deployable document templates
- Examples of over 70 common work products required for support of process improvement activities
- Examples of organizational delineation of process documentation

This book provides a set of IEEE software engineering standards-based templates that support the process definition and work-product documentation required for all activities associated with software development projects. The goal is to provide practical support for individuals responsible for the development and documentation of software processes and procedures. The objective is to present the reader with an integrated set of documents that support the initial requirements of a Lean Six Sigma program.

This book will provide specific support for organizations pursuing Lean Six Sigma guided software process definition and improvement. It is hoped that this knowledge provides the essential foundation on which to build toward the evolution of software process definition, documentation, and improvement, and should be an integral part of every software engineering organization.

The IEEE Computer Society Software and Systems Engineering Standards Committee (S2ESC) is the governing body responsible for the development of software and systems engineering standards. S2ESC has conducted several standards users' surveys; the results of these surveys reflected that standards users found the most value in the guides and standards that provided the specific detail that they needed for the development of their process documentation. Users consistently responded that they used the guides in support of software process definition and improvement but that these standards and guides required considerable adaptation when applied as an integrated set of software process documentation.

This book was written to support software engineering practitioners who are responsible for producing the process documentation, work products, or artifacts associated with support of software process definition and improvement. This book will be most useful to organizations with multiple products and having business customer relationships. In addition to members of project development and test teams working on products with multiple versions, this book will also be useful to members of organizations supporting software project development and testing who work in such areas as project management, configuration management, risk management, human resources, and information technology.

It is the hope of the authors that this book will help members of organizations who are responsible for developing or maintaining their software processes in order to support Lean Six Sigma requirements.

SUSAN K. LAND
DOUGLAS B. SMITH
JOHN W. WALZ

ACKNOWLEDGMENTS

I acknowledge my associates on the America's Army project (www.americasarmy.com) for supplying me with the vision for this book and for also providing me with the ideal environment for the deployment of Lean Six Sigma principles and practices. I would also like to acknowledge my colleagues within the volunteer organizations of the IEEE Computer Society and thank them for their constant encouragement, their dedication to quality, and their friendship. Special thanks again to my family for their continuing and unfailing support.

S. K. L.

I thank my family—Becky, Tim, and Kaitlin—for their tolerance and support. I am deeply indebted to many present and former colleagues from Northrop Grumman: Dr. Ralph Young, who leads and teaches by example in so many ways; Dan Baker, who set me on the Lean Six Sigma path and whose counsel has guided me to personal and professional growth; Craig Hollenbach, font of knowledge, faith, and friendship; my Master Black Belts—Sy Sherman and Jerry Stefanko—whose support has been statistically significant ($p < 0.05$); and Beth Barnickel, whose commitment to One Northrop Grumman and Lean Six Sigma has been crucial to our success and is greatly appreciated.

D. B. S.

Without the support of Ann, my wife, this book would not have been a reality; thank you. I thank my previous managers who guided my professional development and their continued support of my IEEE Computer Society volunteer activities: Thomas J. Scurlock, Jr and Terry L. Welsher, both retired from Lucent Technologies, and James R. McDonnell, AT&T. For my long involvement in software engineering standards, I would like to thank the leadership of Helen M. Wood.

J. W. W.

The authors thank the editing staff at John Wiley & Sons and Angela Burgess of the IEEE Computer Society for their publication support.

1

INTRODUCTION AND OVERVIEW

INTRODUCTION

Software is generated from thousands of lines of code which are developed by a number of people with a variety of skill sets using a multitude of development methods, standards, and rules. These lines of code have to come together at the right time, in the right order, at the right cost, and with the required quality. In order to try to confront this software development challenge, organizations have been employing:

- Software engineering practitioners with better software engineering skills
- Software engineering tools that better fit practitioners and their methods
- Software process improvement methodologies and controls that help to guarantee quality and delivery

Moving an organization from the chaotic environment of free-form development toward an environment that is managed for success using control and communication processes can be confusing to those tasked with making it happen. When those same individuals are also challenged to additionally produce "lean" and "efficient" processes, the task can be overwhelming. It is imperative that the people responsible for the development of software processes and practices understand the foundations of software engineering.

If an organization, and its employees, do not practice or understand the tenants of basic software engineering, trying to employ Lean Six Sigma methodologies is not advisable. Organizations must first have sound software engineering processes and practices. These processes and practices should be based upon standards and industry best practices. Employees must embrace, understand, and should be well trained in these supporting practices before any attempt can be made at increasing software development speed. It should be noted that significant quality improvement gains can be made by simply defining processes and practices in organizations where none exist.

This book is written to support software engineering practitioners who are responsible for producing the process documentation and work products or artifacts associated with support of software process definition and improvement. This book specifically addresses how IEEE standards may be used to facilitate the development of processes, internal plans, and procedures in support of managed and defined software and systems engineering processes in support of Lean Six Sigma implementation [81].

The IEEE Computer Society Software and Systems Engineering Standards Committee (S2ESC) offers an integrated suite of standards based on using IEEE 12207 [40] as a set of reference processes. S2ESC recommends that organizations should use 12207 and related standards to assist them in defining their organizational processes in support of the implementation of the Lean Six Sigma process methodology. However, this book is based on the premise that the reader's processes may not have been defined in this fashion and that the reader seeks help in improving their processes. Therefore, the advice provided in this book does not make the assumption that the user's processes conform to 12207. In general, that advice will be consistent with 12207, but its intent is to assist with process documentation and improvement in the context of Lean Six Sigma. It is the premise of the information provided within these pages that IEEE software engineering standards [42] can be used to provide the basic beginning framework for this type of process improvement.

IEEE standards can be used as instruments to help with the process definition and documentation (e.g., work products) required in support of the process improvement of software and systems development efforts. Many of the IEEE software engineering standards provide detailed procedural explanations; they offer section-by-section guidance on building the necessary support material, and most importantly, they provide best-practice guidance in support of process definition as described by those from academia and industry who sit on the panels of standards reviewers.

In using this book, the key is to use the templates provided to develop simple processes that support your development efforts; team member involvement is critical. The Lean Six Sigma approach combines general quality guidance with a process-based management approach, describing the criteria that the processes should support. IEEE standards are prescriptive. These standards describe how to fulfill the requirements associated with all activities of effective software and systems project life cycle.

IEEE standards are highly specific. They are documented agreements containing technical specifications or other precise criteria to be used consistently as rules, guidelines, or definitions of characteristics to ensure that materials, products, processes, and services are fit for their purpose. In contrast, Lean Six Sigma has been called an improvement method, or improvement engine, using data to identify and eliminate process problems [80]. This book is an attempt to bridge the gap between defined methods and the end goal of a practicing Lean Six Sigma organization.

It is often hard to separate the details associated with product development from the practices required to manage the effort. Simply providing a description of the Six Sigma methodology and lean manufacturing processes to a project lead or manager provides them with a description of an end results model. Pairing this with IEEE standards provides them with a way to work toward this desired end. IEEE standards do not offer a "cookie cutter" approach to management; rather, they support the definition of the management processes in use by describing what is required. IEEE standards provide much needed guidance to organizations working to define their organizational processes and practices.

STIMULUS FOR LEAN SIX SIGMA

Organizations are motivated to use Lean Six Sigma from two directions: external and internal. External motivations may be customers, investors, or competitors. Lean Six Sigma requirements are now showing up in the form of Dept. of Defense (DoD) requirements. The Office of the Secretary of Defense (OSD) and all of the Services are embarked on multiyear continuous process improvement initiatives based on Lean Six Sigma, including the Air Force's Smart Operations for the 21 Century (AFSO21) and the Navy's AIRSpeed. DoD requests for proposals are increasingly asking for Lean Six Sigma qualifications. The Air Force in 2007 awarded a 5-year contract worth up to $99 million for lean expertise and training [123, 124].

Internal motivations for Lean Six Sigma implementation are improving company profitability through improved revenue growth, reduced costs, improved delivery times, and increased customer satisfaction [80]. Lean Six Sigma implementation normally is instigated by top management and enjoys on-going support due to its successes across many industries.

Many software engineering organizations have implemented Lean Six Sigma in minimal form but have not realized its full potential. This may be due to the general and abstract nature of Lean Six Sigma as contrasted with the inherent complexity of software engineering processes. However, it is much more likely that software development organizations may not be realizing the full benefits of Lean Six Sigma because they are trying to apply it to incomplete or nonrobust software engineering development activities. An organization must have sound software engineering practices before it can look to improve them using a methodology like Lean Six Sigma.

Successful software engineering organizations should create well-formed internal work products, or artifacts, during development phases leading to a finished product or as byproducts of their software process or practice definition. Although Lean Six Sigma does not describe these artifacts in detail, this book will describe and provide examples of the important artifacts that are aligned with the IEEE software engineering standards. The usage of IEEE related artifacts allows customers and investors additional opportunities for interaction during the various project phases leading to refined requirements, feedback, and expectations. IEEE-related artifacts provide increased clarity among project personnel, which promotes easier personnel movement between teams and organizations, better alignment between project/program management and software engineering, and better alignment between company goals, objectives, and project/program success factors.

For organizations that do not wish to pursue an implementation of the Lean Six Sigma techniques, this text will show how the application of IEEE standards, and their use as reference material, can facilitate the development of sound software and systems engineering practices. This book is geared for the Lean Six Sigma novice, the project manager, and software engineering practitioners and managers who want a one-stop source, a helpful document, that provides the details and implementation support required when pursuing process definition and improvement.

The plans and artifacts described in this book may be used as an integrated set in support of software and systems process definition and improvement. The plans and artifacts are meant to elicit thought and help guide organizations and development teams through the definition of their own unique software development processes and practices. It should be noted that many of the plans offer examples; these examples are meant to show intent and offer only narrow perspectives on possible plan content.

In addition to Lean Six Sigma, there are also other well-known software process improvement models that have widespread popularity. Two of these models are the Capability Maturity Model Integration (CMMI(r)) and ISO 9001. The authors of this book have taken the realistic view that many software development organizations may have multiple motivators for software process improvement. Organizations may want lean processes, but they may also have dual requirements for CMMI or ISO 9001 certification.

Although these topics will not be covered in this book, they are covered extensively in related texts by the same authors* and deliberate efforts have been made to keep the foundation materials and artifacts provided for the creation of baseline software engineering processes and practices consistent. This is done so that organizations that may have a dual requirement—not only being a Lean organization, but also one meeting the capability requirements of a Level 2 or 3 CMMI organization, or ISO 9001 certified software developing entity. If employed, these plans and artifacts will help organizations establish the sound software engineering foundation required to help them then move on their way toward sound software process improvement.

The authors recommend attendance in the following, or similar, courses: Six Sigma Green Belt Training (5 days) and Lean Practitioner Training (4 days). The authors also recommend the acquisition of the IEEE Software Engineering Collection, which fully integrates over 40 of the most current IEEE software engineering standards onto one CD-ROM [42]. This book provides a list of the IEEE standards abstracts in Appendix C and the mapping IEEE 12207 clause numbers to both the relevant IEEE standards and the IEEE 12207 2008 version clauses in Appendix E.

USING THIS BOOK

The purpose of this book is to assist users in selecting IEEE software engineering standards and their related artifacts appropriate to their needs for process changes leading to Lean Six Sigma implementation. It is important that readers understand that process improvement begins when organizations have a strong foundation of software engineering process methodology. Processes must exist before they can be improved. This book provides two components: a set of templates to establish a core set of software engineering practices and processes, and a description of Lean Six Sigma techniques to accomplish software process improvement.

A key question is, where is your organization? What is its current state? Is your organization currently using

- Six Sigma for software engineering
- Lean software engineering

*Land, Susan K.; *Using IEEE Software Engineering Standards to Jumpstart CMM/CMMI Software Process Improvement,* IEEE/Wiley, 2005; Land, Susan K., and Walz, John, *Practical Support for CMMI-SW Software Project Documentation: Using IEEE Software Engineering Standards,* IEEE/Wiley, 2006; Land, Susan K., and Walz, John, *Practical Support for ISO 9001 Software Project Documentation: Using IEEE Software Engineering Standards,* IEEE/Wiley, 2006.

- Measurements of software engineering processes
- Process frameworks such as CMMI-DEV®
- Formalized management systems such as ISO 9001
- Some of the above on selected projects, but not all projects
- None of the above

A typical journey for newly formed software engineering organizations is to reverse the seven bulleted items above by taking steps, one at a time, starting on a pilot project and applying the learning to all projects in the organization.

Lean manufacturing initiatives are becoming relatively commonplace in the world of manufacturing. Lean manufacturing helps eliminate production waste, introduce value-added measurements, and push for continuous improvements. The Six Sigma methodology uses data and statistical analysis tools to identify, track, and reduce problem areas and defects in products and services.

As Six Sigma is complementary to lean processes, the two are combined into Lean Six Sigma, which strives to present the same types of gains for software engineering companies. Once the waste is removed from software development and production is truly "lean," then the Six Sigma tool set is applied to reduce defects in the value-adding parts of the process.

The subsequent application of Six Sigma methodology supports risk management and problem identification and resolution. In addition to the lean tools to reduce waste and Six Sigma tools to reduce defects, some organizations also include Goldratt's Theory of Constraints [83] in their continuous improvement toolkit to use on bottlenecks. Although we do not cover Theory of Constraints in detail in this book, it is a very compatible approach that, at a high level, consists of a five step process:

1. *Identify* the system's constraint.
2. *Exploit* (maximize throughput of) the system's constraint.
3. *Subordinate* other processes to the constraint; reduce suboptimzation.
4. *Elevate* the constraint (make it more capable).
5. *Repeat* the cycle (what limits the system now that you have broken its previous constraint?).

The Theory of Constraints and its application to scheduling (also called the Critical Chain) are similar to the lean principle of *flow,* and lean's emphasis on reducing multitasking [83].

The Lean Six Sigma approach combines general quality guidance with a process-based management approach, describing the criteria that the processes should support. Just as lean implementation should generally precede the application of Six Sigma techniques, the combined Lean Six Sigma requires a robust management and process measurement system foundation such as ISO 9001 or the CMMI-DEV framework prior to the application of its improvement methodologies. This book is oriented toward organizations whose current state is any of the seven bulleted levels above.

This book provides the coherent software life cycle processes as defined in IEEE 12207 and related software engineering standards to assist organizations in defining their organizational processes and artifacts in support of the implementation of the Lean Six

Sigma methodology. The software engineering standards are used as instruments to help with the process definition and documentation (e.g., work products or artifacts) required in support of the software process improvement toward the desired organizational state. These desired states can be reached in steps: first achieving the robust management system foundation, followed by primary life cycle software processes measurements, and, finally, implementation of the Lean Six Sigma methodology. This journey requires adaptation and tailoring of standards and their artifacts as "one size does not fit all." Projects must design and use their objective experiences to determine what works best when using selected best practices.

Many organizations with a robust management system foundation are pursuing process improvement by establishing a set of organizationally adopted processes to be applied by all of their projects, and then improving them on the basis of their experience. This "experience" is factored by both the lean principles and the Six Sigma principles, resulting in the use of specific Lean Six Sigma improvement techniques that validate specific process changes to be deployed across the organization. Whenever objective experiences demonstrate the need for changes to processes and artifacts, then this book will show the appropriate IEEE software engineering standards to be selected and analyzed so that their best practices can be incorporated into processes, internal plans, procedures, and other artifacts, to support Lean Six Sigma implementation.

Change management is not solely a technical endeavor of engineering process experts. Whether the Lean Six Sigma motivation is from external or internal sources, top management support is required for organizational change management. Top management has the responsibility of providing overall direction, resources, and the control framework for managing the change process. The organizational changes result in changes to participant's skills, their tools, and their work processes and artifacts. This book focuses on the software engineering and improvement work processes and artifacts that support Lean Six Sigma implementation. Although process engineers have major roles, team member involvement in organizational and process changes is critical.

This book will guide your organization to use the provided software engineering standards and artifacts such as document templates to develop or enhance processes that support your development efforts with Lean Six Sigma.

2

STANDARDS AND SOFTWARE PROCESS IMPROVEMENT

WHAT ARE STANDARDS?

Standards are documented sets of rules and guidelines that provide a common framework for communication and are expressed expectations for the performance of work. These rules and guidelines set out what are widely accepted as good principles or practices in a given area. They provide a basis for determining consistent and acceptable minimum levels of quality, performance, safety (low risk) and reliability.

Standards are consensus-based documents that codify best practice [43]. Consensus-based standards have seven essential attributes that aid in process engineering. They

1. Represent the collected experience of others who have been down the same road
2. Describe in detail what it means to perform a certain activity
3. Can be attached to or referenced by contracts
4. Help to assure that two parties define an engineering activity in the same way
5. Increase professional discipline
6. Protect the business and the buyer
7. Improve the product

IEEE software engineering standards provide a framework for documenting software engineering activities. The "soft structure" of the standards set lends well to the instantiation of software engineering processes and practices. The structure of the IEEE software engineering standards set also provides for tailoring. Each standard describes recommended best practices detailing required activities. These standards documents provide a common basis for documenting organizationally unique software process activities.

When trying to understand exactly what the IEEE software and systems engineering standards collection is, and what this body of work represents, the following statement (taken from the Synopses of Standards section in the *IEEE Standards Collection, Software Engineering,* 1994 Edition) summarizes it best:

> The main motivation behind the creation of these IEEE Standards has been to provide recommendations reflecting the state-of-practice in development and maintenance of software. For those who are new to software engineering, these standards are an invaluable source of carefully considered advice, brewed in the caldron of a consensus process of professional discussion and debate. For those who are on the cutting edge of the field, these standards serve as a baseline against which advances can be communicated and evaluated. [42]

IEEE software and systems engineering standards provide a framework for defining and documenting software and systems engineering activities. These standards documents provide a common basis for documenting organizationally unique process activities. The structure of the standards set supports the Lean Six Sigma process methodology while also providing for organizational adaptation. Each standard describes recommended best practices detailing required activities.

IEEE S2ESC Software and Systems Engineering Standards Committee

In 1976, this arm of the IEEE Computer Society was chartered to develop the first standards for software engineering. The mission of the S2ESC is:

1. To develop and maintain a family of software and systems engineering standards that is relevant, coherent, comprehensive, and effective in use. These standards are for use by practitioners, organizations, and educators to improve the effectiveness and efficiency of their software engineering processes, to improve communications between acquirers and suppliers, and to improve the quality of delivered software and systems containing software.
2. To develop supporting knowledge products that aid practitioners, organizations, and educators in understanding and applying our standards.

From a narrow perspective, S2ESC manages the scope and direction of IEEE software and systems engineering and standards. S2ESC is the standards partner of the IEEE Technical Council on Software Engineering (TCSE), and has worked to provide the standards collection that

- Provides a consistent view of the state of the practice
- Is aligned with the Software Engineering Body of Knowledge (SWEBOK)
- Addresses practitioner concerns
- Is affordable

From a broader perspective, in addition to the development of standards, S2ESC develops supporting knowledge products and sponsors or cooperates in annual conferences and workshops in its subject area. S2ESC also participates in international standards making as a member of the U.S. Technical Advisory Group (TAG) to ISO/IEC JTC1/SC7 and as a direct liaison to SC7 itself. Table 2-1 provides a sample of some of the types of software engineering activities supporting by the S2ESC standards.

S2ESC also supports and promotes the Software Engineering Body of Knowledge,

Table 2-1. Some examples of what IEEE standards do

IEEE 982.1 Measures for Reliable Software	Specify techniques to develop software faster, cheaper, better
IEEE 1008 Unit Testing	Describes "best practices"
IEEE 1061 SW Quality Metrics	Provide consensus validity for techniques that cannot be scientifically validated
IEEE/EIA 12207 SW Life Cycle Processes	Provide a framework for communication between buyer and seller
IEEE 1028 SW Reviews	Give succinct, precise names to concepts that are otherwise fuzzy, complex, detailed and multidimensional [159]

certification mechanisms for software engineering professionals, and other products contributing to the profession of software engineering.

SWEBOK

On May 21, 1993, the IEEE Computer Society Board of Governors approved a motion to "establish a steering committee for evaluating, planning, and coordinating actions related to establishing software engineering as a profession" [45]. Many professions are based on a body of knowledge that supports purposes such as accreditation of academic programs, development of education and training programs, certification of specialists, or professional licensing. One of the goals of this motion was to establish a body of knowledge in support of software engineering. In 1996, the initial Strawman version of this body of knowledge was published as the Software Engineering Body of Knowledge (SWEBOK®*). The 2004 version of the SWEBOK® has been released and is available at www.swebok.org. ISO and IEC adopted the SWEBOK in 2005 as a Technical Report, ISO/IEC TR 19759:2005 Software Engineering—Guide to the Software Engineering Body of Knowledge (SWEBOK).

The SWEBOK® outlines the knowledge associated with what has become the consensus body of software engineering knowledge. The goal of the SWEBOK® is to define the core of what the software engineering discipline should contain. The acknowledgement of the SWEBOK® by educational institutions would mean a shift from technology-specific programming courses, like C++ and Java, to the knowledge and practices required in support of software development and software project management.

The purpose of the SWEBOK is to provide a consensually validated characterization of the bounds of the software engineering discipline and to provide a topical access to the body of knowledge supporting that discipline.

Software engineering should be balanced among the following guiding principles:

- *Rigor and formality.* Repeatable methods and processes promote better estimates and less waste (e.g., defects). This channels creativity to the software development product or service instead of the methods.

*SWEBOK is a registered trademark of the IEEE.

- *Separation of concerns.* Supports the management of complexity and parallelization. Design techniques include modularity and decomposition, cohesion and coupling, and abstraction.
- *Anticipation of change.* Customers expect change and sometimes do not know all of their requirements.
- *Generality.* Allows reuse to quickly solve problems or implement similar features.
- *Incrementally.* Frequent customer interaction to confirm and prioritize requirements.
- *Scalability.* Supports quick deployment or addresses larger customer needs.
- *Compositionality.* Allows for precise requirements under various conditions or understanding different user/actor roles.
- *Heterogeneity.* Understanding the variability or differences between studies in the estimates of effects [158].

The SWEBOK® is divided into ten knowledge areas (refer to Table 2-2) that contain descriptions and references to topically supportive materials that reflect the knowledge required to support the software engineering discipline. Appendix C of the SWEBOK *Guide* provides an Allocation of Standards to Knowledge Area, supplying an annotated table of the most relevant standards, mostly from the IEEE and the ISO, allocated to the SWEBOK *Guide* Knowledge Areas [45]. Process teams can use the SWEBOK® as a reference tool supporting their requirements for process development and definition.

The SWEBOK® will continue to evolve, as do all the bodies of knowledge associated with other professional disciplines. Areas will be redefined and refined over time to address all of the factors that affect the creation or maintenance of a software product. The SWEBOK® offers a challenge to all of those involved with software—broaden the knowledge base and move from computer science to software engineering. The requirement is for software engineering practitioners to not only know how to write code, but to understand all aspects of the processes that support the creation of their products.

Table 2-2. The ten SWEBOK Knowledge Areas [45]

SWEBOK® Knowledge Area	Supports
Software Requirements	Software Requirements Fundamentals Requirements Process Requirements Elicitation Requirements Analysis Requirements Specification Requirements Validation Practical Considerations
Software Design	Software Design Fundamentals Key Issues in Software Design Software Structure and Architecture Software Design Quality Analysis and Evaluation Software Design Notations Software Design Strategies and Methods
Software Construction	Software Construction Fundamentals Managing Construction Practical Considerations

Table 2-2. *Continued*

SWEBOK® Knowledge Area	Supports
Software Testing	Software Testing Fundamentals Test Levels Test Techniques Test-Related Measures Test Process
Software Maintenance	Software Maintenance Fundamentals Key Issues in Software Maintenance Maintenance Process Techniques for Maintenance
Software Configuration Management	Management of the SCM Process Software Configuration Identification Software Configuration Control Software Configuration Status Accounting Software Configuration Auditing Software Release Management and Delivery
Software Engineering Management	Initiation and Scope Definition Software Project Planning Software Project Enactment Review and Evaluation Closure Software Engineering Measurement
Software Engineering Process	Process Implementation and Change Process Definition Process Assessment Product and Process Measurement
Software Engineering Tools and Methods	Software Tools Software Requirements Tools Software Design Tools Software Construction Tools Software Testing Tools Software Maintenance Tools Software Engineering Process Tools Software Quality Tools Software Configuration Management Tools Software Engineering Management Tools Infrastructure Support Tools Miscellaneous Tool Issues Software Engineering Methods
Software Quality	Software Quality Fundamentals Software Quality Management Processes Practical Considerations

IEEE 12207

The SWEBOK Knowledge Area breakdown is broadly compatible with the sections of IEEE 12207 Standard for Software Life Cycle Processes, and refers to its various requirements activities. The mapping table in Table 2-3 allows process teams to use the SWEBOK® as a reference to additional process development and definition.

Table 2-3. High-level mapping of the SWEBOK to IEEE 12207

SWEBOK® Area	IEEE 12207
Software Requirements	5.1 Acquisition 6. Supporting Life Cycle Processes: 6.1 Documentation
Software Design	5.2 Supply 6. Supporting Life Cycle Processes: 6.1 Documentation
Software Construction	5.3 Development 6. Supporting Life Cycle Processes: 6.1 Documentation
Software Testing	6. Supporting Life Cycle Processes: 6.1 Documentation 6.4 Verification 6.5 Validation 6.8 Problem Resolution
Software Maintenance	5.4 Operation 5.5 Maintenance 6. Supporting Life Cycle Processes: 6.1 Documentation
Software Configuration Management	6. Supporting Life Cycle Processes: 6.1 Documentation 6.2 Configuration Management
Software Engineering Management	7. Organizational Life Cycle Processes: 7.1 Management 6. Supporting Life Cycle Processes: 6.1 Documentation 6.6 Joint Review
Software Engineering Process	6. Supporting Life Cycle Processes: 6.1 Documentation 7. Organizational Life Cycle Processes: 7.2 Infrastructure 7.3 Training 7.4 Improvement
Software Engineering Tools and Methods	6. Supporting Life Cycle Processes: 6.1 Documentation
Software Quality	6. Supporting Life Cycle Processes: 6.1 Documentation 6.3 Quality Assurance 6.7 Audit

SOFTWARE PROCESSES MUST EXIST BEFORE THEY CAN BE REDESIGNED

The basic software engineering foundations must be in place prior to the initiation of any Lean Six Sigma software process initiative. Only then can organizations move forward to trim the waste and increase their overall production value and product quality. The alternative is improving an undocumented "fuzzy" process, which means making assumptions about what the process might be. The improvements are likely to be based on ideas about the process that are to some extent erroneous. In order to reduce risk, begin with sound software engineering foundations and then incrementally improve these using software process improvement methods.

Software Engineering and Training

Process group members need software engineering skills, as they must understand the domain that they are working to improve. Implementing process improvement can be very time-consuming, depending upon the scope and complexity of the process. Expectations for the process owner's time commitments and job responsibilities must be modified accordingly to reflect the new responsibilities. This commitment should reflect time budgeted for process definition and improvement and any required refresher training.

In order for software process improvement to be successful, everyone involved should know how to effectively perform his or her roles. Many times, key individuals have not been effectively trained to perform in support of the software engineering process. IEEE software and system engineering standards provide process and practice support to the software engineering knowledge as defined by the SWEBOK®. Each standard contains additional reference material that may be used in support of software engineering training activities. Many of the specific questions that practitioners have may be answered by turning to the *IEEE Software Engineering Standards Collection* [42]. Table 2-4 provides a list of knowledge requirements and the IEEE software engineering standard that may be used to support the requirement.

Table 2-4. IEEE standards and training

Requirements	
How to document requirements?	IEEE Std 830, Recommended Practice for Software Requirements Specification
Test	
How to classify software anomalies?	IEEE Std 1044, Standard Classification for Software Anomalies
How to select and apply software measures?	IEEE Std 982.1, Standard Dictionary of Measures to Produce Reliable Software
How to define a test unit?	IEEE Std 1008, Standard for Software Unit Testing
What documentation is required in support of the testing process?	IEEE Std 829, Standard for Software Test Documentation
Maintenance	
What maintenance activities are required prior to product delivery?	IEEE Std 1219, Standard for Software Maintenance

(continued)

Table 2-4   *(continued)*

Communication	
How to communicate using consistent terminology?	IEEE Std 610.12, Standard Glossary of Software Engineering Terminology
Configuration Management	
What describes the requirements and categories of information in support of configuration management planning?	IEEE Std 828, Standard for Software Configuration Management Plans
Reviews and Audits	
Where to find information describing software audit procedures?	IEEE Std 1028, Standard for Software Reviews

Software Engineering and Organizational Support

For the business improvement project to implement any type of software process improvement initiative, all stakeholders need to be identified and committed to their roles. Several stakeholders are identified as follows:

- **Senior Management** initiates the implementation/improvement project, commits to the investment, monitors the various stages toward completion, and guides the organization through is periodic review and planning cycles.
- **Customers, Investors,** or their representatives have a variety of interfacing roles, depending on their maturity, previous experiences, and types of contracting agreements.
- **Software Engineering Team Members** have technical and supporting roles as defined in their business improvement model and their software development life cycle model.

Charter Infrastructure

The definition of the Lean Six Sigma (LSS) implementation project must be defined. The project charter should be described in terms of time frame, membership, organizational and project scope, and level of conformity (self-declared or external assessment). The degree of formality, required infrastructure investments and communication channels should also be defined.

Establish Steering Committee and Process Group

Process improvement normally works best with two levels of committees or groups. The first is the Steering Committee, which includes the sponsors and meets infrequently through the year. The second is the Process Group, a process improvement team made up of people who are given responsibility and authority for improving a selected process in an organization; this team must have the backing of senior management represented by the Steering Committee. Process owners in the Process Group are responsible for the process design, not for the performance, of their associated process areas. The process owner is further responsible for the process measurement and feedback systems, the process documentation, and the training of the process performers in its structure and conduct. In essence, the process owner is the person ultimately responsible for improving a process. IEEE software engineering standards provide valuable support to the process team and each individual process owner. The standards can be used to help define and document the initial baseline of recommended processes and practices.

Software Engineering Processes

Software engineering processes have generic descriptions; the following is provided as an example:

> A software engineering process is an implementation of a combined set of supporting practices. Subsequently, software engineering practices are comprised of various sets of activities. A process can be documented to show how the set of activities performed can achieve a given purpose. A process may have any number of elements or subprocesses [168].

Software engineering processes should be defined and then executed by trained practitioners. During and following their execution process, measurements must be collected. One measure important to lean processes is cycle time. Another measure important to Six Sigma is defects found or prevented. Both software engineering product and software engineering process have "qualities." Any software engineering process can be improved according to several qualities or factors:

- Reliability
- Robustness
- Performance
- Evolvability
- Reusability

Although there is some slight variation among the most widely accepted process improvement methodologies, the authors have chosen to turn to the *Guide to The Software Engineering Body of Knowledge (SWEBOK®)*. The SWEBOK® is directly supported by the IEEE Software and Systems Engineering Standards (S2ESC) Collection.

Software Engineering Artifacts

All software engineering processes create work products or artifacts. Some are just records as simple as process start and end dates. Other artifacts, such as requirement specifications, are important to share with the customer for confirmation. Most artifacts are placed under configuration control. The major artifacts are called out in the development plan. Just as the development plan can be tracked to completion, software engineering processes can be assessed according to any associated artifacts.

3

12207 PRIMARY LIFE CYCLE PROCESSES

THE IEEE 12207 PROCESS FRAMEWORK

IEEE 12207 standards support a framework for the life cycle of software. The life cycle begins with an idea or need that can be satisfied wholly or partly by software and ends with the retirement of that software. The framework architecture is built with a set of processes as defined by IEEE 11207. Interrelationships exist among these processes and each process is placed under the responsibility of an organization or a party in the software life cycle. An organization may employ an organizational process to establish, control, and improve a life cycle process.

The selection and order of software engineering processes determine the development life cycle. IEEE 12207 does not prescribe a specific life cycle model or software development method. The users (e.g., process experts) of the standard are responsible for selecting a life cycle model for the software project and mapping the processes, activities, and tasks in this standard onto that model. They are also responsible for selecting and applying the software development methods and for performing the activities and tasks suitable for the software project. Figure 3-1 provides an overview of these processes.

The IEEE 12207 standards group the activities that may be performed during the life cycle of software into five primary processes, eight supporting processes, and four organizational processes. Each life cycle process is divided into a set of activities; each activity is further divided into a set of tasks (Figure 3.2).

The five primary processes serve primary parties during the life cycle of the software. A primary party is one that initiates or performs the development, operation, or maintenance of software products. These primary parties are the acquirer, the supplier, the developer, the operator, and the maintainer of software products.

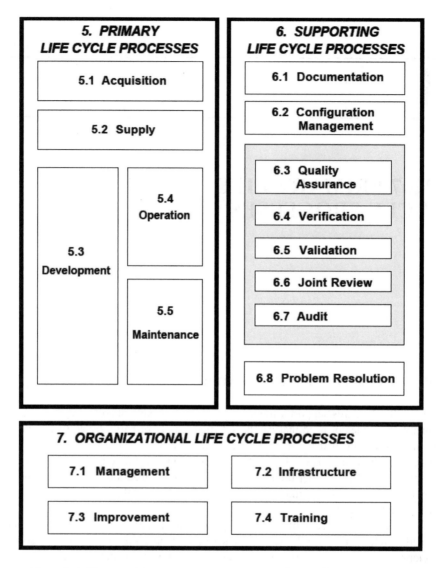

Figure 3-1. Diagram of the primary, supporting, and organizational processes [40].

The eight supporting life cycle processes support other processes as an integral part with a distinct purpose and contribute to the success and quality of the software project. A supporting process is employed and executed, as needed, by another process. The four organizational life cycle processes are employed by an organization to establish and implement an underlying structure made up of associated life cycle processes and personnel and continuously improve the structure and processes. They are typically employed outside the realm of specific projects and contracts; however, lessons from such projects and contracts contribute to the improvement of the organization.

IEEE Std. 1074 Standard for Developing a Software Life Cycle Process describes an approach that uses IEEE 12207 processes to create the software project life cycle process (SPLCP). The five 12207 primary processes, shown in Table 3-1, are performed by dis-

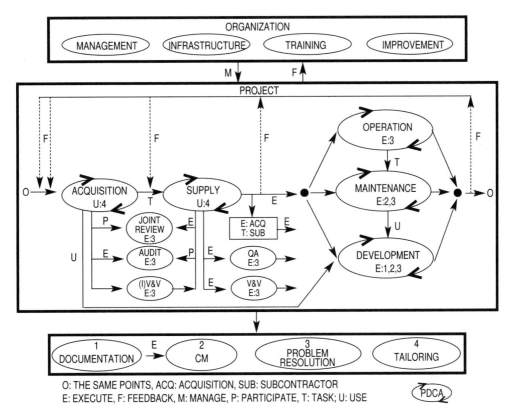

Figure 3-2. Primary, supporting, and organizational SLC processes [40].

Table 3-1. The Five 12207 primary processes [40]

Process	Description
Acquisition	Defines the activities of the acquirer, the organization that acquires a system, software product, or software service.
Development	Defines the activities of the developer, the organization that defines and develops the software product.
Operation	Defines the activities of the operator, the organization that provides the service of operating a computer system in its live environment for its users.
Maintenance	Defines the activities of the maintainer, the organization that provides the service of maintaining the software product; that is, managing modifications to the software product to keep it current and in operational fitness. This process includes the migration and retirement of the software product.
Supply	The supply process contains the activities and tasks of the supplier.

tinct parties: the acquirer, the supplier, the developer, the operator, and the maintainer of software products.

ACQUISITION

The acquisition process defines the activities of the acquirer. The process begins with the definition of the need to acquire a system, software product, or software service. The process continues with the preparation and issue of a request for proposal, selection of a supplier, and management of the acquisition process through to the acceptance of the system, software product, or software service. Table 3-2 provides a list of the acquisition process objectives.

IEEE standards support the acquisition process requirements very directly; the project documentation in support of supply and acquisition activities described in this section are derived from IEEE Std 1062, Software Acquisition Plan.

Table 3-2. Acquisition process objectives [4]

a) Develop a contract, including tailoring of the standard, that clearly expresses the expectation, responsibilities, and liabilities of both the acquirer and the supplier.

b) Obtain products and/or services that satisfy the customer need.

c) Manage the acquisition so that specified constraints (e.g., cost, schedule, and quality) and goals (e.g., degree of software reuse) are met.

d) Establish a statement of work to be performed under contract.

e) Qualify potential suppliers through an assessment of their capability to perform the required software.

f) Select qualified suppliers to perform defined portions of the contract.

g) Establish and manage commitments to and from the supplier.

h) Regularly exchange progress information with the supplier.

i) Assess compliance of the supplier against the agreed upon plans, standards, and procedures.

j) Assess the quality of the supplier's delivered products and services.

k) Establish and execute acceptance strategy and conditions (criteria) for the software product or service being acquired.

l) Establish a means by which the acquirer will assume responsibility for the acquired software product or service.

Software Acquisition Plan

The modification of the recommended Software Acquisition Plan (SAP) table of contents to support the goals of Lean Six Sigma more directly is shown in Table 3-3. Additional information is provided in Appendix A of this text, relating to the work products associated with software acquisition activities. These include a discussion of the nine steps of acquisition, an organizational acquisition strategy checklist, supplier evaluation criteria, and supplier performance standards.

Table 3-3. Software Acquisition Plan document outline [4]

Title Page
Revision Page
Table of Contents
1. Introduction
2. References
3. Definitions and Acronyms
4. Software Acquisition Overview
    4.1   Organization
    4.2   Schedule
    4.3   Resource Summary
    4.4   Responsibilities
    4.5   Tools, Techniques, and Methods
5. Software Acquisition Process
    5.1   Planning Organizational Strategy
    5.2   Implementing the Organization's Process
    5.3   Determining the Software Requirements
    5.4   Identifying Potential Suppliers
    5.5   Preparing Contract Documents
    5.6   Evaluating Proposals and Selecting the Suppliers
    5.7   Managing Supplier Performance
    5.8   Accepting the Software
    5.9   Using the Software
6. Software Acquisition Reporting Requirements
7. Software Acquisition Management Requirements
    7.1   Anomaly Resolution and Reporting
    7.2   Deviation Policy
    7.3   Control Procedures
    7.4   Standards, Practices, and Conventions
    7.5   Performance Tracking
    7.6   Quality Control of Plan
8. Software Acquisition Documentation Requirements

Software Acquisition Plan Document Guidance

The following provides section-by-section guidance in support of the creation of a SAP. The SAP should be considered to be a living document and should change to reflect any process improvement activity. This guidance should be used to help define a software acquisition process and should reflect the actual processes and procedures of the implementing organization. Additional information is provided in the document template, *Software Acquisition Plan.doc,* which is located on the companion CD-ROM provided with this book.

Introduction. This section of the SAP should describe the specific purpose, goals, and scope of the software acquisition effort. All associated requirements and plans should be identified. The type of contract and support concept should be identified.

References. This section should list all supporting documents supplementing or implementing the SAP, including other plans, processes, or task descriptions that elaborate details of this plan.

Definitions and Acronyms. This section of the SAP should define or reference all terms unique to the development and understanding of the SAP. All abbreviations and notations used in the SAP should also be described.

Software Acquisition Overview. This section should describe the acquiring organization, acquisition schedule, and associated resources. All responsibilities, tools, techniques, and methods necessary to perform the software acquisition process should also be described.

Organization. This subsection of the Software Acquisition Overview should describe the organization of the acquisition effort. This description of the organization should include the approval hierarchy and points of contact.

Schedule. This subsection of the Software Acquisition Overview should describe the schedule of the acquisition, including all milestones. If this schedule is included as part of the acquiring organization's software project management plan, this plan may be referenced.

Resource Summary. This subsection of the Software Acquisition Overview should describe all staffing, facilities, tools, finances, and any special procedural requirements that are needed in support of the software acquisition effort.

Responsibilities. This subsection of the Software Acquisition Overview should provide an overview of acquisition responsibilities.

Tools, Techniques, and Methods. This subsection of the Software Acquisition Overview should describe all documentation, tools, techniques, methods, and environments to be used during the acquisition process. Acquisition, training, support, and qualification information for each should be provided. The SAP should document the measures to be used by the acquisition process and should describe how these measures support the acquisition process.

Software Acquisition Process. This section should identify all actions to be performed for each of the software acquisition steps shown:

1. Planning organizational strategy
2. Implementing the organization's process
3. Determining the software requirements
4. Identifying potential suppliers
5. Preparing contract documents
6. Evaluating proposals and selecting the suppliers
7. Managing supplier performance
8. Accepting the software
9. Using the software

For further descriptions, refer to the table in Appendix A (Software Process Work Products) entitled Recommendations for Software Acquisition. Any additional information required in support of the software acquisition process should be included as deemed necessary.

Software Acquisition Reporting Requirements. This section should describe all reporting requirements. This should include a description of the reporting in support of acquisition status and risk.

Software Acquisition Management Requirements. This section should describe all procedures and processes in support of anomaly resolution and reporting. References should be provided to other plans describing the management of the software acquisition process.

Anomaly Resolution and Reporting. This subsection of the Software Acquisition Management Requirements describes the method of reporting and resolving anomalies, including all reporting and resolution criteria.

Deviation Policy. This subsection of the Software Acquisition Management Requirements should describe all procedures and forms used if deviation is required. All deviation approval authorities should also be identified.

Control Procedures. This subsection of the Software Acquisition Management Requirements should describe all procedures and processes supporting the configuration, protection, and storage of associated software products.

Standards, Practices, and Conventions. This subsection of the Software Acquisition Management Requirements should describe all standards, practices, and conventions used in the development of this plan or in support of the acquisition process.

Performance Tracking. This subsection of the Software Acquisition Management Requirements should describe the processes supporting performance monitoring. All items tracked should be identified. All reporting procedures should also be described, including reporting format.

Quality Control of the Plan. This subsection of the Software Acquisition Management Requirements should describe the processes supporting the development and maintenance of this plan. Other project plans may be referenced in support of this section (e.g., Software Quality Assurance Plan).

Software Acquisition Documentation Requirements. All documentation requirements required in support of the acquisition process should be described here. An appendix providing examples of all required document formats could be included.

See Appendix A for templates useful in selecting the best suppliers: Make/Buy Decision Matrix; Alternative Solution Screening Criteria Matrix; and Cost–Benefit Ratio.

Concept of Operations

The following information is based on IEEE Std 1362, IEEE Guide for Information Technology-System Definition—Concept of Operations (ConOps) document. The following provides section-by-section guidance in support of the creation of a ConOps document. This guidance should be used to help define a requirements management process and should reflect the actual processes and procedures of the implementing organization.

IEEE Std 12207.0 provides information in support of the development of a concept of operations description. The information provided in this section is in conformance with IEEE Std 12207.

Additional information is provided in the document template, *ConOpsDocument.doc,* which is located on the companion CD-ROM. Table 3-4 provides an example of a suggested document outline for a concept of operations document.

Table 3-4. Concept of Operations (ConOps) document outline

Title page
Revision Page
Table of Contents
1. Scope
    1.1   Document Overview
    1.2   System Overview
2. Referenced Documents
3. Definitions and Acronyms
4. Operating Procedures
    4.1   Background
    4.2   Operational Policies and Constraints
    4.3   Current Operating Procedures
    4.4   Modes of Operation
    4.5   User Classes
    4.6   Support Environment
5. Change Justification
    5.1   Changes Considered
    5.2   Assumptions and Constraints
6. Proposed System
7. Operational Scenarios
8. Impact Summary
    8.1   Operational Impact
    8.2   Organizational Impact
9. System Analysis
    9.1   Summary of Improvements
    9.2   Disadvantages and Limitations
    9.3   Alternatives
Appendices

Concept of Operations (ConOps) Document Guidance

A ConOps document is used to represent the user viewpoint of the system or software. The ConOps should effectively describe the system characteristics to all participants and how these characteristics meet the mission and objectives of the organization. The ConOps should also present a summary of all assumptions and constraints, perceived impact to existing systems, disadvantages or limitations, and alternatives examined as options to the solution being proposed.

Document Overview. A description of the document, the target audience, and the rationale for its use should be presented in this section.

System Overview. This section of the document should provide an overview of the system and the intended audience and purpose. A diagram of the system, providing an overview of the functionality, is often useful when trying to convey this type of information.

Referenced Documents. This section should provide all supporting documentation used in the development of this document. All documentation referenced by the ConOps should also be listed in this section.

Definitions and Acronyms. This section should provide a list of all definitions and acronyms unique to this document and critical to understanding its content.

Operating Procedures. This section should describe the current operating procedures. These procedures can be based upon an existing system or may describe manual procedures. If there are no current operating procedures, then the ideal procedures should be described.

Background. This subsection of the Operating Procedures should provide the reader with information in support of the project background, objectives, and scope. The information provided here should describe the problem set and proposed solution.

Operational Policies and Constraints. This subsection of the Operating Procedures should provide information in support of all operational policies and constraints as they apply to the current system or situation. As defined by IEEE Standard 1362, IEEE Guide for Information Technology—System Definition—Concept of Operations (ConOps) Document, "Policies limit decision-making freedom but do allow for some discretion. Operational constraints are limitations placed on the operations of the current system. Examples of operational constraints include the following:

- A constraint on the hours of operation of the system, perhaps limited by access to secure terminals.
- A constraint on the number of personnel available to operate the system.
- A constraint on the computer hardware (for example, must operate on computer X).
- A constraint on the operational facilities, such as office space."

Current Operating Procedures. This subsection of the Operating Procedures should provide a detailed description of the current system or situation. This should include a description of the operating environment, major system components, required interfaces, performance requirements, and desired features. Presenting this information to the system user in graphical format is useful when attempting to describe a system or situation. It may be useful to include items such as schedules, charts, functional and/or data flow, or workflow diagrams.

This subsection should also provide a description of all operational requirements in language that the system user would understand. All required facilities, material, hardware and software, and personnel should be described.

Modes of Operation. This subsection of the Operating Procedures should provide detailed information describing all required modes of operation for the system or situation.

User Classes. This subsection of the Operating Proceduresshould provide a description of all user classes. This should include a description of the organizational structure, user profiles with associated roles and required skill sets, and required interaction between user classes. This section should also describe all key personnel associated with the project. Behavior diagrams can include the use case diagram (used by some methodologies during requirements gathering), sequence diagram, activity diagram, collaboration diagram, and statechart diagram.

Support Environment. As applicable, this subsection of the Operating Procedures should describe all support activity required for the maintenance of the system or situation.

Change Justification. This section should provide a description of all deficiencies of the current system or situation. This section should provide a description of all identified problems in terms that can be easily understood by the user. This section should provide a description of all known issues, proposed changes, and a justification for the proposed change. It is helpful to also provide an Appendix to this document that lists all proposed changes in order of importance. A common method used is to categorize items as essential (must have), desired (nice to have), or optional. This ranking is helpful during the translation the items described in the ConOps document to a more formal requirements specification.

Changes Considered. This subsection of the Change Justification should provide information regarding all features (changes) considered but not included in the proposed software application. The impact of not including these features should be described along with any plans for future adoption.

Assumptions and Constraints. This subsection of the Change Justification should address all assumptions and constraints that will affect users during development and operation of the software application. It is important to tie assumptions to assessments of impact. Describe what the new system will provide in terms of performance gains, interfaces to external systems, or schedule impact.

Proposed System. This section should describe the proposed system, providing a high-level, broad system description. This description should be solution oriented. It should focus on the user needs and how the proposed system will support the needs of the user community. Any description of the proposed system should address the characteristics of the operating environment, the performance characteristics, all interface requirements, the capabilities and functions, a description of the data flow requirements, associated cost and risk, and quality requirements.

It is important to keep in mind that the ConOps should be written using common language. Avoid computer-related jargon and use graphics where possible. All items described in Section 3 (Definitions and Acronyms) of this document should be addressed in this section.

Operational Scenarios. This section should provide a description, or series of descriptions, of how the proposed system will operate. Scenarios may also be used to describe what the system will not do. Scenarios should be used to help readers understand how system functionality will support operational requirements.

Impact Summary. This section should describe the positive and negative perceived impact of the proposed system. All predeployment, deployment, and training activity should be described, with any accompanying impact description. Providing this type of information will help organizations prepare for any disruptions caused during system deployment.

Operational Impact. This subsection of the Impact Summary should describe all anticipated impact to the system user during system operation. System deployment may require a change in current policy or procedure and these should be addressed.

Organizational Impacts. This subsection of the Impact Summary should describe all anticipated organizational impact. All impacts associated with users, system development, and system deployment during the operation of the proposed system should be addressed.

System Analysis. This section should provide an analysis of the proposed system. A summary of all improvements, disadvantages and limitations, and alternatives considered as relating to the proposed system should be described.

Summary of Improvements. This subsection of the System Analysis should provide an evaluation of all improvements to existing processes or practices to be provided by the proposed system. This summary should include a description of any new or enhanced capabilities.

Disadvantages and Limitations. This subsectionof the System Analysis should describe any perceived disadvantages or limitations presented by the deployment of the proposed system.

Alternatives. This subsection of the System Analysis should present a description of any alternatives considered. All alternatives considered, but rejected, should be documented with a rationale for nonacceptance.

Appendices. To facilitate ease of use and maintenance of the ConOps document, some information may be placed in appendices to the document. Each appendix should be referenced in the main body of the document where that information would normally have been provided.

Decision Tree Analysis

Decision analysis tools and techniques may be used to support the technical decisions that must be made during the life cycle of a software project. Some of these decisions include the type of architecture, determining whether to build or buy, product design, platform type, life cycle selection, and testing approaches. Table 3-5 provides the steps typical in the decision process and associated sample questions.

Decision trees are often useful when attempting to choose between several courses of action. They provide a highly effective structure for the evaluation of options, associated outcomes, and associated risk and benefit. A decision tree begins with a decision that is

Table 3-5. Basic decision analysis and resolution process [162]

Step in Process	Sample Questions
1. Draft Decision Statement	On what situation do we need to take action? What are we trying to achieve?
2. Establish Decision Objectives	What are the anticipated results? What resources do we have to work with?
3. Objectives: Required or Desired?	What is critical to the success of the decision and can this be measured?
4. Value the Desired Objectives	What is the value of each objective? What is the value scale?
5. Develop the Alternatives	Are there possible alternative solutions?
6. Test Alternatives against Required Objectives	How do the alternatives compare against the objectives?
7. Score Alternatives against Desired Objectives	What is the value of each alternative? What is the value scale?
8. Determine Risks	What are the associated probabilities and impact of the risk?
9. Select Best Alternative	What alternative provides the most benefit for the least risk?

represented by a small square toward the left of the drawing area. Lines are drawn out to the right from this initial decision box, each line representing an alternative solution.

Consider the results of each decision path. If the alternative presented results in uncertainty, draw a circle and note the uncertainty above the circle. In a decision tree, squares represent decision and circles represent uncertain outcomes. If the alternative results in another decision draw another square noting the decision above the square. This process should be repeated until all possibilities are exhausted.

The decision tree must now be evaluated. Begin by identifying a percentage value for each of the possible outcomes. Also assign a dollar amount or value to each alternative outcome. Next, perform the calculations associated with each node of the decision tree and record the result. When calculations are based upon uncertainties (circled items), multiply the value of the outcome by the probability shown. After the estimated outcome or benefit is calculated it is important to then subtract the estimate for all associated costs. The decision tree will then reflect net benefit values that may be used in the decision making process. An example of a basic decision tree is presented in Figure 3-3. An example of the recorded outcomes, or benefit, for a new product through development is provided in Table 3-6.

Decision trees can help provide objective insight when analyzing information for decision making. They help to clarify the problem by laying out all the options. All probabilities are factored into the various alternatives. The key to an effective decision-making tool is that it should help individuals make objective decisions.

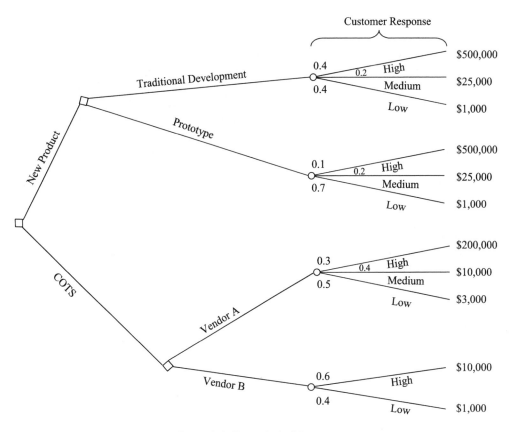

Figure 3-3. Example decision tree.

Table 3-6. Example decision tree results

Node	Benefit Level	Benefit Calculations	Benefit Subtotal	Outcome Total	Estimated Cost	Net Benefit
Traditional Development	High	0.4 × 500,000	200,000	210,000	75,000	135,000
	Medium	0.4 × 25,000	10,000			
	Low	0.2 × 1,000	200			
Prototype	High	0.1 × 500,000	50,000	55,700	40,000	15,700
	Medium	0.2 × 25,000	5,000			
	Low	0.7 × 1,000	700			
Vendor A	High	0.3 × 200,000	60,000	65,500	15,000	50,000
	Medium	0.4 × 10,000	4,000			
	Low	0.5 × 3,000	1,500			
Vendor B	High	0.6 × 10,000	6,000	6,400	0	6,400
	Low	0.4 × 1,000	400			

SUPPLY

The supply process contains the activities and tasks of the supplier. The process may be initiated either by a decision to prepare a proposal to answer an acquirer's request for proposal or by signing and entering into a contract with the acquirer to provide the system, software product, or software service. The process continues with the determination of procedures and resources needed to manage and assure the project, including development of project plans and execution of the plans through delivery of the system, software product, or software service to the acquirer. Table 3-7 provides a list of the supply process objectives.

Table 3-7. Supply process objectives

a) Establish clear and ongoing communication with the customer.
b) Define documented and agreed-on customer requirements, with managed changes.
c) Establish a mechanism for ongoing monitoring of customer needs.
d) Establish a mechanism for ensuring that customers can easily determine the status and disposition of their requests.
e) Determine requirements for replication, distribution, installation, and testing of the system containing the software or stand-alone software product.
f) Package the system containing the software or the stand-alone software product in a way that facilitates its efficient and effective replication, distribution, installation, testing, and operation.
g) Deliver a quality system containing the software or stand-alone software product to the customer, as defined by the requirements, and install in accordance with the identified requirements.

Request for Proposal

The Customers Request for Proposal (RFP) is usually the first document to trigger the supply process. Table 3-8 provides an outline for a RFP.

Table 3-8. Request for proposal outline

Project Title
Company Background
Project Description
Functional Requirements
Design Requirements
Technical and Infrastructure Requirements
Estimated Project Duration
Assumptions and Agreements
Submission Information
For Additional Information or Clarification
Basis for Award of Contract
Bidder's General Information
Anticipated Selection Schedule
Terms and Conditions

Request for Proposal (RFP) Guidance

The information presented here is based upon general recommendations provided in IEEE Std 12207.0, IEEE Std 1220, and IEEE Std 12207.2.

Project Title. A carefully crafted "Request for Proposal" (RFP) is the key to getting the best quality services you require for your project. The RFP is your "official" statement to vendors about the products you require. Vendors typically try to respond, point by point, to your RFP when they make their proposals.

Company Background. Insert a concise paragraph outlining your company's background.

Project Description. Insert a summary of your project, including the problem/ opportunity, goals/objectives, and any information that will help the vendors understand the need for the project. Include the context of this RFP. Be sure not to outline specific requirements in this section.

Functional Requirements. Insert an outline of all the functionality you would like your project to have, along with a short description. Differentiate between mandatory requirements and desirable requirements.

Design Requirements. Insert an outline of any requirements that pertain to the design of the project. Include any compatibility issues.

Technical and Infrastructure Requirements. Insert any technical or infrastructure-related requirements, such as a server or database configuration.

Estimated Project Duration. Insert the project start and estimated duration of the project or the required completion date. Also include a listing of expected milestones and deliverables.

Assumptions and Agreements. Insert a list of any assumptions or agreements the product or the vendors must meet, such as on-going communication channels or expected collaboration. Include the expected usage of the proposed product and prime contractor responsibility.

Submission Information. Insert the deadline for RFP submission, assumptions, and the address to submit the proposal to. Also state the confidentiality of evaluation and correction of errors.

For Additional Information or Clarification. Insert a list of contacts that are available to clarify any questions regarding the RFP. Decide whether to allow vendor suggestions or creativity and, thus, RFP resubmissions.

Basis for Award of Contract. Insert the outline of the list of the evaluation criteria that will be used to determine the best vendor proposal. It is not uncommon to list the weight that each criterion holds in relation to the others. Include here descriptions of required development process and revision cycles, standards of quality, quality measures, and so on.

Bidder's General Information. Insert bidder's profile, customer base, financial details, technical capability (CMMI®), quality certification (ISO 9001), conformance to industry standards, tender's qualifications, relevant services, any specific skills and experience, details of management and key personnel, and reference projects.

Anticipated Selection Schedule. Insert the schedule for your RFP selection process, including the closing date for receipt of proposal.

Terms and Conditions. Attach a blank contract or spell out special conditions or requirements that vendors must meet. Also include buyer's responsibilities such as payment schedule.

Joint Customer Technical Reviews

The information provided here is based upon the recommendations provided in IEEE Std 1028, IEEE Standard for Software Reviews. Technical reviews are an effective way to detect and identify problems early in the software development process. The criteria for technical reviews are presented here to help organizations define their review practices.

Introduction. Technical reviews are an effective way for a team of qualified personnel to evaluate a software product to identify discrepancies from specifications and standards. Whereas technical reviews identify anomalies, they may also provide the recommendation and examination of various alternatives. The examination need not address all aspects of the product and may only focus on a selected aspect of a software product.

RFP, software requirements specifications, design descriptions, test and user documentation, installation and maintenance procedures, and build processes are all examples of items subject to technical reviews.

Responsibilities. The roles in support of a technical review are:

Decision maker. The decision maker is the individual who requests the review and determines whether objectives have been met.

Review leader. The review leader is responsible for the review and must perform all administrative tasks (to include summary reporting), ensure that the review is conducted in an orderly manner, and that the review meets its objectives.

Recorder. The recorder is responsible for the documentation of all anomalies, action items, decisions, and recommendations made by the review team.

Technical staff. The technical staff should actively participate in the review and evaluation of the software product.

The following roles are optional and may also be established for the technical review:

Management staff. Management staff may participate in the technical review for the purpose of identifying issues that will require management resolution.

Customer or user representative. The role of the customer or user representative should be determined by the review leader prior to the review.

Input. Input to the technical review should include a statement of objectives, the software product being examined, existing anomalies and review reports, review procedures, and any standard against which the product is to be examined. Anomaly categories should be defined and available during the technical review. For additional information in support of the categorization of software product anomalies, refer to IEEE Std 1044 [7].

Authorization. All technical reviews should be defined in the software project management plan (SPMP). The plan should describe the review schedule and all allocation of resources. In addition to those technical reviews required by the SPMP, other technical reviews may be scheduled.

Preconditions. A technical review shall be conducted only when the objectives have been established and the required review inputs are available.

Procedures. These are described in the following subsections.

Management Preparation. Managers are responsible for ensuring that all reviews are planned, that all team members are trained and knowledgeable, and that adequate resources are provided. They are also responsible for ensuring that all review procedures are followed.

Planning the Review. The review leader is responsible for the identification of the review team and their assignment of responsibility. The leader should schedule and announce the meeting, prepare participants for the review by providing them the required material and collecting all comments.

Overview of Review Procedures. The team should be presented with an overview of the review procedures. This overview may occur as a part of the review meeting or as a separate meeting.

Overview of the Software Product. The team should receive an overview of the software product. This overview may occur either as a part of the review meeting or as a separate meeting.

Preparation. All team members are responsible for reviewing the product prior to the review meeting. All anomalies identified during this prereview process should be presented to the review leader. Prior to the review meeting, the leader should classify all anomalies and forward these to the author of the software product for disposition. The review leader is also responsible for the collection of all individual preparation times to determine the total preparation time associated with the review.

Examination. The review meeting should have a defined agenda that should be based upon the premeeting anomaly summary. Based upon the information presented, the team should determine whether the product is suitable for its intended use, whether it conforms to appropriate standards, and is ready for the next project activity. All anomalies should be identified and documented.

Rework/Follow-up. The review leader is responsible for verifying that the action items assigned in the meeting are closed.

Exit Criteria. A technical review shall be considered complete when all follow-up activities have been completed and the review report has been published.

Output. The output from the technical review should consist of the project being reviewed, a list of the review team members, a description of the review objectives, and a list of resolved and unresolved software product anomalies. The output should also include a list of management issues, all action items and their status, and any recommendations made by the review team.

Software Project Management Plan

IEEE Std 1058, IEEE Standard for Software Project Management Plans [13], and IEEE 12207.0, IEEE Standard for Life Cycle Processes [40], provide effective support of software project planning activities. However, information regarding Lean Six Sigma related measurement activities needs to be added to and stated explicitly for the project plan. Table 3-9 provides a document outline for a software project management plan.

Table 3-9. Software project management plan document outline

Title Page
Revision Page
Table of Contents
1. Introduction
    1.1   Project Overview
    1.2   Project Deliverables
    1.3   Document Overview
    1.4   Acronyms and Definitions
2. References
3. Project Organization
    3.1   Organizational Policies
    3.2   Process Model
    3.3   Organizational Structure
    3.4   Organizational Boundaries and Interfaces
    3.5   Project Responsibilities
4. Managerial Process
    4.1   Management Objectives and Priorities
    4.2   Assumptions, Dependencies, and Constraints
    4.3   Risk Management
    4.4   Monitoring and Controlling Mechanisms
    4.5   Staffing Plan
5. Technical Process
    5.1   Tools, Techniques, and Methods
    5.1   Software Documentation
    5.3   Project Support Functions
6. Work Packages
    6.1   Work Packages
    6.2   Dependencies
    6.3   Resource Requirements
    6.4   Budget and Resource Allocation
    6.5   Schedule
7. Additional Components

Software Project Management Plan Document Guidance

The following provides section-by-section guidance in support of the creation of a SPMP. The SPMP should be considered to be a living document. The SPMP should change, in particular any associated schedules, to reflect any required change during the life cycle of a project. This guidance should be used to help define a management process and should reflect the actual processes and procedures of the implementing organization. Additional information is provided in the document template, *Software Project Management Plan.doc,* that is located on the companion CD-ROM. Additional information is provided in Appendix A, Software Process Work Products, which describes the work breakdown structure, workflow diagram, and stakeholder involvement matrix work products.

Introduction. This section contains the following subsections.

Project Overview. This subsection should briefly state the purpose, scope, and objectives of the system and the software to which this document applies. It should describe the general nature of the system and software; summarize the history of system development, operation, and maintenance; identify the project sponsor, acquirer, user, developer, and support agencies; and identify current and planned operating sites.

The project overview should also describe the relationship of this project to other projects, as appropriate, addressing any assumptions and constraints. This subsection should also provide a brief schedule and budget summary. This overview should not be construed as an official statement of product requirements. Reference to the official statement of product requirements should be provided in this subsection of the SPMP.

Project Deliverables. This subsection of the SPMP should list all of the items to be delivered to the customer, the delivery dates, delivery locations, and quantities required to satisfy the terms of the project agreement. This list of project deliverables should not be construed as an official statement of project requirements.

Document Overview. This subsection should summarize the purpose and contents of this document and describe any security or privacy considerations that should be considered associated with its use. This subsection of the SPMP should also specify the plans for producing both scheduled and unscheduled updates to the SPMP. Methods of disseminating the updates should be specified. This subsection should also specify the mechanisms used to place the initial version of the SPMP under change control and to control subsequent changes to the SPMP.

Acronyms and Definitions. This subsection should identify acronyms and definitions used within the project SPMP. The project SPMP should only list acronyms and definitions used within the SPMP.

References. This section should identify the specific references used within the project SPMP. The project SPMP should only contain references used within the SPMP.

Project Organization. This section contains the following subsections.

Organizational Policies. This subsection of the SPMP should identify all organizational policies relative to the software project.

Process Model. This subsection of the SPMP should specify the (life cycle) software development process model for the project, describe the project organizational structure, identify organizational boundaries and interfaces, and define individual or stakeholder responsibilities for the various software development elements.

Organizational Structure. This subsection should describe the makeup of the team to be used for the project. All project roles, and stakeholders, should be identified as well as a description of the internal management structure of the project. Diagrams may be used to depict the lines of authority, responsibility, and communication within the project.

Organizational Boundaries and Interfaces. This subsection should describe the limits of the project including any interfaces with other projects or programs, the application of the program's software configuration management (SCM) and software quality assurance (SQA) (including any divergence from those plans), and the interface with the project's customer. This section should describe the administrative and managerial boundaries between the project and each of the following entities: the parent organization, the customer organization, subcontracted organizations, or any other organizational entities that interact with the project. In addition, the administrative and managerial interfaces of the project-support functions, such as configuration management, quality assurance, and verification, should be specified in this subsection.

Project Responsibilities. This subsection should describe the project's approach through a description of the tasks required to complete the project (e.g., requirements → design → implementation → test) and any efforts (update documentation, etc.) required to successfully complete the project. It should state the nature of each major project function and activity, and identify the individuals, or stakeholders, who are responsible for those functions and activities.

Managerial Process. This subsection should specify management objectives and priorities; project assumptions, dependencies, and constraints; risk management techniques; monitoring and controlling mechanisms to be used; and the staffing plan.

Management Objectives and Priorities. This subsection should describe the philosophy, goals, and priorities for management activities during the project. Topics to be specified may include, but are not limited to, the frequency and mechanisms of reporting to be used; the relative priorities among requirements, schedule, and budget for this project; risk management procedures to be followed; and a statement of intent to acquire, modify, or use existing software.

Assumptions, Dependencies, and Constraints. This subsection should state the assumptions on which the project is based, the external events the project is dependent upon, and the constraints under which the project is to be conducted.

Risk Management. This subsection should identify the risks for the project. Completed risk management forms should be maintained and tracked by the project leader with associated project information. These forms should be reviewed at weekly staff meetings. Risk factors that should be considered include contractual risks, technological risks, risks due to size and complexity of the project, risks in personnel acquisition and retention, and risks in achieving customer acceptance of the product.

Monitoring and Controlling Mechanisms. This subsection should define the reporting mechanisms, report formats, information flows, review and audit mechanisms, and other tools and techniques to be used in monitoring and controlling adherence to the SPMP. A typical set of software reviews is listed in Appendix A. Project monitoring should occur at the level of work packages. The relationship of monitoring and controlling mechanisms to the project support functions should be delineated in this subsection. This subsection should also describe the approach to be followed for providing the acquirer or its authorized representative access to developer and subcontractor facilities for review of software products and activities.

Staffing Plan. This subsection should specify the numbers and types of personnel required to conduct the project. Required skill levels, start times, duration of need, and methods for obtaining, training, retaining, and phasing out of personnel should be specified.

Technical Process. This subsection should specify the technical methods, tools, and techniques to be used on the project. In addition, the plan for software documentation should be specified, and plans for project support functions such as quality assurance, configuration management, and verification and validation may be specified.

Tools, Techniques, and Methods. This subsection of the SPMP should specify the computing system(s), development methodology(s), team structures(s), programming language(s), and other notations, tools, techniques, and methods to be used to specify, design, build, test, integrate, document, deliver, modify or maintain or both (as appropriate) the project deliverables.

This subsection should also describe any tools (compilers, CASE tools, and project management tools), any techniques (review, walk-through, inspection, and prototyping) and the methods (object-oriented design or rapid prototyping) to be used during the project.

Software Documentation. This subsection should contain, either directly or by reference, the documentation plan for the software project. The documentation plan should specify the documentation requirements and the milestones, baselines, reviews, and sign-offs for software documentation. The documentation plan may also contain a style guide, naming conventions, and documentation formats. The documentation plan should provide a summary of the schedule and resource requirements for the documentation effort. IEEE Std for Software Test Documentation (IEEE Std 829) [15] provides a standard for software test documentation.

Project Support Functions. This subsection should contain, either directly or by reference, plans for the supporting functions for the software project. These functions may include, but are not limited to, configuration management, software quality assurance, and verification and validation.

Work Packages. This section contains the following subsections.

Work Packages. This subsection of the SPMP should specify the work packages, identify the dependency relationships among them, state the resource requirements, provide the allocation of budget and resources to work packages, and establish a project schedule.

The work packages for the activities and tasks that must be completed in order to satisfy the project agreement must be described in this section. Each work package should be uniquely identified; identification may be based on a numbering scheme and descriptive titles. A diagram depicting the breakdown of activities into subactivities and tasks may be used to depict hierarchical relationships among work packages.

Dependencies. This subsection should specify the ordering relations among work packages to account for interdependencies among them and dependencies on external events. Techniques such as dependency lists, activity networks, and the critical path may be used to depict dependencies.

Resource Requirements. This subsection should provide, as a function of time, estimates of the total resources required to complete the project. Numbers and types of personnel, computer time, support software, computer hardware, office and laboratory facilities, travel, and maintenance requirements for the project resources are typical resources that should be specified.

Budget and Resource Allocation. This subsection should specify the allocation of budget and resources to the various project functions, activities, and tasks. Defined resources should be tracked.

Schedule. This subsection should be used to capture the project's schedule, including all milestones and critical paths. Options include Gantt charts (*Milestones Etc.*™ or *Microsoft Project*™), Pert charts, or simple time lines.

Additional Components. This subsection should address additional items of importance on any particular project. This may include subcontractor management plans, security plans, independent verification and validation plans, training plans, hardware procurement plans, facilities plans, installation plans, data conversion plans, system transition plans, or product maintenance plans.

DEVELOPMENT

The development process contains the activities and tasks of the developer. The developer manages the development process at the project level following the management process, infrastructure process, and tailoring process. Also, the developer manages the process at the organizational level following the improvement process and the training process. Finally, the developer performs the supply process if it is the supplier of developed software products. Table 3-10 provides a list of the development process objectives.

Table 3-10. Development process objectives

a) Develop requirements of the system that match the customer's stated and implied needs.
b) Propose an effective solution that identifies the main elements of the system.
c) Allocate the defined requirements to each of those main elements.
d) Develop a system release strategy.
e) Communicate the requirements, proposed solution, and their relationships to all affected parties.

(continued)

Table 3-10. *(continued)*

f) Define the requirements allocated to software components of the system and their interfaces to match the customer's stated and implied needs.

g) Develop software requirements that are analyzed, correct, and testable.

h) Understand the impact of software requirements on the operating environment.

i) Develop a software release strategy.

j) Approve and update the software requirements, as needed.

k) Communicate the software requirements to all affected parties.

l) Develop an architectural design.

m) Define internal and external interfaces of each software component.

n) Establish traceability between system requirements and design and software requirements, between software requirements and software design, and between software requirements and tests.

o) Define verification criteria for all software units against the software requirements.

p) Produce software units defined by the design.

q) Accomplish verification of the software units against the design.

r) Develop an integration strategy for software units consistent with the release strategy.

s) Develop acceptance criteria for software unit aggregates that verify compliance with the software requirements allocated to the units.

t) Verify software aggregates using the defined acceptance criteria.

u) Verify integrated software using the defined acceptance criteria.

v) Record the results of the software tests.

w) Develop a regression strategy for retesting aggregates, or the integrated software, should a change in components be made.

x) Develop an integration plan to build system unit aggregates according to the release strategy.

y) Define acceptance criteria for each aggregate to verify compliance with the system requirements allocated to the units.

z) Verify system aggregates using the defined acceptance criteria.

aa) Construct an integrated system demonstrating compliance with the system requirements (functional, nonfunctional, operations and maintenance).

ab) Record the results of the system tests.

ac) Develop a regression strategy for retesting aggregates or the integrated system should a change in components be made.

ad) Identify transition concerns, such as availability of work products, availability of system resources to resolve problems and adequately test before fielding corrections, maintainability, and assessment of transitioned work products.

The development process is the largest of the 17 processes in IEEE 12207. The development activities are:

- Process implementation
- System requirements analysis
- System architectural design
- Software requirements analysis
- Software architectural design
- Software detailed design
- Software coding and testing

- Software integration
- Software qualification testing
- System integration;
- System qualification testing
- Software installation
- Software acceptance support

· Depending upon the type of contract, the development process begins with process implementation and continues through to the customer acceptance.

The Development Process leverages several IEEE standards:

- IEEE Std 829, Standard for Software Test Documentation
- IEEE Std. 830, Recommended Practice for Software Requirements Specifications
- IEEE Std 1008, Standard for Software Unit Testing
- IEEE Std 1012, Standard for Software Verification and Validation Plans
- IEEE Std. 1016, Recommended Practice for Software Design Descriptions
- IEEE Std 1063, Standard for Software User Documentation
- IEEE Std. 1074 Standard for Developing a Software Project Life Cycle Process
- IEEE Std 1220, Standard for Application and Management of the Systems Engineering Process
- IEEE Std 1233, Guide to Developing System Requirements Specifications
- IEEE Std. 1320.1, Standard for Functional Modeling Language—Syntax and Semantics for IDEF0
- IEEE Stds1420.1, 1420.1a, and 1420.1b Software Reuse—Data Model for Reuse Library Interoperability
- IEEE Std. 1471, Recommended Practice for Architectural Description of Software Intensive Systems
- IEEE/EIA Std. 12207.0, Standard for Information Technology—Software Life Cycle Processes
- IEEE/EIA Std. 12207.1, Standard for Information Technology—Software life cycle processes—Life cycle data,

The next sections will address each development activity and the related project documentation supported by the IEEE Standards. Unless stipulated in the contract, the developer should define the software life cycle model for the project. IEEE Std. 1074, the standard for developing a software project life cycle process, is specifically constructed for this development activity of process implementation.

System Requirements Analysis

When the software product is part of a system, then the software life cycle (SLC) will specify the system requirements analysis activity. This analysis will result in the system requirements specification, which describes: functions and capabilities of the system; business, organizational, and user requirements; safety, security, human-factors engineering (ergonomics), interface requirements, operational, and maintenance requirements; design constraints; and qualification requirements.

System Requirements Specification

The information provided here in support of requirements management is designed to facilitate the definition of a system requirements specification. This information was developed using IEEE Std 1233, IEEE Guide for Developing System Requirements Specifications [25]. Table 3-11 provides an example system requirements specification document outline.

Table 3-11. System requirements specification document outline

Title Page
Revision Page
Table of Contents
1. Introduction
    1.1 System Purpose
    1.2 System Scope
    1.3 Definitions, Acronyms, and Abbreviations
    1.4 References
    1.5 System Overview
2. General System Description
    2.1 System Context
    2.2 System Modes and States
        2.2.1 Configurations
    2.3 Major System Capabilities
    2.4 Major System Conditions
    2.5 Major System Constraints
    2.6 User Characteristics
    2.7 Assumptions and Dependencies
    2.8 Operational Scenarios
3. System Capabilities, Conditions, and Constraints
    3.1 Physical
        3.1.1 Physical Construction
        3.1.2 Physical Dependability
        3.1.3 Physical Adaptability
        3.1.4 Environmental Conditions
    3.2 System Performance Characteristics
        3.2.1 Load
        3.2.2 Stress
        3.2.3 Contention
        3.2.4 Availability
    3.3 System Security and Safety
    3.4 Information Technology Management
    3.5 System Operations
        3.5.1 System Human Factors
        3.5.2 System Usability
        3.5.3 Internationalization
        3.5.4 System Maintainability
        3.5.5 System Reliability
    3.6 Policy and Regulation
    3.7 System Life Cycle Sustainment
4. System Interfaces
5. Specific Requirements
6. Traceability Matrix

System Requirements Specification Document Guidance

The following provides section-by-section guidance in support of the creation of a system requirements specification (SysRS). This guidance should be used to help establish a requirements baseline and should reflect the actual processes and procedures of the implementing organization. Additional information is provided in the document template, *System Requirements Specification.doc,* which is located on the companion CD-ROM.

Introduction. Explain the purpose and scope of the project system requirements specification (SysRS), as well as provide clarification of definitions, acronyms, and references. This section should also provide an overview of the project.

System Purpose. Explain the purpose for writing the SysRS for this project and describe the intended audience for the SysRS. (Note this maybe aligned with ConOps document System Overview.)

System Scope. Identify the System products to be produced, by name, explain what the System products will, and will not do, and describe the application of the System being specified including all relevant goals, objectives, and benefits from producing the System. (Note this maybe aligned with ConOps document System Overview.)

Definitions, Acronyms, and Abbreviations. Provide the definitions of all terms, acronyms, and abbreviations required to properly interpret the SysRS. (Note this maybe aligned with ConOps document Definitions and Acronyms.)

References. List all references used within the SysRS.

System Overview. Describe what the rest of the SysRS contains as it relates to the systems and software components effort. (Note: this may be aligned with ConOps document, System Overview.) It should also explain how the document is organized.

General System Description. Describe the general factors that affect the product and its requirements. (Note: this may be aligned with ConOps document, Proposed System.)

System Context. Include diagrams and narrative to provide an overview of the context of the system, defining all significant interfaces crossing the system's boundaries.

System Modes and States. Include diagrams and narrative to provide usage/operation modes and transition states (Note: this may be aligned with ConOps document, Modes of Operation.) This subsection contains the Configurations subsection, which describes typical system configurations that meet various customers' needs.

Major System Capabilities. Provide diagrams and accompanying narrative to show major capability groupings of the requirements.

Major System Conditions. Show major conditions, relative to the attributes of the major capability groupings.

Major System Constraints. Show major constraints, relative to the boundaries of the major capability groupings.

User Characteristics. Identify each type of user of the system (by function, location, or type of device), the number in each group, and the nature of their use of the system. (Note: this maybe aligned with ConOps document, User Classes.)

Assumptions and Dependencies. Address all assumptions and dependencies that impact the system resulting from the SysRS. It is important to tie assumptions to assessments of impact. This subsection should be the source for recognizing the impact of any changes to the assumptions about or dependencies on the SysRS and resulting system. This subsection can highlight unresolved requirement issues and should be recorded in the Project Manager's Open Issues List.

Operational Scenarios. Provide descriptive examples of how the system will be used. Scenarios may also be used to describe what the system will not do. Scenarios should be used to help readers understand how system functionality will support operational requirements. (Note: this maybe aligned with ConOps document, Operational Scenarios.)

System Capabilities, Conditions, and Constraints. Describe the next levels of system structure. System behavior, exception handling, manufacturability, and deployment should be covered under each capability, condition, and constraint.

Physical. This subsection to system capabilities, conditions, and constraints contains the following subsections:

   Physical Construction. Include the environmental (mechanical, electrical, chemical) characteristics of where the system will be installed.
   Physical Dependability. Include the degree to which an system component is operable and capable of performing its required function at any (random) time, given its suitability for the mission and whether the system will be available and operate—when, as many times, and as long as needed. Examples of dependability measures are availability, interoperability, compatibility, reliability, repeatability, usage rates, vulnerability, survivability, penetrability, durability, mobility, flexibility, and reparability.
   Physical Adaptability. Include how to address growth, expansion, capability, and contraction.
   Environmental Conditions. Include environmental conditions to be encountered by the system. The following subjects should be considered for coverage: natural environment (wind, rain, temperature); induced environment (motion, shock, noise); and electromagnetic signal environment.

System Performance Characteristics. Include the critical performance conditions and their associated capabilities. Consider performance requirements for the operational phases and modes. Also consider endurance capabilities, minimum total life expectancy, operational session duration, planned utilization rate, and dynamic actions or changes. This subsection contains the following subsections:

   Load. Describe the range of work or transaction loads the system shall operate on or process in one time period.
   Stress. Describe the extreme but valid conditions for system work or transaction loads.

Contention. Identify any sharing of computer resources whereby two or more processes can access the computer's processor simultaneously.

Availability. Describe the range of operational availability or the actual availability of a system under logistics constraints.

System Security and Safety. Cover both the facility that houses the system and operational security requirements, including requirements of privacy, protection factors, and safety.

Information Technology Management. Describe the various stages of information processing from production to storage and retrieval to dissemination for the better working of the user organization.

System Operations. Describe how the system supports user needs and the needs of the user community. This subsection contains the following subsections:

System Human Factors. Include references to applicable documents and specify any special or unique requirements for personnel and communications and personnel/equipment interactions.

System Usability. Describe the special or unique usability requirements for personnel and communications and personnel/equipment interactions.

Internationalization. For each major global customer segment, describe the special or unique internationalization requirements for personnel and communications and personnel/equipment interactions.

System Maintainability. Describe, in quantitative terms, the requirements for the planned maintenance, support environment, and continued enhancements.

System Reliability. Specify, in quantitative terms, the reliability conditions to be met. Consider including the reliability apportionment model to support allocation of reliability values assigned to system functions for their share in achieving desired system reliability.

Policy and Regulation. Describe relevant organizational policies that will affect the operation or performance of the system. Include any external regulatory requirements, or constraints imposed by normal business practices.

System Life Cycle Sustainment. Describe quality activities and measurement collection and analysis to help sustain the system through its expected life cycle.

System Interfaces. Include diagrams and narrative to describe the interfaces among different system components and their external capabilities, including all users, both human and other systems. Include interface interdependencies or constraints.

Specific Requirements. This section should contain all details that the system/software developer needs to create a design. The details within this section should be defined as individual, specific requirements with sufficient cross-referencing back to related discussions in the Introduction and General Description sections above. The ordering of subsections that will be shown in this section is one of several possible combinations that may be selected. This ordering will normally be used but others may be selected provided they are justified. Each specific requirement should be stated such that its achievement can be objectively verified by a prescribed method. Each requirement should be enumerated for tracking. The method of verification will be shown in a requirements traceability matrix in Appendix A of the SySRS.

Traceability Matrix. This is a comprehensive baseline of all system requirements described within the SysRS. This list should be used as the basis for all requirements management, system and allocated software requirements, design, development, and test.

Software Requirements Analysis

The resulting software requirements specifications should include the quality characteristics, functional and capability requirements, external interfaces, qualification, safety, security, human-factors engineering, data definitions, installation and acceptance, user documentation, user operations, and user maintenance.

Software Requirements Specification

The information provided here in support of requirements management is designed to facilitate the definition of a software requirements specification. This information was developed using IEEE Std 830, IEEE Recommended Practice for Software Requirements Specifications [6]. Table 3-12 provides an example software requirements specification document outline.

Table 3-12. Software requirements specification document outline

Title Page
Revision Page
Table of Contents
1. Introduction
    1.1 Software Purpose
    1.2 Software Scope
    1.3 Definitions, Acronyms, and Abbreviations
    1.4 References
    1.5 Software Overview
2. Overall Description
    2.1 Product Perspective
    2.2 Product Functions
    2.3 Environmental Conditions
    2.4 User Characteristics
    2.5 External Interfaces
    2.6 Constraints
        2.6.1 Safety
        2.6.2 Security
        2.6.3 Human Factors
    2.7 Assumptions and Dependencies
3. Requirements Management

(continued)

Table 3-12. *(continued)*

3.1 Resources and Funding
3.2 Reporting Procedures
3.3 Training
4. Specific Requirements
5. Acceptance
6. Documentation
7. Maintenance
Appendixes
Traceability Matrix
Index

Software Requirements Specification Document Guidance

The following provides section-by-section guidance in support of the creation of a software requirements specification. This guidance should be used to help establish a requirements baseline and should reflect the actual processes and procedures of the implementing organization. Additional information is provided in the document template, *Software Requirements Specification.doc,* which is located on the companion CD-ROM.

Title Page. The title page should include the names and titles of all document approval authorities.

Revision Sheet. The revision sheet should provide the reader with a list of cumulative document revisions. The minimum information recorded should include the revision number, the document version number, the revision date, and a brief summary of changes.

Introduction. This section should explain the purpose and scope of the project software requirements specification (SRS), as well as provide clarification of definitions, acronyms, and references. This section should also provide an overview of the project.

Software Purpose. This subsection should explain the purpose for writing an SRS for this project and describe the intended audience for the SRS.

Software Scope. This subsection should identify the software products to be produced, by name, and explain what the software products will and, if necessary, will not do, and describe the application of the software being specified, including all relevant goals of, objectives for, and benefits from producing the software.

Definitions, Acronyms, and Abbreviations. This subsection should provide the definitions of all terms, acronyms, and abbreviations required to properly interpret the SRS.

References. This subsection should list all references used within the SRS.

Software Overview. This subsection should describe what the rest of the SRS contains as it relates to the software effort. It should also explain how the document is organized.

Overall Description. This section should describe the general factors that affect the product and its requirements.

Product Perspective. This subsection should put the product into perspective vis-à-vis other related products or projects. If the product to be produced from this SRS is totally independent, it should be clearly stated here. If the product to be produced from this SRS is part of a larger system, then this subsection should describe the functions of each component of the larger system or project and identify the interfaces between this product and the remainder of the system or project. This subsection should identify all principle external interfaces for this software product (Note: descriptions of the interfaces will be contained in another part of the SRS.) The product perspective requirements maybe documented in Product Packaging Information, shown in Appendix A.

Product Functions. This subsection should provide a summary of the functions to be performed by the software produced as a result of this SRS. Functions listed in this subsection should be organized in a way that will be understandable to the intended audience of the SRS.

Environmental Conditions. This subsection should provide a summary of the environment in which the software must operate. (Note: this subsection is an overview; details of the specific requirements will be contained in section 4 of the SRS.)

User Characteristics. This subsection should describe the general characteristics of the eventual users of the product, which will affect the specific requirements. Eventual users of the product will include end-product customers, operators, maintainers, and systems people as appropriate. For any users who impact the requirements, characteristics such as education, skill level, and experience levels will be documented within this subsection, as they impose constraints on the product.

External Interfaces. This subsection should describe all required external interface requirements. This should include references to any existing interface control documentation.

Constraints. This subsection should provide a list of the general constraints imposed on the system that may limit designer's choices. Constraints may come from: regulations or policy, hardware limitations, interfaces to other applications, parallel operations, audit functions, control functions, higher-order language requirements, communication protocols, criticality of the applications or safety, and security restrictions. This section should not describe the implementation of the constraints, but should provide a source for understanding why certain constraints will be expected from the design.

Assumptions and Dependencies. This subsection should list all assumptions and dependencies that impact the product of the software resulting from the SRS. This subsection should be the source for recognizing the impact of any changes to the assumptions or dependencies on the SRS and resulting software. This section can highlight unresolved requirement issues and should be recorded in the Project Manager's Open Issues List.

Requirements Management. This subsection should provide an overview of the requirements management process and any associated procedures. References to existing organizational requirements management plans may be used.

Resources and Funding. This subsection should identify individuals responsible for requirements management. If this information were provided in an existing requirements management plan, a reference to this plan would be provided here.

Reporting Procedures. This subsection should describe all reporting procedures associated with requirements management activities. Details should be provided for the reporting procedures and reporting schedule. If this information is provided in an organizational software requirements management plan, or if this information is provided in the software project management plan, then a reference to these plans should be provided.

Training. This subsection should either reference an organizational training plan, which describes the method and frequency of requirements elicitation, documentation, and management training, or provide the details for all associated training.

Specific Requirements. This subsection should contain all details that the software developer needs to create a design. The details within this section should be defined as individual, specific requirements with sufficient cross-referencing back to related discussions in the Introduction and General Description sections above. The ordering of subsections that will be shown in this section is one of several possible combinations that may be selected. This ordering should normally be used, but others may be selected provided they are justified. Each specific requirement should be stated such that its achievement can be objectively verified by a prescribed method. The method of verification will be shown in a requirements traceability matrix of the SRS.

Acceptance. This subsection should describe the acceptance process, referencing any associated documentation (e.g., software test plan).

Documentation. This section should list any additional supporting process or project documentation not previously referenced.

Maintenance. This section is required if the system must meet requirements for continued operation and enhancement.

Traceability Matrix. This is a comprehensive baseline of all software requirements described within the SRS. This list should be used as the basis for all requirements management, software design, development, and testing.

Software Design Document

The following information is primarily based on IEEE Std 1016, IEEE Recommended Practice for Software Design Descriptions [5] and IEEE 12207.1, Standard for Information Technology—Software Life Cycle Processes [40]. IEEE Std 1471, IEEE Recommended Practice for Architectural Description of Software Intensive Systems [34]; IEEE

Std 1320.1, IEEE Standard for Functional Modeling Language—Syntax and Semantics for IDEF0 [26]; and IEEE Stds1420.1, 1420.1a, and 1420.1b, Software Reuse—Data Model for Reuse Library Interoperability [29–31] were also used as source documents for the development of this information.

The resulting software design document describes the detailed design for each software component of the software item. The software components must be refined into lower levels containing software units that can be coded, compiled, and tested. It shall be ensured that all the software requirements are allocated from the software components to software units. Table 3-13 provides an example software design document outline.

Table 3-13. Software design document outline

Title Page
Revision Page
Table of Contents
  1. Introduction
      1.1   Purpose
      1.2   Scope
      1.3   Definitions, Acronyms, and Abbreviations
      1.4   References
  2. Design Overview
      2.1   Background Information
      2.2   Alternatives
  3. User Characteristics
  4. Requirements and Constraints
      4.1   Performance Requirements
      4.2   Security Requirements
      4.3   Design Constraints
  5. System Architecture
  6. Detailed Design
      6.1   Description for Component N
      6.2   Processing Narrative for Component N
      6.3   Component N Interface Description
      6.4   Component N Processing Detail
  7. Data Architecture
      7.1   Data Analysis
      7.2   Output Specifications
      7.3   Logical Database Model
      7.4   Data Conversion
  8. Interface Requirements
      8.1   Required Interfaces
      8.2   External System Dependencies
  9. User Interface
      9.1   Module [X] Interface Design
      9.2   Functionality
  10. NonFunctional Requirements
  11. Traceability Matrix

Software Design Document Guidance

The following provides section-by-section guidance in support of the creation of a software design document (SDD). This guidance should be used to help define associated design decisions and should reflect the actual requirements of the implementing organization. Additional information is provided in the document template, *Software Design Document.doc,* which is located on the companion CD-ROM.

Introduction. As stated in IEEE 1016, "The SDD shows how the software system will be structured to satisfy the requirements identified in the software requirements specification. It is a translation of requirements into a description of the software structure, software components, interfaces, and data necessary for implementation. In essence, the SDD becomes a detailed blueprint for the implementation activity."

This section should explain the purpose and scope of project software design, as well as provide clarification of definitions, acronyms, and references. This section should also provide an overview of the project.

Purpose. This subsection should explain the purpose for writing an SDD for this project and describe the intended audience for the SDD.

Scope. This subsection should:

- Identify the software products to be produced, by name.
- Explain what the software products will, and if necessary, will not do.
- Describe the application of the software being specified including all relevant goals, objectives, and benefits of producing the software.

Provide a description of the dominant design methodology. Provide a brief overview of the product architecture. Briefly describe the external systems with which this system must interface. Also explain how this document might evolve throughout the project life cycle.

Definitions, Acronyms, and Abbreviations. This subsection should provide the definitions of all terms, acronyms, and abbreviations required to properly interpret the SDD.

References. This subsection should list all references used within the SDD. All relationships to other plans and policies should be described, as well as existing design standards.

Design Overview. This section contains the following subsections.

Background Information. This subsection should briefly present background information relevant to the development of the system design. All stakeholders should be identified and their contact information provided. A description of associated project risks and issues may also be presented along with assumptions and dependencies critical to project success. Describe the business processes that will be modeled by the system.

Alternatives. All design alternatives considered, and the rationale for nonacceptance, should be briefly addressed in this subsection. See Appendix A for Make/Buy/Mine/ Commission Decision Matrix and Alternative Solution Screening Criteria Matrix.

User Characteristics. Identify the potential system users. Specify the levels of expertise needed by the various users and indicate how each user will interact with the system. Describe how the system design will meet specific user requirements.

Requirements and Constraints. This section contains the following subsections.

Performance Requirements. This subsection should describe how the proposed design will ensure that all associated performance requirements will be met.

Security Requirements. This subsection should describe how the proposed design will ensure that all associated security requirements will be met. List any access restrictions for the various types of system users. Describe any access code systems used in the software. Identify any safeguards that protect the system and its data. Specify communications security requirements.

Design Constraints. This subsection should describe how requirements that place constraints on the system design would be addressed. List all dependencies and limitations that may affect the software. Examples include budget and schedule constraints, staffing issues, availability of components, and so on.

System Architecture. This section should provide a description of the architectural design. See Appendix A for "Architecture Design Success Factors and Pitfalls." All entities should be described, as well as their interdependent relationships. A top-level diagram may be provided. See Appendix A for a "Unified Modeling Language (UML)" example of showing inheritance, aggregation, and reference relationships.

Detailed Design. Decompose the system into design components that will interact with and transform data to perform the system objectives. Assign a unique name to each component, and group these components by type, for example, class, object, procedure. Describe how each component satisfies system requirements. In user terminology, specify the inputs, outputs, and transformation rules for each component. Depict how the components depend on one another. Each component should be described as shown by the following list of suggested subsections:

- Description for Component N
- Processing Narrative for Component N
- Component N Interface Description
- Component N Processing Detail

Data Architecture. This section should describe the data structures to be used in support of the implementation. If these include databases, define the table structure, including full field descriptions, relationships, and critical database objects. Graphical languages are appropriate. This information is often provided in a separate Database Design Document; if this is the case, simply refer to this document and omit the remainder of this section.

Data Analysis. A brief description of the procedures used in support of data analysis activities should be described in this subsection. Any analysis of the data that resulted in a change to the system design, or that impacted system design, should be noted.

Output Specifications. All designs supporting requirements for system outputs should be described in this subsection. These may include designs to support reporting, printing, e-mail, and so on.

Logical Database Model. Identify specific data elements and logical data groupings that are stored and processed by the design components in the Detailed Design section. Outline data dependencies, relationships, and integrity rules in a data dictionary. Specify the format and attributes of all data elements or data groupings. A logical model of data flow, depicting how design elements transform input data into outputs should be developed and presented here.

Data Conversion. This subsection should describe all design requirements in support of data conversion activities. This may be covered in a separate data migration plan, but if it is not, it should be documented in the design documentation. The migration and validation of any converted legacy data should be described.

Interface Requirements. This section contains the following subsections.

Required Interfaces. This subsection should describe all interfaces required in support of hardware and software communications. If an Interface Control Document was developed along with the Software Requirements Specification it should be referenced here. The effectiveness of how the system design addresses relevant interface issues should be discussed. Specify how the product will interface with other systems. For each interface, describe the inputs and outputs for the interacting systems. Explain how data is formatted for transmission and validated upon arrival. Note the frequency of data exchange.

External System Dependencies. This subsection should provide a description of all external system dependencies. A diagram may be used to provide a description of the system with each processor and device indicated.

User Interface. Describe the user interface and the operating environment, including the menu hierarchy, data entry screens, display screens, online help, and system messages. Specify where in this environment the necessary inputs are made, and list the methods of data outputs, for example, printer, screen, or file. This section contains the following subsections.

Module [X] Interface Design. This subsection should contain screen Images and a description of all associated design rules.

Functionality. All objects and actions should be described in this subsection. Any reuse of existing components should be identified.

Nonfunctional Requirements. This section should address design items associated with nonfunctional requirements relating to system performance, security, licensing, language, or other related items.

Traceability Matrix. It is necessary to show traceability throughout the product life cycle. All design items should be traceable to original system requirements. This can be annotated in the Requirements Traceability Matrix, the Design Review document, or in the Configuration Management System.

Interface Control Document

The following information is based on IEEE Std 830, IEEE Recommended Practice for Software Requirements Specifications [6] and The Dept. of Justice Systems Development Life Cycle Guidance Document, Interface Control Document (ICD) Template.* Additional information is provided in the document template, *Interface Control Document.doc,* which is located on the companion CD-ROM. Table 3-14 provides an example document outline. Additional information in support of associated development work products is provided in Appendix A.

Table 3-14. Interface control document outline

Title Page
Revision Page
Table of Contents
1. Introduction
    1.1   System Identification
    1.2   Document Overview
    1.3   References
    1.4   Definitions and Acronyms
2. Description
    2.1   System Overview
    2.2   Interface Overview
    2.3   Functional Allocation
    2.4   Data Transfer
    2.5   Transactions
    2.6   Security and Safety
3. Detailed Interface Requirements
    3.1   Interface 1 Requirements
    3.2   Interface Processing Time Requirements
    3.3   Message (or File) Requirements
    3.4   Communication Methods
    3.5   Security Requirements
    3.5   Interface 2 Requirements
4. Qualification Methods

Interface Control Document Guidance

The following provides section-by-section guidance in support of the creation of an ICD. This guidance should be used to help define associated interface control requirements and should reflect the actual requirements of the implementing organization.

    The Introduction contains the following subsections.

System Identification. This subsection should contain a full identification of the participat-

*Department of Justice Systems Development Life Cycle Guidance, Interface Control Document Template, Appendix C-17; http://www.usdoj.gov/jmd/irm/lifecycle/table.htm.

ing systems, the developing organizations, responsible points of contact, and the interfaces to which this document applies, including, as applicable, identification numbers(s), title(s), abbreviation(s), version number(s), release number(s), or any version descriptors used. A separate paragraph should be included for each system that comprises the interface.

Document Overview. This subsection should provide an overview of the document, including a description of all sections.

References. This subsection should list the number, title, revision, and date of all documents referenced or used in the preparation of this document.

Definitions and Acronyms. This subsection should describe all terms and abbreviations used in support of the development of this document that are critical to the comprehension of its content.

Description. A description of the interfaces between the associated systems should be described in the following subsections.

System Overview. This subsection should describe each interface and the data exchanged between the interfaces. Each system should be briefly summarized, with special emphasis on functionality relating to the interface. The hardware and software components of each system should be identified.

Interface Overview. This subsection should describe the functionality and architecture of the interfacing system(s) as they relate to the proposed interface. Briefly summarize each system, placing special emphasis on functionality. Identify all key hardware and software components as they relate to the interface.

Functional Allocation. This subsection should describe the operations are performed on each system involved in the interface. It should also describe how the end user would interact with the interface being defined. If the end user does not interact directly with the interface being defined, a description of the events that trigger the movement of information should be defined.

Data Transfer. This subsection should describe how data would be moved among all component systems of the interface. Diagrams illustrating the connectivity among the systems are often helpful communication tools in support of this type of information.

Transactions. This subsection should describe the types of transactions that move data among the component systems of the interface being defined. If multiple types of transactions are utilized for different portions of the interface, a separate section may be included for each interface.

Security and Safety. If the interface defined has security and safety requirements, briefly describe how access security will be implemented and how data transmission security and safety requirements will be implemented for the interface being defined.

Detailed Interface Requirements. This section should provide a detailed description of all requirements in support of all interfaces between associated systems. This should include definitions of the content and format of every message or file that may pass between the two systems and the conditions under which each message or file is to be sent. The information presented in the subsection Interface 1 Requirements should be replicated as needed to support the description of all interface requirements.

Interface 1 Requirements. Briefly describe the interface, indicating data protocol, communication method(s), and processing priority.

Interface Processing Time Requirements. If the interface requires that data be formatted and communicated as the data is created, as a batch of data is created by operator action, or in accordance with some periodic schedule, indicate processing priority. Priority should be stated as measurable performance requirements defining how quickly data requests must be processed by the interfacing system(s).

Message (or File) Requirements. This subsection should describe the transmission requirements. The definition, characteristics, and attributes of the requirements should be described.

Data Assembly Characteristics. This subsection should define all associated data elements that the interfacing entities must provide as well as required access requirements.

Field/Element Definition. All characteristics of individual data elements should be described.

Communication Methods. This subsection should address all communication requirements, including a description of all connectivity and availability requirements. All aspects of the flow of communication should be described.

Security Requirements. This subsection should address all security features that are required in support of the interface process.

Note: When more than one interface between two systems is being defined in a single ICD, each should be defined separately, including all of the characteristics described in Interface 1 Requirements for each. There is no limit on the number of unique interfaces that can be defined in a single Interface Control Document. In general, all interfaces defined should involve the same two systems.

Qualification Methods. This section should describe all qualification methods to be used to verify that the requirements for the interfaces have been met. Qualification methods may include:

- *Demonstration*–The operation of interfacing entities that relies on observable functional operation not requiring the use of instrumentation, special test equipment, or subsequent analysis.
- *Test*–The operation of interfacing entities using instrumentation or special test equipment to collect data for later analysis.
- *Analysis*–The processing of accumulated data obtained from other qualification methods. Examples are reduction, interpretation, or extrapolation of test results.

- *Inspection*–The visual examination of interfacing entities, documentation, and so on.

- *Special qualification methods*–Any special qualification methods for the interfacing entities, such as special tools, techniques, procedures, facilities, and acceptance limits.

If a separate test plan exists, then a reference to this document should be provided.

OPERATION

The operations process contains the activities and tasks of the operator. This process covers the operation of the software product and operational support to users. The operator manages the operation process at the project level, following the management process, which is instantiated in this process; establishes an infrastructure under the process following the infrastructure process, tailors the process for the project following the tailoring process, and manages the process at the organizational level following the improvement process and the training process. Table 3-15 describes the operation process objectives.

Table 3-15. Operation process objectives

a) Identify and mitigate operational risks for the software introduction and operation.
b) Operate the software in its intended environment according to documented procedures.
c) Provide operational support by resolving operational problems and handling user inquires and requests.
d) Provide assurance that software (and host system) capacities are adequate to meet user needs.
e) Identify customer support service needs on an ongoing basis.
f) Assess customer satisfaction with both the support services being provided and the product itself on an ongoing basis.
g) Deliver needed customer services.

This document guidance is provided because the documentation associated with product delivery and the user's manual is critical in support of the successful transition, or delivery, of any software product. If such user's information is deemed a requirement, the development of user's documentation supports the requirement of customer communication and also supports requirements associated with customer-related processes.

User's Manual

The following provides section-by-section guidance in support of the creation of a software user's manual or operator's manual. IEEE Std 1063, IEEE Standards for Software User Documentation [18] was the primary reference document for the development of this material. Table 3-16 provides an example user's manual document outline.

Table 3-16. User's manual document outline

Title Page
Revision Page
Table of Contents
1. Introduction
    1.1   Document Use
    1.2   Definitions and Acronyms
    1.3   References
2. Concept of Operations
3. General Use
4. Procedures and Tutorials
5. Software Commands
6. Navigational Features
7. Error Messages and Problem Resolution
Index

User's Manual Document Guidance

This guidance should be used to help define a management process and should reflect the actual processes and procedures of the implementing organization. Additional information is provided in the document template, *Software Users Manual.doc,* which is located on the companion CD-ROM.

Introduction. This section should provide information on document use, all definitions and acronyms, and references.

Document Use. This subsection should describe the intended use of the software user's manual. The organization of the user's document should effectively support its use. If the user's manual is going to contain both instructional and reference material, each type should be clearly separated into different chapters or topics. Task-oriented documentation (instructional) should include procedures that are structured according to user's tasks. Documentation used as reference material should be arranged to provide access to individual units of information. This section can provide an overview of the type of information provided, its intended use, and the organization of the user's manual.

Definitions and Acronyms. This section should identify all definitions and acronyms specific to this software user's manual. This should be an alphabetical list of application-specific terminology. All terminology used with the user's manual should be consistently applied.

References. This section should provide a list of all references used in support of the development of the software user's manual. Include a listing of all the documentation related to the product that is to be transitioned to the operations area, which includes any security or privacy protection consideration associated with its use. Also include as a part of this any licensing information for the product.

Concept of Operations. This section should provide an overview of the software, including its intended use. Descriptions of any relevant business processes or workflow activities should be included. Any items required in support of the understanding of the software product should be included. This may require a description of theory, method, or algorithm critical to the effective use and understanding of the product.

General Use. Information should be provided in support of routine user activities. It is important to identify actions that will be performed repetitively to avoid redundancy within the user's manual. For example, describing how to cancel, or interrupt, an operation while using the software would be in this section. Other task-oriented routine documentation could include software installation and deinstallation procedures, how to log on and off the application, and the identification of basic items/actions that are common across the applications' user interface.

Procedures and Tutorials. Information of a tutorial (i.e., procedural) nature should be provided in the user's documentation as clearly as possible. A consistent approach to the presentation of the material is important when trying to clearly communicate a concept to the user.

Describe the purpose and concept for the tutorial information presented in the user's manual. Include a list of all activities that must be completed prior to the initiation of the procedure or tutorial. Identify any material that should be used as reference in support of the task. List all cautions and other supporting information that are relevant in supporting the performance of the task.

It is important to list all instructional steps in the order that they should be performed, with any optional steps clearly identified. The steps should be consecutively numbered and the initial and last steps of the task should be clearly identified. It is important that the user understand how to successfully initiate and complete the procedure or tutorial.

Warnings and cautions should be distinguishable from instructional steps, and should be preceded by a word and graphic symbol alerting the user to the item. For example, warning (graphic) would precede a warning to the user. The use of the following format for warnings and cautions is suggested: word and graphic, brief description, instructional text, description of consequences, and proposed solution or workaround.

Software Commands. The user's manual should describe all software commands, including required and optional parameters, defaults, precedence, and syntax. All reserved words and commands should be listed. This section should not only provide the commands, but should also provide examples of their use. Documentation should include a visual representation of the element, a description of its purpose, and an explanation of intended action. A quick reference card may be included in the users documentation, providing the user with the ability to rapidly refer to commonly used commands.

Navigational Features. The document should describe all methods of navigation related to the software application. All function keys, graphical user interface items, and commands used in support of application navigation should be described and supported with examples.

Error Messages and Problem Resolution. Information in support of problem resolution (i.e., references) should address all known problems or error codes present in the software application. Users should be provided information that will either help them recover from known problems, report unknown issues, or suggest application enhancements.

Index. An index provides an effective way for users to access documented information. It is important to remember that for an index to be useful, it should contain words that users are most likely to look up and should list all topics in the user's documentation. Pay special attention to the granularity and presentation of the index topics. Place minor keywords under major ones, for example instead of using *files* with 30 pages listed, use *files, saving,* and *files, deleting,* with their associated specific pages listed.

MAINTENANCE

The maintenance process contains the activities and tasks of the maintainer. The objective is to modify an existing software product while preserving its integrity. The maintainer manages the maintenance process at the project level following the management process, which is instantiated in this process; establishes an infrastructure under the process following the Infrastructure process; tailors the process for the project following the tailoring process; and manages the process at the organizational level following the improvement process and the training process. Table 3-17 describes the maintenance process objectives.

Table 3-17. Operation process objectives

a) Define the impact of organization, operations, and interfaces on the existing system in operation.
b) Identify and update appropriate life cycle data.
c) Develop modified system components with associated documentation and tests that demonstrate that the system requirements are not compromised.
d) Migrate system and software upgrades to the user's environment.
e) Ensure that fielding of new systems or versions does not adversely affect ongoing operations.
f) Maintain the capability to resume processing with prior versions.

This process consists of the following activities:

1. Process implementation
2. Problem and modification analysis
3. Modification implementation
4. Maintenance review/acceptance
5. Migration
6. Software retirement [48]

Transition Plan

The purpose of transition planning is to lay out the tasks and activities that need to take place to efficiently move a product (i.e., specify name brand or in-house developed software or COTS software, middleware or component software/hardware) from the development or pilot environment to the production, operations, and maintenance environment. The transition planning steps apply whether the product is being transitioned within an agency (i.e. from agency development staff to agency network or operations staff) or to an outside agency (i.e., information technology systems).

The transition plan is designed to facilitate migration of an application system from development to production (i.e., maintenance). The information provided here in support of transition plan development is based upon IEEE Std-1219, IEEE Standard for Software Maintenance [22]; IEEE Std 12207.0, Standard for Information Technology—Life Cycle Processes [40]; and US DOD Data Item Description DI-IPSC-81429, Software Transition Plan [139] as primary reference material. Additional information has been incorporated as "lessons learned" from multiple production application systems and transition opportunities. Table 3-18 provides a proposed document outline in support of Lean Six Sigma requirements.

Table 3-18. Transition plan document outline

Title Page
Revision Page
Table of Contents
1. Introduction
    1.1 Overview
    1.2 Scope
    1.3 Definitions and Acronyms
        1.3.1 Key acronyms
        1.3.2 Key terms
    1.4 References
2. Product
    2.1 Relationships
3. Strategies
    3.1 Identify Strategy
    3.2 Select Strategy
4. Transition Schedules, Tasks, and Activities
    4.1 Installation
    4.2 Operations and Support
    4.3 Conversion
    4.4 Maintenance
5. Resource Requirements
    5.1 Software Resources
    5.2 Hardware Resources
    5.3 Facilities
    5.4 Personnel
    5.5 Other Resources
6. Acceptance Criteria
7. Management Controls
8. Reporting Procedures
9. Risks and Contingencies
10. Transition Team Information
11. Transition Impact Statement
12. Plan Review Process
13. Configuration Control

Transition Plan Document Guidance

The following provides section-by-section guidance in support of the creation of a software transition plan. The development of a software transition plan is critical to the successful transition of software from development to deployment. This guidance should be used to help develop and define the transition process and should reflect the actual processes and procedures of the implementing organization. Additional information is provided in the document template, *Software Transition Plan.doc,* which is located on the companion CD-ROM.

The Introduction contains the following subsections.

Overview. The transition plan should include an introduction addressing background information on the project. Show the relationship of the project to other projects and/or organizations or agencies, address maintenance resources required, identify the transition team's organization and responsibilities, as well as the tools, techniques, and methodologies that are needed to perform an efficient and effective transition.

The transition plan should include deployment schedules, resource estimates, identification of special resources, and staffing. The transition plan should also define management controls and reporting procedures, as well as the risks and contingencies. Special attention must be given to minimizing operational risks. An impact statement should be produced outlining the potential impact of the transition on the existing infrastructure, operations, support staff, and user community.

Scope. This subsection should include a statement of the scope of the transition plan. Include a full identification of the product to which this document applies, including (as applicable) identification numbers, titles, abbreviations, version numbers, and release numbers. Include product overview, overview of the supporting documentation, and description of the relationship of the product to other related projects and agencies.

Definitions and Acronyms. This subsection should identify all definitions and acronyms specific to this software configuration management plan.

References. This subsection should provide a list of all references used in support of the development of the software configuration management plan. Include a listing of all the documentation related to the product that is to be transitioned to the operations area, including any security or privacy protection consideration associated with its use. Include as a part of this any licensing information for the product.

Product. This section should include a brief statement of the purpose of the product to which this document applies. It should also describe the general nature of the product; summarize the history of development, operation, and maintenance; identify the project sponsor, acquirer, user, developer, vendor, and maintenance organizations; identify current and planned operating sites; and list other relevant documents.

Relationships. This subsection should describe the relationship(s) of the product being transitioned to any other projects and agencies. The inclusion of a diagram or flowchart to help indicate these relationships is often helpful.

Strategies. This section includes the following subsections.

Identify Strategies. This subsection should identify the transition strategies and tools to be used as part of the transition plan. Identify all the options for moving the product from its present state into production/operations. These options could include:

- Incremental implementation or phased approach
- Parallel execution
- One-time conversion and switchover
- Any combinations of the above

Each option should also identify the advantages and disadvantages, risks, estimated time frames, and estimated resources.

Select Strategy. This subsection should include an evaluation of each of the transition options, comparing them to the transition requirements, and selecting the one that is most appropriate for the project. Once a transition strategy has been selected, then the justification is documented and approved.

Transition Schedules, Tasks and Activities. This section should include, or provide reference to, detailed schedules for the selected transition strategy. These schedules should include equipment installation, training, conversion, deployment, and or retirement of the existing system (if applicable), as well as any transition activities required to turn over the product from developers or vendors to operational staff. The schedules should reflect all milestones for conducting transition activities.

Also include an installation schedule for equipment (new or existing), software, databases, and so on. Include provisions for training personnel to use the operational software and target computer(s), as well as any maintenance software and/or host system(s). Describe the developer's plans for transitioning the deliverable product to the maintenance organization. This should address the following:

- Planning/coordination of meetings
- Preparation of items to be delivered to the maintenance organization
- Packaging, shipment, installation, and checkout of the product maintenance environment
- Packaging, shipment, installation, and checkout of the operational software
- Training of maintenance/operational personnel

Installation. Installation consists of the transportation and installation of the product from the development environment to the target environment(s). This subsection should describe any required modifications to the product, checkout in the target environment(s), and customer acceptance procedures. If problems arise, these should be identified and reported; corrective procedures should be addressed here as well. If known, any temporary "work-around(s)" should also be described.

Operations and Support. This subsection should address user operations and all required ongoing support activity. Support includes providing technical assistance, consulting with the user, and recording user support requests by maintaining a Support Request Log. The operations and support activities can trigger maintenance activities via the ongoing project monitoring and controlling activities or problem and change logs, and this process should be described or a reference to associated documentation should be provided.

Conversion. This section should address any data or database transfers to the product and its underlying components that would occur during the transition.

Maintenance. Maintenance activities are concerned with the identification of enhancements and the resolution of product errors, faults, and failures. The requirements for software maintenance initiate "service level changes" or "product modification requests" by using defined problem and change management reporting procedures. This section should describe all issues and activities associated with product maintenance during the transition.

Resource Requirements. All estimates for resources (hardware, software, and facility) as well as any special resources (i.e., service and maintenance contracts) and staffing for the selected transition strategy should be described in this section. The assignment of staff, agency, and vendor responsibility for each task identified should be documented. This allows managers and project team members to plan and coordinate the work of this project with other assignments. If specific individuals cannot be identified when the transition plan is developed, generic names may be used and replaced with individual names as soon as the resources are identified.

Software Resources. A description of any software and associated documentation needed to maintain the deliverable product should be included in this subsection. The description should include specific names, identification numbers, version numbers, release numbers, and configurations as applicable. References to user/operator manuals or instructions for each item should be included. Identify where each product item is to come from—acquirer-furnished, currently owned by the organization, or to be purchased. Include information about vendor support, licensing, and usage and ownership rights, whether the item is currently supported by the vendor, whether it is expected to be supported at the time of delivery, whether licenses will be assigned to the maintenance organization, and the terms of such licenses. Include any required service and maintenance contract costs as well as payment responsibility.

Hardware Resources. This subsection should include a description of all hardware and associated documentation needed to maintain the deliverable product. This hardware may include computers, peripheral equipment, simulators, emulators, diagnostic equipment, and noncomputer equipment. The description shall include specific models, versions, and configurations. References to user/operator manuals or instructions for each item should also be included.

Include information about manufacturer support, licensing, usage and ownership rights, whether the items are currently supported by the manufacturer, or will be in the future, whether licenses will be assigned to the maintenance organization, and the terms of such licenses.

Facilities. Describe any facilities needed to maintain the deliverable product in this subsection. These facilities may include special buildings, rooms, mock-ups, building features such as raised flooring or cabling, building features to support security and privacy protection requirements, building features to support safety requirements, special power requirements, and so on. Include any diagrams that may be applicable.

Personnel. This section should include a description of all personnel needed to maintain the deliverable product, include anticipated number of personnel, types of support personnel (job descriptions), skill levels and expertise requirements, and security clearance.

Other Resources. Identify any other consumables (i.e. technology, supplies, and materials) required to support the product. Provide the names, identification numbers, version numbers, and release numbers. Identify if the document or consumable is acquirer-furnished, an item that will be delivered to the maintenance organization, or an item the organization current owns or needs to acquire. If the maintenance/operational organization needs to acquire it, indicate whether or not the budget will cover the expense.

Acceptance Criteria. It is important to establish the exit or acceptance criteria for transitioning the product. These criteria will determine the acceptability of the deliverable work products and should be specified in this section. Representatives of the transitioning organization and the acquiring organization should sign a formal agreement, such as a service level agreement, that outlines the acceptance criteria. Any technical processes methods, or tools, as well as performance benchmarks required for product acceptance, should be specified in the agreement. Also include an estimation of the operational budget for the product and how these expenses will be covered.

Management Controls. Describe all management controls to ensure that each task is successfully executed and completed based on the approved acceptance criteria. This should include procedures for progress control, quality control, change control, version control, and issue management during the transition process.

Reporting Procedures. This section should define the reporting procedures for the transition period. Include such things as type of evaluations (review, audit, or test) as well as how anomalies that are identified during the performance of these evaluations should be reported.

Risks and Contingencies. This section should identify all known risks and contingencies encountered during the transition process, with special attention given to minimizing operational risks. Reference to an organization risk management plan should be provided here. A description of risk mitigation, tracking, and reporting should be presented.

Transition Team Information. This section should include all transition team information should include the transition team's organization, roles and responsibilities for each activity, as well as the tools, techniques, and methodologies and/or procedures that are needed to perform the transition.

Transition Impact Statement. This section should contain a transition impact statement that describes any anticipated impact on existing network infrastructure, support staff, and user community during the system transition. The impact statement should include descriptions for the performance requirements, availability, security requirements, expected response times, system backups, expected transaction rates, initial storage requirements with expected growth rate, as well as help desk support requirements.

Plan Review Process. This section should describe the review procedures/process in support of this document. A review should be held to identify and remove any defects from the transition plan before it is distributed. This is a content review and should be conducted by the appropriate members of the project team or an independent third party. The results should be recorded in an Inspection Report and Inspection Log Defect Summary, shown in Appendix A.

Configuration Control. The transition plan information should be subject to the configuration control process for the project. Subsequent changes are tracked to ensure that the configuration of the transition plan information is known at all times. Changes should be allowed only with the approval of the responsible authority. Transition plan changes should follow the same criteria established for the project's change control procedures and reference to any associated project-level configuration management plan should be provided here.

4

12207 SUPPORTING LIFE CYCLE PROCESSES

IEEE 12207 SUPPORTING PROCESSES

IEEE 12207 Supporting Life Cycle Processes provide a specific focused set of activities to assist the Primary Life Cycle Processes as an integrated part with a distinct purpose, contributing to the success and quality of the project. These processes may also provide services to other processes, for example, the Agreement Processes, System Qualification Testing, Software Acceptance Support, Operation, and the Maintenance Process. The eight supporting processes that support the project from the organizational level are:

1. Documentation
2. Configuration management
3. Quality assurance
4. Verification
5. Validation
6. Joint review
7. Audit
8. Problem resolution

The organization employing and performing a supporting process

- Manages it at the project level following the management process
- Establishes an infrastructure under it following the infrastructure process
- Tailors it for the project following the tailoring process

- Manages it at the organizational level following the improvement process and the training process
- Assures its quality through joint reviews, audits, verification, and validation

DOCUMENTATION

The documentation process is a process for recording information produced by a life cycle process or activity. This process contains the set of activities to plan, design, develop, produce, edit, distribute, and maintain those documents needed by all concerned, such as managers, engineers, and users of the software product. Table 4-1 provides a list of the document process objectives.

Table 4-1. Document process objectives

a) Identify all documents to be produced by the process or project
b) Specify the content and purpose of all documents and plan and schedule their production
c) Identify the standards to be applied for development of documents
d) Develop and publish all documents in accordance with identified standards and in accordance with nominated plans
e) Maintain all documents in accordance with specified criteria

Configuration Management Record

Many existing software configuration management systems can be expanded to plan and track all organizational and project documents and related records. An example of a configuration management record is shown in Table 4-2.

Table 4-2. Configuration management record

Configuration Item (CI) identifier
CI Type (e.g., Policy, Procedures, Plan, Project Artifact, Report, Record)
CI Entity Control Level (e.g., Organization, Project, System, Subsystem, Customer)
CI Full title
CI Version
CI Status (e.g., V&V, CR#)
CI Owner name
CI Owner Contact Information
CI Date Last Updated
CI References
CI Location
CI List of Keywords
CI Links to Other Information Sources
Change Request (CR) Identifier
CR Requester Name
CR Requester Contact Information
CR Date

(continued)

Table 4-2.   *(continued)*

CR Title
CR Description
CR Justification
CR Suggested Effective Date
CR Acceptance Disposition
CR Implementer Information
CR Targeted Effective Date
CR Brief Discussion
CR Date of Last Data Entry
CR Detailed Summary of the Results, Including Objectives and
   Procedures
CR Validation Tracking (e.g., Test Phases)
CR Validation Contact Information

CONFIGURATION MANAGEMENT

The configuration management process works throughout the software life cycle to

- Identify, define, and baseline software configuration items (CI) in a system
- Control modifications and releases of the CI (e.g., updating of multiple products in one or more locations)
- Record and report the status of the CI and modification requests (e.g., including all actions and changes resulting from a change request or problem)
- Ensure the completeness, consistency, and correctness of the CI (e.g., verification and validation activities)
- Control storage, handling, and delivery of the CI (e.g., release management and delivery)

The CI is defined within a configuration that satisfies an end use function and can be uniquely identified at a given reference point. Table 4-3 provides a list of the configuration management process objectives.

Table 4-3. Configuration management process objectives

a) Identify, define, and control all relevant items of the project
b) Control modifications of the items
c) Record and report the status of items and modification requests
d) Ensure the completeness of the items
e) Control storage, handling, release, and delivery of the items

Software Configuration Management Plan

IEEE Std 828, IEEE Standard for Software Configuration Management Plans (SCMP), can be used to help support this requirement. Appendix A also provides examples of supporting CM work products, which include a Configuration Control Board (CCB) Letter of Authorization, CCB Charter, and Software Change Request Procedures. Table 4-4 pro-

Table 4-4. Software configuration management plan document outline

Title Page
Revision Page
Table of Contents
1.0 Introduction
    1.1 Purpose
    1.2 Scope
    1.3 Definitions/Acronyms
    1.4 References
2.0 Software Configuration Management
    2.1 SCM Organization
    2.2 SCM Responsibilities
    2.3 Relationship of CM to the Software Process Life Cycle
        2.3.1 Interfaces to Other Organizations on the Project
        2.3.2 Other Project Organizations CM Responsibilities
    2.4 SCM Resources
3.0 Software Configuration Management Activities
    3.1 Configuration Identification
        3.1.1 Specification Identification
        3.1.2 Change Control Form Identification
        3.1.3 Project Baselines
        3.1.4 Library
    3.2 Configuration Control
        3.2.1 Procedures for Changing Baselines
        3.2.2 Procedures for Processing Change Requests and Approvals
        3.2.3 Organizations Assigned Responsibilities for Change Control
        3.2.4 Change Control Boards (CCBs)
        3.2.5 Interfaces
        3.2.6 Level of Control
        3.2.7 Document Revisions
        3.2.8 Automated Tools Used to Perform Change Control
    3.3 Configuration Status Accounting
        3.3.1 Storage, Handling and Release of Project Media
        3.3.2 Information and Control
        3.3.3 Reporting
        3.3.4 Release Process
        3.3.5 Document Status Accounting
        3.3.6 Change Management Status Accounting
    3.4 Configuration Audits and Reviews
4.0 Configuration Management Milestones
5.0 Training
6.0 Subcontractor Vendor Support
7.0 Appendices

vides a proposed software configuration management plan document outline in support of Lean Six Sigma requirements.

Software Configuration Management Plan Document Guidance

The following provides section-by-section guidance in support of the creation of a Software Configuration Management Plan (SCMP). The SCMP should be considered to be a living document and should change to reflect any process improvement activity. This guidance should be used to help define a software configuration management process and should reflect the actual processes and procedures of the implementing organization. Additional information is provided in the document template, *Software Configuration Management Plan.doc,* which is located on the companion CD-ROM.

Introduction. The introduction information provides a simplified overview of the software configuration management (SCM) activities so that those approving, those performing, and those interacting with SCM can obtain a clear understanding of the plan. The introduction should include four topics: the purpose of the plan, the scope, the definition of key terms, and references.

Purpose. The purpose should address why the plan exists and the intended audience.

Scope. The scope should address SCM applicability, limitations, and assumptions on which the plan is based. The scope should provide an overview description of the software development project, identification of the software CI(s) to which SCM will be applied, identification of other software to be included as part of the plan (e.g., support or test software), and relationship of SCM to the hardware or system configuration management activities for the project. The scope should address the degree of formality, depth of control, and portion of the software life cycle for applying SCM on this project, including any limitations, such as time constraints, that apply to the plan. Any assumptions that might have an impact on the cost, schedule, or ability to perform defined SCM activities (e.g., assumptions of the degree of customer participation in SCM activities or the availability of automated aids) must also be addressed. The following is an example of scope:

> This document defines CM activities for all software and data produced during the development of the [Project Name] ([Project Abbreviation]) software. This document applies to all module products, end-user products, and data developed and maintained for the [Project Abbreviation] Program. CM activities as defined herein will be applied to all future [Project Abbreviation] projects.
>
> This document conforms to [Company Name]'s Software Configuration Management Policy [Policy #] and IEEE standards for software configuration management, and will change as needed to maintain conformance.

Software Configuration Management (SCM). Appropriately documented SCM information describes the allocation of responsibilities and authorities for SCM activities to organizations and individuals within the project structure. SCM management information should include three topics: the project organization(s) within which SCM is to apply, the SCM responsibilities of these organizations, and references to the SCM policies and directives that apply to this project.

SCM Organization. The SCM organizational context must be described. The plan should identify all participants, or those responsible for any SCM activity on the project. All functional roles should be described, as well as any relationships to external organizations. Organization charts, supplemented by statements of function and relationships, can be an effective way of presenting this information. Figure 4-1 provides an overview of an example SCM organization.

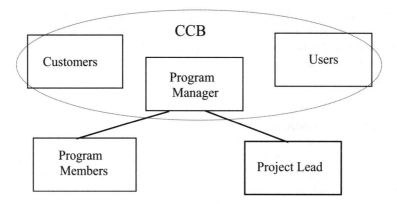

Figure 4-1. Example of SCM organization.

SCM Responsibilities. All those responsible for SCM implementation and performance should be described in this subsection. A matrix describing SCM functions, activities, and tasks can be useful for documenting the SCM responsibilities. For any review board or special organization established for performing SCM activities on this project, the plan should describe its

1. Purpose and objectives
2. Membership and affiliations
3. Period of effectivity
4. Scope of authority
5. Operational procedures

The following provides an example of typical responsibilities:

The Program Manager (PM) is responsible for ensuring that the SCM process is developed, maintained, and implemented. The PM is responsible for ensuring that project leads and team members are adequately trained in SCM policy and procedure. The [Project Abbreviation] Program Manager and Project Leads work together to develop, maintain, and implement effective software configuration management. Please refer to Figure [number].

The [Project Abbreviation] Program Manager (PM) has the authority to ensure development of an appropriate configuration management process. Since CCB responsibility exists with the customer (See Figure [number]), team member(s) of the [Project Abbreviation] Project will be identified as members of this CCB and will represent our interests there.

The Configuration Control Board (CCB) is responsible for the review and approval of all Software Change Requests (SCRs), all baseline items, and all changes to baseline items. The CCB considers the cost and impact of all proposed changes, and considers the impact to all interfacing items as part of their approval action. The CCB for [Project Abbreviation] consists of the PM, the PL, and two or more members from the government Program Management team.

The associated Configuration Control Board (CCB) has the authority to approve a baseline version of this process, as well as any changes to the process as recommended by any associated CCB representative. The same CCB has the authority to define the baseline for each software module, and end-user products, and to approve start of work and acceptance of completed changes to the baseline products. This CCB also has the authority to approve reversion to a previous version of the product if warranted.

The [Project Abbreviation] Program SCM is responsible for implementing the actions of this process and for the implementation of [Company Name] process improvement recommendations.

The Software Configuration Manager (SCM) ensures that the process is implemented once approved and that proposed changes to the process are reviewed by [Project Abbreviation] team members and approved by the CCB as required.

Project Leads, or a designee, are responsible for adhering to the [Project Abbreviation] CM process and for ensuring that only approved changes are entered into the module product baselines. Project Leads, or a designee, are also responsible for identifying the module version to be included in the end-user products. The Project Lead, or a designee, is responsible for controlling the software product(s) development and/or maintenance baseline to ensure that only approved changes are entered into the baseline(s) system. The Project Lead maintains the software product, with all changes to be approved by the associated CCB.

The Project Lead (PL) develops and proposes a baseline version of the software configuration management product, ensuring that the SCM process is followed, once approved, and proposes process changes to the [Project Abbreviation] SCM and CCB, as required.

Each Project Lead (PL) is responsible for ensuring that project configuration management processes are followed. The PL also ensures all members of the project team are trained in SCM policies and procedures that apply to their function on the team. This consists of initial training for new members and refresher training, as required, due to process or personnel changes.

Relationship of CM to the Software Process Life Cycle. This subsection should describe CM as it relates to other elements of the software process life cycle. A diagram showing the relationship between CM elements and the overall process is often used here.

SCM Resources. This subsection should describe the minimum recommended resources (eg., time, tools, and materials) dedicated to software configuration management activities.

Software Configuration Management Activities. This section contains the following subsections.

Configuration Identification. All technical and managerial SCM activities should be described, as well as all functions and tasks required to manage the configuration of the software. General project activities that have SCM implications should also be described from the SCM perspective. SCM activities are traditionally grouped into four functions: configuration identification, configuration control, status accounting, and configuration audits and reviews. An example of the type of information provided in this subsection follows.

All [Project Abbreviation] modules and baselines are to be approved by the CCB. Once a module has been baselined, all changes made will be identified in their respective Version Description Documents (VDDs).

The [Project Abbreviation] is currently in development; product baseline to be approved upon delivery.

Product baselines are to be maintained by the [Project Abbreviation] SCM. The baseline will consist of the following: source files, libraries, executable files, data files, make files, link files, VDDs, and user's manuals.

Specification Identification. SCM activities should identify, name, and describe all code, specifications, design, and data elements to be controlled for the project. Controlled items may include executable code, source code, user documentation, program listings, databases, test cases, test plans, specifications, management plans, and elements of the support environment (such as compilers, operating systems, programming tools, and test beds).

Change Control Form Identification. The plan should specify the procedures for requesting a change. As a minimum, the information recorded for a proposed change should contain the following:

1. The name(s) and version(s) of the CIs in which the problem appears
2. Originator's name and organization
3. Date of request
4. Indication of urgency
5. The need for the change
6. Description of the requested change

Additional information, such as priority or classification, may be included to assist in analysis and evaluation. Other information, such as change request number, status, and disposition, may be recorded for change tracking. An example of the type of information provided in this section follows:

Software Change Request (SCR) procedures for each development team are defined in Appendix A. These procedures are used to add to, change, or remove items from the baselines. The identification and tracking of change requests is accomplished through Change Enhancement Requests (CERs).

Project Baselines. This subsection should describe the configuration control activities required implement changes to baselined CIs. The plan should define the following sequence of specifc steps:

1. Identification and documentation of the need for a change
2. Analysis and evaluation of a change request
3. Approval or disapproval of a request
4. Verification, implementation, and release of a change

The plan should identify the records to be used for tracking and documenting this sequence of steps for each change.

Library. This subsection should identify the software configuration management repository, if one exists. The following provides an example of the type of typical information:

[Project Abbreviation] is composed of end-user products, module products, and data. The organization of these computer software units is given in Appendix [letter].

Configuration Control. This subsection contains the following subsections.

Procedures for Changing Baseline. This subsection should describe the procedures for making controlled changes to the development baseline. The following provides an example of this type of information.

Changes to each software product baseline are made according to the change process.

Change Tracking. Changes to each software product baseline are tracked using an automated CER and SCR tracking system. These systems report current status for all CERs and SCRs for a specified version and product.

Change Release. The PL provides reports containing the following information for review at each CCB meeting for:

a)   Number of new CERs
b)   Type
c)   Priority
d)   Source (Internal, Field, FOT&E)
e)   Action required
f) ˙ Number of CERs in each status

Each development team Project Lead, or designated Change Request Coordinator (CRC), is responsible for keeping the information in the CER tracking system up to date so the information is available to the PL and PM prior to CCB meetings.

Procedures for Processing Change Requests and Approvals. This subsection should describe the procedures for processing changes requests and the associated approval process. An example of a change request process is provided in Appendix A, Software Change Request Procedures, of this book.

Organizations Assigned Responsibilities for Change Control. This subsection should describe all organizations associated with the change control process and their associated responsibilities.

Change Control Boards (CCBs). This subsection should describe the role of the CCB and refer to any associated documentation (e.g., CCB Charter).

Interfaces. This subsection should describe any required software interfaces and point to the source of their SCM documentation.

Level of Control. This subsection should describe the levels of control and implementing authorities. The following is an example.

Evaluating Changes. Changes to existing software baselines are evaluated by the software product's associated CCB.

Approving or Disapproving Changes. Changes to existing software baselines are either approved or disapproved by the associated project CCB.

Implementing Changes. Approved changes are incorporated into the product baseline by the development team as described by the change control process documented in Appendix [letter].

Document Revisions. This subsection should describe SCM plan maintenance information. All activities and responsibilities necessary to ensure continued SCM planning during the life cycle of the project should be described. The plan should describe:

1. Who is responsible for monitoring the plan
2. How frequently updates are to be performed
3. How changes to the Plan are to be evaluated and approved
4. How changes to the Plan are to be made and communicated

Automated Tools Used To Perform Change Control. This subsection should identify all software tools, techniques, equipment, personnel, and training necessary for the implementation of SCM activities. SCM can be performed by a combination of software tools and manual procedures. For each software tool, whether developed within the project or brought in from outside the project, the plan should describe or reference its functions and identify all configuration controls to be placed on the tool.

Configuration Status Accounting. Configuration status accounting activities record and report the status of a project's CIs. The SCM plan should describe what will be tracked and reported, describe the types of reporting and their frequency, and how the information will be processed and controlled.

If an automated system is used for any status accounting activity, its function should be described or referenced. The following minimum data elements should be tracked and reported for each CI: its initial approved version, the status of requested changes, and the implementation status of approved changes. The level of detail and specific data required may vary according to the information needs of the project and the customer.

Storage, Handling, and Release of Project Media. This subsection should describe the storage, handling, and release of SCM products. The following is provided as an example:

> The initial approved version of a [Project Abbreviation] software product is separately maintained by the PL until the approval of subsequent version of the product. Upon approval of the subsequent version, the initial version is held in the project SCM library. This library is under configuration control.

Information and Control. Implementation of internal and customer controls, e.g., the point at which changes to baselines will be controlled by change control forms.

Reporting. This subsection should describe the format, frequency, and process for reporting SCM product baseline status. Providing configuration items to testing may require a test item transmittal report. This test item report should identify the person responsible for each item, the physical location of the item, and item status. The following is an example of status report of requested changes:

> The status report of requested changes is available to the PM at any time. A schedule status report may be provided to the CCB weekly and reported to [Company Name] senior management on a monthly basis. Please refer to the [Project Abbreviation] Software Development Plan, Section 6, for a status report schedule example.

Release Process. This subsection should describe the release process, including all required approvals. A summary chart is sometimes used to describe the release process effectively.

Document Status Accounting. This subsection should describe how changes to the SCM plan are managed and reported.

Change Management Status Accounting. This subsection should describe the format, frequency, and process for reporting SCM change management status.

Configuration Audits and Reviews. This section should provide information in support of required configuration audits and reviews. The following is provided as an example:

> Each team PL reviews the software baseline semiannually. The [Project Abbreviation] PM and [Project Abbreviation] SCM review the baseline of each software product annually. This baseline review consists of checking for CER implementation. The CERs are randomly selected. Discrepancies will be annotated and reported to the PL. Corrective action is taken if required.

The [Project Abbreviation] CCB reviews the SCM process annually. Randomly selected problem reports are identified and walked through the change process. Discrepancies are annotated and reported to the [Project Abbreviation] PM and [Project Abbreviation] SCM. Corrective action is taken if required.

A software baseline, the associated status reporting, and associated documentation for each software product are available for SQA review at all times.

Configuration Management Milestones. This section should provide a description of the minimum set of SCM-related project milestones that are acceptable for compliance. These milestones should be reflected in the software project management plan schedule and resource allocation.

Training. This section can refer to an independent training plan. However, this training plan must include information relating specifically to SCM training. If an independent plan does not exist, then information regarding the type and frequency of SCM training should be identified here.

Subcontractor/Vendor Support. This section should describe the SCM activities required by subcontractors. The following information is provided as a supporting example:

[Project Abbreviation] subcontractor support will be identified by the [Project Abbreviation] Program Manager. Subcontractor support will be required to follow the process defined in this document. If a wavier is requested, the subcontractor must provide evidence of comparable configuration management procedures. These procedures will follow the same audit and control procedures described in Section xxx.

Appendix A. Software Change Request Procedures. Refer to Appendix A. Software Change Request Procedures, of this book for example software change request procedures.

Appendix B. [Program Abbreviation] Software Organization. This appendix should provide an overview of all data products. The following example is provided:

[Program Abbreviation] software elements can be categorized into three groups: module products, data products, and end-user products.

Module products and data products are used by software developers, not the users, and are components of [Program Abbreviation].

Module products—Module products are software libraries that perform a specific set of functions. Module products provide reusable code ensuring consistency in function performance, eliminating duplication of effort in developing like functions, and reducing the amount of code. Module products are used in our end-user products and may be available for use by other companies/agencies.

Data products—Data products are database data files. Data products allow a single development, maintenance, and testing source and provides consistent data and format. Data products are included as part of our end-user products and may be available for use by other companies/agencies. Data products may be sent to users as a product itself.

End-user products—End-user products are executable programs for users.

Configuration management of module products is accomplished according to this document. Each module product has an individual baseline and version numbers. Module product managers specify the version of the module product to be used in end-user products.

B.1. [Program Abbreviation] Module Products. A directory structure for all module products exists for each software product version released should be provided.

B.1 [Program Abbreviation] Data Products. A listing of all data products should be provided for each software product version released.

B.1 [Program Abbreviation] Deliverables. A description of each software deliverable should be provided.

B.1 [Program Abbreviation] Support Software. A support software should be described.

QUALITY ASSURANCE

The quality assurance (QA) process provides adequate assurance that the software products and processes in the project life cycle conform to their specified requirements and adhere to their established plans. The QA process must assure that each process, activity, and task required by the contract or described in plans are being performed in accordance with the contract and with those plans. QA also provides assurances that each software product has undergone software product evaluation, testing, and problem resolution. The quality assurance process may make use of the results of other supporting processes, such as verification, validation, joint review, audit, and problem resolution. Table 4-5 provides a list of the quality assurance process objectives.

Table 4-5. Quality assurance process objectives

a) Identify, plan, and schedule quality assurance activities for the process or product
b) Identify quality standards, methodologies, procedures, and tools for performing quality assurance activities and tailor them to the project
c) Identify resources and responsibilities for the performance of quality assurance activities
d) Establish and guarantee the independence of those responsible for performing quality assurance activities
e) Perform the identified quality assurance activities in line with the relevant plans, procedures, and schedules
f) Apply organizational quality management systems to the project

Software Quality Assurance Plan (SQAP)

The information provided here is designed to facilitate the definition of processes and procedures relating to software quality assurance activities. This guidance was developed using IEEE Std 12207.0, Guide to Life Cycle Processes [40], and IEEE Std 730-2002, IEEE Standard for Software Quality Assurance Plans. Additional information is provided in Appendix A, Software Process Work Products, which presents a recommended minimum set of software reviews and an example SQA inspection log. Table 4-6 provides a proposed software quality assurance plan document outline in support of Lean Six Sigma requirements.

Table 4-6. software quality assurance plan document outline

Title Page
Revision Page
Table of Contents
1. Introduction
    1.1 Purpose
    1.2 Scope
    1.3 Definitions, Acronyms, and Abbreviations
    1.4 References
2. SQA Management
    2.1 Organization
    2.2 Tasks
    2.3 Responsibilities
3. SQA Documentation
    3.1 Development, Verification and Validation, Use, and
        Maintenance
    3.2 Control
4. Standards and Practices
    4.1 Coding/Design Language Standards
    4.2 Documentation Standards
5. Reviews and Audits
6. Configuration Management
7. Testing
8. Software Measures
9. Problem Tracking
10. Records Collection, Maintenance, and Retention
11. Training
12. Risk Management

Software Quality Assurance Plan Document Guidance

The following provides section-by-section guidance in support of the creation of a SQAP. The SQAP should be considered to be a living document. The SQAP should change to reflect any process improvement activity. This guidance should be used to help define a software quality process and should reflect the actual processes and procedures of the implementing organization. All information is provided for illustrative purposes only. Additional information is provided in the document template, *Software Quality Assurance Plan.doc,* which is located on the companion CD-ROM.

Introduction. This section should provide the reader with a basic description of the project, including associated contract information and any standards used in the creation of the document. The following is provided as an example.

> The [Project Name] ([Proj Abbrev]) software products are developed under contract # [contract number], subcontract # [sub number], controlled by [customer office designator] at [customer location]. This software quality assurance plan (SQAP) has been developed to ensure that [Project Abbreviation] software products conform to established technical and contractual requirements. These may include, but are not limited to [company name] standards for software quality assurance plans.

As described by IEEE Std 610.12, Software Quality Assurance (SQA) is "a planned and systematic pattern of all actions necessary to provide adequate confidence that an item or product conforms to established technical requirements"*. This definition could be interpreted somewhat narrowly. The CMMI supports the notion that quality assurance is performed on both processes and products, and the 610.12 definition mentions only an item or product. This definition should be expanded to read, "a planned and systematic pattern of all actions necessary to provide adequate confidence that an item, product, or process conforms to established technical requirements."

Purpose. This subsection shall describe the purpose of the SQAP. It shall list the name(s) of the software items covered by the SQAP and the intended use of the software. An example is provided:

> The purpose of this SQAP is to establish, document, and maintain a consistent approach for controlling the quality of all software products developed and maintained by the [Project Abbreviation] Group. This plan defines the standardized set of techniques used to evaluate and report upon the process by which software products and documentation are developed and maintained.
> The SQA process described applies to all development, deliverables, and documentation maintained for the [Project Abbreviation] project.

Scope. This subsection shall describe the specific products and the portion of the software life cycle covered by the SQAP for each software item specified. The following is an example of scope:

> The specific products developed and maintained by the [Project Abbreviation] development team may be found in the [Project Abbreviation] Software Configuration Management Plan.
> In general, the software development products covered by this SQA Plan include, but are not limited to, the following: source files, libraries, executable files, makefiles, link files, and documentation.
> The software deliverables covered by this SQA Plan are: [project name] system software, version description documents (VDDs),† user's manuals, programmer's manuals, software design documents, and software requirements specifications.
> [Project Abbreviation] software development tasks cover the full spectrum of software life cycle stages from requirements to system testing. As a result, software quality assurance (SQA) activities must be adapted as needed for each specific software task.
> Because of the cumulative and diverse nature of [Project Abbreviation] software, this life cycle often includes the use of programs or subprograms that are already in use. These programs are included in the preliminary design, test documentation, and testing. They are included in the detailed design strictly in the design or modification of their interfaces with the remainder of the newly developed code.
> In many of the [Project Abbreviation] software products, changes are required to existing code, either as corrections or upgrades. The life cycle of software change is the same as the life cycle of new software, although certain steps may be significantly shorter. For example, system testing may omit cases that do not exercise the altered code.
> The decision as to whether an individual task involves a new software development or a change to an existing program will be the responsibility of the project lead, program manager, customer, associated CCB, or a combination thereof.

---

*IEEE Std. 610.12-1990 Standard Glossary of Software Engineering (ANSI).
†A version description document template is provided on the companion CD-ROM.

The life cycle for a specific project would be described in the associated software project management plan (SPMP). An example of a project life cycle and information in support of SQA activities is provided in Appendix A, Example Life Cycle, of this text.

SQA Management. This section shall describe the project's quality assurance organization, its tasks, and its roles and responsibilities, which will be used to effectively implement process and product quality assurance. A reference to any associated software project management plan should be provided here. The following is an example paragraph:

> The [Project Abbreviation] Program Manager, each [Project Abbreviation] Project Team Lead, and the [Project Abbreviation] Software Quality Assurance Manager (SQAM) will work together with the [company name] Software Engineering Process Group to develop, maintain, and implement effective software quality assurance management.

Organization. This subsection shall depict the organizational structure that influences and controls the quality of the software. This shall include a description of each major element of the organization together with the roles and delegated responsibilities. The amount of organizational freedom and objectivity to evaluate and monitor the quality of the software, and to verify problem resolutions, shall be clearly described and documented. In addition, the organization responsible for preparing and maintaining the SQAP should be identified. A typical organizational breakdown follows:

> *Program Manager.* The [Project Abbreviation] Program Manager (PM) is ultimately responsible for ensuring the SQA process is developed, implemented, and maintained. In this capacity, the PM will work with the [Company Name] SQAM and the [Project Abbreviation] SQAM in establishing standards and implementing the corporate SQAP as it applies to the [Project Abbreviation] program.
>
> The [Project Abbreviation] PM is ultimately responsible for ensuring that the Project Lead(s) and their associated project team(s) are trained in SQA policies and procedures that apply to their function on each team. This will consist of initial training for new members and refresher training, as required, due to process or personnel changes.
>
> *Project Lead.* The [Project Abbreviation] Project Lead (PL) shall provide direct oversight of SQA activities associated with project software development, modification, or maintenance tasks assigned within project functional areas.
>
> The Project Lead(s) are responsible for ensuring that project software quality assurance processes are followed. Similarly, the PL will ensure that all members of their project team regularly implement the SQA policies and procedures that apply to their specific function on the team.
>
> The PL is responsible for implementing the actions of this SQA process for all [Project Abbreviation] software products, including end-user products, module products and data. The PL will be responsible for controlling the product, ensuring that only approved changes are implemented. Each PL will maintain this process, with all changes to the process approved by the PM and associated.

Tasks. This subsection shall describe the portion of the software life cycle covered by the SQAP. All tasks to be performed, the entry and exit criteria for each task, and relationships between these tasks and the planned major checkpoints should also be described. The sequence and relationships of tasks, and their relationship to the project management plan master schedule, shall also be indicated. The following provides a content example:

The [Project Abbreviation] PM shall ensure that a planned and systematic process exists for all actions necessary to provide software products that conform to established technical and contractual requirements.

The following SQA checkpoints will be placed in the software development life cycle:

    a. At the end of the Software Requirements and Software System Test Planning activities, there shall be a Requirements Review in accordance with Section xxx of this document.
    b. At the end of the Preliminary Design activity, there shall be a Preliminary Design Review in accordance with Section xxx of this document.
    c. At the end of Detailed Design, there shall be a Critical Design Review in accordance with Section xxx of this document.
    d. At the end of Unit Testing, there shall be a walk-through in accordance with Section xxx of this document.
    e. At the end of Integration Testing, there shall be a Test Review in accordance with Section xxx of this document.
    f. At the end of System Testing, there shall be a Test Review and an audit in accordance with Section xxx of this document.
    g. At Release, there shall be an SQA review using the checklists found in Appendix xx.

The following SQA checkpoints will be placed in the software maintenance life cycle:

    a. At the end of the Analysis activity there shall be a Requirements Review in accordance with Section xxx of this document.
    b. At the end of Design, there shall be a Preliminary Design Review in accordance with Section xxx of this document.
    c. At the end of Implementation, there will be an walk-through in accordance with procedures identified in the associated System Test Plan and in Section xxx of this document.
    d. At Release, there shall be an SQA review using the release checklists found in Appendix xx.
    e. Acceptance Testing is dependant upon criteria determined by the customer. These criteria are based upon the documentation released with the software (i.e., the SRS, VDD, users documentation, etc.).

Each PL, or a designee, shall evaluate all software products to be delivered to the customer during the course of any project. This evaluation must ensure that the products are properly annotated, are adequately tested, are of acceptable quality, and are updated to reflect changes specified by the customer. Additionally, the PL is responsible for ensuring that both the content and format of the software documentation meet customer requirements.

Responsibilities. This subsection shall identify the specific organizational element that is responsible for performing each task as shown by the following:

Project Lead Responsibilities. The [Project Abbreviation] PM has overall responsibility for [Project Abbreviation] program SQA responsibilities.

The [Project Abbreviation] SQAM has the overall responsibility for monitoring compliance with the project procedures, noting any deviations and findings reporting.

The [Project Abbreviation] PLs have the responsibility for ensuring that individual team members are trained in Project SQA procedures and practices. They may delegate the actual training activities to [Project Abbreviation] SQAM. Project Leads are also responsible or reporting changes in Project procedures to the [Project Abbreviation] SQAM.

Each individual [Project Abbreviation] employee shall be responsible for learning and following the SQA procedures and for suggesting changes to their PL when necessary. Each PL shall be responsible for the implementation and accuracy of all SQA documentation and life cycle tracking.

Configuration Control Board (CCB). Each baseline software product developed by [Project Abbreviation] has an associated Configuration Control Board (CCB). The associated CCB has the authority to approve, disapprove, or revise all software quality assurance reporting procedures.

An example of the delineation of SQA life cycle responsibility should also be provided.

SQA Documentation. This section shall identify the documentation governing the development, verification and validation, use, and maintenance of the software. List all documents that are to be reviewed or audited for adequacy, identifying the reviews or audits to be conducted. Refer to IEEE Std 730 for detailed information supporting SQA documentation requirements.

Development, Verification and Validation, Use, and Maintenance. This subsection should address how project documentation is developed, verified, validated and maintained.

Control. This subsection should describe the responsibilities associated with software project documentation. The following provides an example of the type of information to support this requirement:

The [Project Abbreviation] ([Project Abbreviation]) Program Manager (PM), Project Lead(s), and team members shall share responsibility for SQA documentation. Document reviews and integration of any required documentation changes will occur as specified in Section 2.3.1 of this document. Refer to [Project Abbreviation] Software Configuration Management Process for additional information regarding specific product (i.e., software, documentation . . .) maintenance.

Changes to requirements, software baselines, and completed maintenance items will be reported in the deliverable's Version Description Document. Status reports will be provided to [Customer Name] no less than biweekly.

Standards and Practices. This section shall identify the standards, practices, conventions, quality requirements, and measurements to be applied. Product and process measures shall be included in the measurements used and may be identified in a separate measurements plan. This section should also describe how conformance with these items will be monitored and assured. The following provides an example of typical information:

The following standards shall be applied to all programs and incorporated into existing programs whenever and wherever possible. Retrofitting of existing programs to meet new standards will be avoided when this retrofitting impacts production.

Coding/Design Language Standards. All software developed and maintained by [Project Abbreviation] project teams shall adhere to [Project Abbreviation] *Programming Standards , [Additional Coding Reference].* Deviations from the Standards shall be noted on the SCR form or recorded as a CER.

Documentation Standards. All [Project Abbreviation] documentation shall be adapted from the [Additional Documentation Reference] and IEEE documentation guidelines.

Reviews and Audits. This section shall describe all software reviews to be conducted. These might include managerial reviews, acquirer–supplier reviews, technical reviews, inspections, walk-throughs, and audits. The schedule for software reviews as they relate to the software project's schedule should also be listed, or reference made to a software project management plan containing the information. This section should describe how the software reviews would be accomplished.

Configuration Management. This section shall identify associated configuration management activities necessary to meet the needs of the SQAP. An example follows:

> For information regarding configuration control and status accounting procedures refer to the [Project Abbreviation] *Configuration Management Plan.* Configuration Management Maintenance is the responsibility of each [Project Abbreviation] PL and each project team Configuration Manager.
>
> The Project SQA Manager shall audit the baseline versions of the software product and accompanying documentation prior to software release for validation. (Please refer to Appendix xx, software checklists.) The [Project Abbreviation] PM will verify that a baseline is established and preserved and that documented configuration management policies and [Company Name] corporate policies are implemented.

Testing. This section shall identify all the tests not included in the software verification and validation plan for the software covered by the SQAP and shall state the methods to be used. If a separate test plan exists, it should be referenced. An example follows:

> Testing of software developed, modified, and/or maintained under the [Project Abbreviation] contract must be tailored to meet the requirements of the specific project work request. Typical testing efforts include unit, integration, and system testing. The [Project Abbreviation] Program Manager will ensure that all project software testing is planned and executed properly. Please refer to the [Project Abbreviation] *Configuration Management Process* and program test plans, listed in Section xxx.

The system test program should be broad enough to ensure that all the software requirements are met. The software test plan is to be written at the same time that the software requirements are being finalized. It may be useful to draft a test plan while the requirements are being defined. As each requirement is defined, a way to determine that it is satisfied should be delineated. This method is then translated into a test case for the system testing and is added to the software test plan.

System Testing. The system testing should consist of a list of test cases (descriptions of the data to be entered) and a description of the resulting program output.

Integration Testing. The integration testing should check to ensure that the modules work together as specified in the SDD. Integration testing should be included as part of the software test plan, ensuring that all the modules to be tested are executed and exercised thoroughly.

Unit Testing. Unit testing is the detailed testing of a single module to ensure that it works as it should. Unit testing shall be identified in the software test plan to ensure that every critical function in the module is executed at least once. Critical functionality should be identified in the software test plan.

Software Measures. This section should describe the collection, maintenance, and retention of measures associated with the software quality assurance process. A separate software measurement and measures plan is often used, and, if so, a reference to this plan should be provided. The following is an example of the information supporting this section of the document:

> The [Project Abbreviation] ([Project Abbreviation]) SQA Manager shall ensure that [Project Abbreviation] Software projects follow the guidelines established by the [Project Abbreviation] *Software Measurement and Measures Plan.*
>
> The SQA Manager shall ensure the project measures are appropriate and will provide recommendations for improvements on the project's measurement process, as required.
>
> The following software measures will be recorded and used in support of the [Project Abbreviation] SQA effort:
>
> a) A history of the bugs and enhancements, a list of functions that are changed to fix the bugs or implement the enhancements, and the time and effort required to fix the bugs or implement the enhancements; the [Project Abbreviation] Software Change Request Tracking System will provide such information (CER/SCR System).
>
> b) Total numbers of errors, opened versus closed; errors opened and closed since last report. Errors being defined as:
>
>> 1. Requirements errors—Baseline Change Request (BCR)
>> 2. Coding defects—Test Problem Reports (TPRs)
>> 3. Documentation defects—Interface Change Notice (DCR)
>> 4. Bad Fixes—Internal Test Report (ITR)
>
> c) Total implemented requirements versus planned requirements.

Problem Tracking. This section shall describe the practices and procedures to be followed for reporting, tracking, and resolving problems or issues identified in both software items and the software development and maintenance process. It should also describe organizational responsibilities concerned with their implementation. The following provides some example text:

> Please refer to [Project Abbreviation] *Configuration Management Process.* This document describes the practices and procedures followed for reporting, tracking, and resolving problems identified in software items and the associated development, maintenance, and organizational responsibilities.
>
> Deficiencies in the SQA Plan shall be corrected using the following procedure:
>
>> 1. A deficiency is noted, either by a [Company Name] employee finding the procedure difficult to follow or by the SQA manager noting that the procedure has not been followed.
>> 2. The PM, or a designee, notes the deficiency and develops a way to correct it. In this, he shall have help from the employee who had trouble following the procedure or who did not follow the procedure.
>> 3. A new version of the SQA Plan will be written, printed, and disseminated to [Project Abbreviation] Program personnel and [Company Name] Corporate SQA. The new version shall incorporate the correction.

Records Collection, Maintenance, and Retention. This section shall identify all SQA documentation and describe how the documentation is maintained while designating the retention period. The following is an example:

> The Software Developmental Library (SDL) for [Project Abbreviation] program software may be found in Appendix [letter]. Documents pertaining to developmental configuration, which are also maintained, may be found in Section [x.x]. Access to all documents is controlled by each project team's Project Lead, or designee.
>
> This documentation will be maintained in [Maintenance Location and Address]. All documentation will be maintained on in a shared [Project Abbreviation] subdirectory [Project Abbreviation]/[path]/shared/<document name>. Weekly backups of this system occur and are stored off-site.
>
> All documentation will be maintained throughout the length of the [Project Abbreviation] contract.

Training. This section shall identify the training activities necessary to meet the needs of the SQAP. The following is an example:

> It is the responsibility of the [Project Abbreviation] Program Manager (PM) to allow for the training necessary to accomplish SQA goals and implementation.
>
> Similarly, the PL will ensure that all members of the project team are trained in the SQA policies and procedures that apply to their function on each team. This will consist of initial training for new members and refresher training, as required, due to process or personnel changes.

Risk Management. This section shall specify all methods and procedures employed to identify, assess, monitor, and control areas of risk arising during the portion of the software life cycle covered by the SQAP. Reference to the software project management plan, or a separate risk management plan, would suffice here, as shown by the following text:

> Program risks will be identified either by individual team members or CCB members. The risk tracking form (refer to [Project Abbreviation] Software Development Plan) will be used to document and track all risks identified. All risks will be submitted to the [Project Abbreviation] CCB at weekly staff meetings to determine the severity and probability. Risks determined to be significant will be assigned an owner. The owner will determine the plan for dealing with the risk and propose a plan to the [Project Abbreviation] CCB. The owner will track the risk to closure. Risks that are not deemed significant will be canceled.
>
> The guidelines used by the [Project Abbreviation] CCB to define risk severity and probability is listed in the [Project Abbreviation] Software Development Plan.

Additional information is provided in the *Software Quality Assurance Plan.doc* document template on the companion CD-ROM. Appendix A, Software Process Work Products, of this book also provides several examples of supporting process artifacts.

VERIFICATION

The verification process determines whether the software products of an activity fulfill the requirements or conditions imposed on them in the previous activities. This process may include analysis, review, and test. Verification should be integrated, as early as possible, with the process that employs it. Verification is defined as confirmation by examination and provision of objective evidence that specified requirements have been fulfilled. Simply put, verification ensures that work products meet their specified requirements. Table 4-7 provides a list of the verification process objectives.

Table 4-7. Verification process objectives

---
a) Identify criteria for verification of all required work products
b) Perform requirements verification activities
c) Find and remove defects from products produced by the project

---

Inspections

The information provided here is based upon the recommendations provided in IEEE Std 1028, IEEE Standard for Software Reviews. Inspections are an effective way to detect and identify problems early in the software development process. Inspections should be peer-driven and should follow a defined, predetermined practice. The criteria for inspections are presented here to help organization define their inspection practices. IEEE Std 1044, IEEE Standard Classification for Software Anomalies [7] provides valuable information regarding the categorization of software anomalies. Additional information is provided in Appendix A, which presents a recommended minimum set of software reviews and an example SQA inspection log.

Introduction. Software inspections can be used to verify that a specified software product meets its requirements. These requirements can address functional characteristics, quality attributes, or adherence to regulations or standards. Any inspection practice should collect anomaly data to use identify trends and improve the development process. Use *Classic Anomaly Class Categories* in Appendix A, Software Process Work Products.

Examples of software products subject to inspections are:

1. Software requirements specification
2. Software design description
3. Source code
4. Software test plan documentation
5. Software user documentation
6. Maintenance manual
7. System build procedures
8. Installation procedures
9. Release notes

Responsibilities. Each inspection should have a predefined set of roles and responsibilities; the assignment of these responsibilities should occur on a case-by-case basis and is dependant upon the item subject to inspection.

Participants. Inspections should consist of three to six participants and should be led by an impartial facilitator who is trained in inspection techniques. Determination of remedial or investigative action for an anomaly is a mandatory element of a software inspection, although the resolution should not occur in the inspection meeting. Collection of data for the purpose of analysis and improvement of software engineering procedures (including all review procedures) is strongly recommended but is not a mandatory element of software inspections. Participant roles are defined in the following five subsections.

Inspection Leader. The inspection leader is responsible for administrative tasks pertaining to the inspection and for the planning and preparation of the inspection. The leader is responsible for ensuring that the inspection is conducted in an orderly manner and meets its objectives. The leader is responsible for the collection of inspection data and the issuance of the inspection report.

Recorder. The recorder is responsible for recording all inspection data required for process analysis. The recorder should document all anomalies, action items, decisions, and recommendations made by the inspection team. The inspection leader may also act as recorder.

Reader. The reader should methodically lead the inspection team through the software product, providing a summary, when appropriate, highlighting all items of interest.

Author. The author contributes the product for inspection and participates when a special understanding of the software product is required. The author is responsible for any rework required to make the software product meet inspection exit criteria.

Inspector. All participants in the review are inspectors. The author should not act as inspection leader and should not act as reader or recorder. All other roles may be shared among the team members and participants may act in more than one role. Individuals holding management positions over any member of the inspection team should not participate in the inspection.

Inspectors are responsible for the identification and description of all anomalies found in the software product. Inspectors should represent different viewpoints at the meeting (for example, sponsor, requirements, design, code, safety, test, independent test, project management, quality management, and hardware engineering). Only those viewpoints with any relevance to the product should be represented.

Each inspector should be assigned a specific review topic to ensure effective product coverage. The inspection leader should assign these roles prior to the inspection.

Input. Input to the inspection should include the objectives, identification of the software product, the inspection procedure, all reporting forms, all guidelines or standards, and any known anomalies or issues.

Authorization. Inspections should be identified in any associated software project management plan, software verification and validation plan, and should be reflected in project scheduling and resource allocation.

Preconditions. An inspection should be conducted only when a statement of objectives for the inspection is established and the required inspection inputs are available.

Minimum Entry Criteria. An inspection should not be conducted until the software product is complete and conforms to project standards for content and format, all tools and documentation required in support of the inspection are available, and any related prior milestones have been satisfied. If the inspection is a reinspection, all previously identified problems should have been resolved.

Management Preparation. Managers are responsible for planning the time and resources required for inspections, including support functions, and making sure that these are identified in the associated software project management plan. Managers should also ensure that individuals receive adequate training and orientation on inspection procedures and they possess appropriate levels of expertise and knowledge sufficient to comprehend the software product under inspection.

Planning the Inspection. The author of the software product is responsible for the assembly of all inspection materials. The inspection leader is responsible for the identification of the inspection team, the assignment of specific responsibility, and scheduling and conducting the inspection. As a part of the planning procedure, the inspection team should determine if anomaly resolution is to occur during the inspection meeting or if resolution will occur under other circumstances.

Overview of Inspection Procedures. The author of the software product should present a summary overview for the inspectors. The inspection leader should answer questions about any checklists and assign roles. Expectation should be reviewed, such as the minimum preparation time anticipated and the typical number of anticipated anomalies.

Preparation. Each inspection team member should examine the software product and document all anomalies found. All items found should be sent to the inspection leader. The inspection leader should classify (i.e., consolidate) all anomalies and forward these to the author of the software product for disposition. The inspection leader, or reader, should specify the order of the inspection (e.g., hierarchical). The reader should be prepared to present the software product at the inspection meeting.

Examination. The inspection meeting should follow this agenda:

Introduce the meeting. The inspection leader should introduce all participants and review their roles. The inspection leader should state the purpose of the inspection and should remind the inspectors to focus their efforts toward anomaly detection, direct remarks to the reader and not the author, and to comment only on the software product and not the author. See Classic Anomaly Class Categories in Appendix A.

Establish preparedness. The inspection leader should verify that all participants are prepared. The inspection leader should gather individual preparation times and record the total in associated inspection documentation.

Review general items. General, nonspecific, or pervasive anomalies should be presented first. If problems are discovered that inhibit the further efficient detection of error, the leader may choose to reconvene the inspection following the resolution of all items found.

Review software product and record anomalies. The reader should present the software product to the inspection team. The inspection team should examine the software product and create an anomaly list of items found. The recorder should enter each anomaly, location, description, and classification on the anomaly list.* See Inspection Log Defect Summary Description in Appendix A.

Review the anomaly list. At the end of the inspection meeting, the team should review the anomaly list to ensure its completeness and accuracy. The inspection leader should allow time to discuss every anomaly about which any disagreement occurred.

Make an exit decision. The purpose of the exit decision is to bring an unambiguous close to the inspection meeting. Common exit criteria are listed below:

1. *Accept with no or minor rework.* The software product is accepted as is or with only minor rework (requires no further verification).

2. *Accept with rework verification.* The software product is to be accepted after the inspection leader or a designated member of the inspection team verifies rework.

3. *Reinspect.* Schedule a reinspection to verify rework. At a minimum, a reinspection should examine the software product areas changed to resolve anomalies identified in the last inspection, as well as side effects of those changes.

Exit criteria. An inspection is considered complete when prescribed exit criteria have been met.

Output. The output of the inspection provides documented evidence that the project is being inspected. It provides a list of all team members, the meeting duration, a description of the product inspected, all inspection inputs, objectives, anomaly list and summary, a rework estimate, and total inspection preparation time.

Data Collection Recommendations. Inspection data should contain the identification of the software product, the date and time of the inspection, the inspection leader, the preparation and inspection times, the volume of the materials inspected, and the disposition of the inspected software product. The management of inspection data should provide for the capability to store, enter, access, update, summarize, and report categorized anomalies.

Anomaly Classification. Anomalies may be classified by technical type. IEEE Std 1044 provides effective guidance in support of anomaly classification. Appendix A of IEEE Std 1044 provides sample screens from an anomaly reporting system. The data fields for this system are listed in Appendix A under the Verification Artifacts section. See Classic Anomaly Class Categories in Appendix A.

Anomaly Ranking. Anomalies may be ranked by potential impact on the software product. Three ranking categories are provided as examples:

1. *Category 1.* Major anomaly of the software product, or an observable departure from specification, with no available workaround.

---

*IEEE Std 1044, IEEE Std Classification for Software Anomalies [7] may be used to support anomaly classification.

2. *Category 2.* Minor anomaly of the software product, or departure from specification, with existing workaround.

3. *Category 3.* Cosmetic anomalies that deviate from relevant specifications but do not cause failure of the software product.

The information is captured in the Inspection Report Description and the Inspection Log Defect Summary Description as shown in Appendix A.

Improvement. Inspection data should be analyzed regularly in order to improve the inspection and software development processes. Frequently occurring anomalies may be included in the inspection checklists or role assignments. The checklists themselves should also be evaluated for relevance and effectiveness.

Walk-throughs

The following information is based on IEEE Std 1028, IEEE Standard for Software Reviews; and IEEE Std 1044, IEEE Guide to Classification of Software Anomalies.

The purpose of a systematic walk-through is to evaluate a software product. Walk-throughs effectively identify anomalies and contribute toward the improvement of the software development process. Walk-throughs are also effective training activities, encouraging technical exchanges and information sharing. Examples of software products subject to walk-throughs include software project documentation, source code, and system build or installation procedures.

Responsibilities. The following roles should be established:

Leader. The walk-through leader is responsible for administrative tasks pertaining to the walk-through and is responsible for the planning and preparation of the walk-through. The leader is responsible for ensuring that the walk-through is conducted in an orderly manner and meets its objectives. The leader is responsible for the collection of walk-through data and the issuance of the summary report.

Recorder. The recorder should methodically document all decisions, actions, and anomalies discussed during the walk-through.

Author. The author contributes the product for walk-through and participates when a special understanding of the software product is required. The author is responsible for any rework required to address all anomalies identified.

Team Member. All participants in the walk-through are team members. All other roles may be shared among the team members and participants may act in more than one role. Individuals holding management positions over any member of the walk-through team should not participate in the walk-through.

Input. Input criteria should include the walk-through objectives, the software product, anomaly categories,* and all relevant standards. See Classic Anomaly Class Categories in Appendix A.

---

*IEEE Std 1044, IEEE Std Classification for Software Anomalies [7] may be used to support anomaly classification.

Authorization. Walk-throughs should be identified in any associated software project management plan or software verification and validation plan, and should be reflected in project scheduling and resource allocation.

Preconditions. A walk-through should be conducted only when a statement of objectives for the walk-through is established and the required walk-through inputs are available.

Management Preparation. Managers are responsible for planning the time and re-sources required for walk-through, including support functions and that these are identi-fied in the associated software project management plan. Managers should also ensure that individuals receive adequate training and orientation on walk-through procedures and possess appropriate levels of expertise and knowledge sufficient to evaluate the software product.

Planning the Walk-through. The walk-through leader is responsible for identifying the walk-through team, distributing walk-through materials, and scheduling and conducting the walk-through.

Overview. The author of the software product should present a summary overview for the team.

Preparation. Each walk-through team member should examine the software product and prepare a list of items for discussion. These items should be categorized as either specific or general, where general would apply to the entire product and specific to one part. All anomalies detected by team members should be sent to the walk-through leader for classi-fication. The leader should forward these to the author of the software product for disposi-tion. The walk-through leader should specify the order of the walk-through (e.g., hierar-chical).

Examination. The walk-through leader should introduce all participants, describe their roles, and state the purpose of the walk-through. All team members should be reminded to focus their efforts toward anomaly detection. The walk-through leader should remind the team members to comment only on the software product and not its author. Team mem-bers may pose general questions to the author regarding the software product.

The author should then present an overview of the software product under review. This is followed by a general discussion during which team members raise their general items. After the general discussion, the author serially presents the software product in detail. Team members raise their specific items when the author reaches them in the presenta-tion. It is important to note that new items may be raised during the meeting.

The walk-through leader is responsible for guiding the meeting to a decision or action on each item. The recorder is responsible for noting all recommendations and required ac-tions. Appendix A, Software Process Work Products, contains several walk-through forms: Requirements Walk-through Form, Software Project Plan Walk-through Check-list, Preliminary Design Walk-through Checklist, Detailed Design Walk-through Check-list, Program Code Walk-through Checklist, Test Plan Walk-through Checklist, Walk-through Summary Report, and Classic Anomaly Class Categories.

After the walk-through meeting, the walk-through leader should issue a walk-through report detailing anomalies, decisions, actions, and other information of interest. This re-port should include a list of all team members, description of the software product, statement of walk-through objectives, list of anomalies and associated actions, all due dates and individual responsible, and the identification of follow-up activities.

Data Collection Recommendations. Walk-through data should contain the identification of the software product, the date and time of the walk-through, the walk-through leader, the preparation and walk-through times, the volume of the materials inspected, and the disposition of the software product. The management of walk-through data should provide for the capability to store, enter, access, update, summarize, and report categorized anomalies.

Anomaly Classification. Anomalies may be classified by technical type. IEEE Std 1044-1993 provides effective guidance in support of anomaly classification. Appendix A of IEEE Std 1044 provides sample screens from an anomaly reporting system. The data fields for this system are listed in Appendix A under the Verification Artifacts section.

Anomaly Ranking. Anomalies may be ranked by potential impact on the software product. Three ranking categories are provided as examples:

1. *Category 1.* Major anomaly of the software product, or an observable departure from specification, with no available workaround.
2. *Category 2.* Minor anomaly of the software product, or departure from specification, with existing workaround.
3. *Category 3.* Cosmetic anomalies that deviate from relevant specifications but do not cause failure of the software product.

Improvement. Walk-through data should be analyzed regularly in order to improve the walk-through and software development processes. Frequently occurring anomalies may be included in the walk-through checklists or role assignments. The checklists themselves should also be evaluated for relevance and effectiveness.

VALIDATION

The validation process is a process for determining whether the final system or software product meets the requirements and fulfills its specific intended use. Validation is defined as confirmation by examination and provision of objective evidence that the particular requirements for a specific intended use are fulfilled. Validation is often performed by testing, conducted at several levels or by several approaches. It is not to be confused with Verification, which ensures that the product or artifact satisfies or matches the requirements—that is, you built the right product. Table 4-8 provides a list of the Validation process objectives.

Table 4-8. Validation process objectives

a) Identify criteria for validation of all required work products;
b) Perform required validation activities;
c) Provide evidence that the work products, as developed, are suitable for their intended use.

Software Test Plan

The documentation supporting software testing should define the scope, approach, resources, and schedule of the testing activities. It should identify all items being tested, the features to be tested, the testing tasks to be performed, the personnel responsible for each task, and the risks associated with testing. IEEE Std 829, IEEE Standard for Software Test Documentation [15], and IEEE 12207.1—Standard for Information Technology— Software Life Cycle Processes—Life Cycle Data [40], were used as primary references for the information provided here.

The IEEE Computer Society also publishes IEEE Std 1008, IEEE Standard for Software Unit Testing [1], and IEEE Std 1012, IEEE Standard for Software Verification and Validation [19], as part of their software engineering standards collection. These documents provide additional information in support of the development and definition of software test processes and procedures. Table 4-9 provides an example document outline.

Table 4-9. Software test plan document outline

Title Page
Revision Page
Table of Contents
1. Introduction
2. Scope
    2.1 Identification
    2.2 System Overview
    2.3 Document Overview
    2.4 Acronyms and Definitions
        2.4.1 Acronyms
        2.4.2 Definitions
3. Referenced Documents
4. Software Test Environment
    4.1 Development Test and Evaluation
        4.1.1 Software Items
        4.1.2 Hardware and Firmware Items
        4.1.3 Other Materials
        4.1.4 Proprietary Nature, Acquirer's Rights, and Licensing
        4.1.5 Installation, Testing, and Control
        4.1.6 Participating Organizations
    4.2 Test Sites
        4.2.1 Software Items
        4.2.2 Hardware and Firmware Items
        4.2.3 Other Materials
        4.2.4 Proprietary Nature, Acquirer's Rights, and Licensing
        4.2.5 Installation, Testing, and Control
        4.2.6 Participating organizations
5. Test Identification
    5.1 General Information
        5.1.1 Test Levels
        5.1.2 Test Classes
             5.1.2.1 Check for Correct Handling of Erroneous Inputs
             5.1.2.2 Check for Maximum Capacity
             5.1.2.3 User Interaction Behavior Consistency
             5.1.2.4 Retrieving Data

(continued)

Table 4-9   *(continued)*

Software Test Plan Document Guidance

The following provides section-by-section guidance in support of the creation of a software test plan. It does not identify specific testing methodologies, approaches, techniques, facilities, or tools. As with any defined software engineering process, additional supporting documentation may be required. The development of a software test plan (STP) is integral to the software development process. Additional information is provided in the document template, *Software Test Plan.doc,* which is located on the companion CD-ROM.

Identification. This subsection to the Scope section should provide information that uniquely identifies the software effort and this associated test plan. This can also be provided in the form of a unique test plan identifier. The following text provides an example:

> This software test plan is to detail the testing planned for the [Project Name] ([Project Abbreviation]) Version [xx], Statement of Work, [date], Task Order [to number], Contract No. [Contract #], and Amendments.
>
> The goal of [Project Acronym] development is to [goal]. This software will allow [purpose].
>
> Specifics regarding the implementation of these modules are identified in the [Project Abbreviation] Software Requirements Specification (SRS) with line item descriptions in the accompanying Requirements Traceability Matrix (RTM).

Document Overview. This subsection to the Scope section should provide a summary of all software items and software features to be tested. The need for each item and its history may be addressed here as well. References to associated project documents should be cited here. The following provides an example:

This document describes the Software Test Plan (STP) for the [Project Abbreviation] software. [Project Abbreviation] documentation will include this STP, the [Project Abbreviation] Software Requirements Specification (SRS), the Requirements Traceability Matrix (RTM), the [Project Abbreviation] Software Development Plan (SDP), the Software Design Document (SDD), the [Project Abbreviation] User's Manual, the [Project Abbreviation] System Administrator's Manual, and the [Project Abbreviation] Data Dictionary.

This STP describes the process to be used and the deliverables to be generated in the testing of the [Project Abbreviation]. This plan, and the items defined herein, will comply with the procedures defined in the [Project Abbreviation] Software Configuration Management (SCM) Plan, and the [Project Abbreviation] Software Quality Assurance (SQA) Plan. Any nonconformance to these plans will be documented as such in this STP.

This document is based on the [Project Abbreviation] Software Test Plan template with tailoring appropriate to the processes associated with the creation of the [Project Abbreviation]. The information contained in this STP has been created for [Customer Name] and is to be considered "For Official Use Only."

Please refer to section "6. Schedule" in the SDP for information regarding documentation releases.

Acronyms and Definitions. All relevant acronyms and definitions should be included in this subsection.

Referenced Documents. This section should include all material referenced during the creation of the test plan.

Development Test and Evaluation. This section should provide a high level summary of all key players, facilities required, and site of test performance. The following is provided as an example:

Qualification, integration, and module level tests are to be performed at [Company Name], [Site Location]. All testing will be conducted in the development center [Room Number]. The following individuals must be in attendance: [provide list of performers and observers].

Software Items. This subsection should provide a complete list of all software items used in support of testing, including versions. If the software is kept in an online repository, reference to the storage location may be cited instead of listing all items here. The following is an example:

Software used in the testing of [Project Abbreviation]:
    [Software List]
    For details of client and server software specifications see the [Project Abbreviation] Software Requirements Specifications (SRS).

Hardware and Firmware Items. This subsection should provide a complete list of all hardware and firmware items used in support of testing. If this information is available in another project document it may be referenced instead of listing all equipment. The following is an example:

[Hardware Items] are to be used during testing, connected together in a client–server relationship through a TCP/IP compatible network. For details of client and server hardware specifications, see the [Project Abbreviation] Software Requirements Specifications (SRS). For integration and qualification level testing, a representative hardware set as specified in the SRS is to be used.

Other Materials. This subsection should list all other materials required to support testing.

Proprietary Nature, Acquirer's Rights, and Licensing. Any of the issues associated with the potential proprietary nature of the software, acquirer's rights, or licensing should be addressed in this subsection. An example is provided:

> Licensing of commercial software is one purchased copy for each PC it is to be used on with the exception of [Group Software], which also requires licensing for the number of users/client systems. Some of the information in the dataset is covered by the Privacy Act and will have to be protected in some manner.

Installation, Testing, and Control. This section should address environment requirements or actions required to support the installation of the application for testing. The following detail is provided:

> [Company Name] will acquire the software and hardware needed as per the [Project Abbreviation] Software Development Plan (SDP) in order to run [Project Abbreviation] in accordance with contract directions and limitations.
>
> [Company Name] will install the supporting software and test it per the procedures in the [Project Abbreviation] Installation Software Test Description.
>
> The [Project Abbreviation] Change Enhancement Request (CER) tracking system will be used to determine eligibility for testing at any level. See the [Project Abbreviation] Software Configuration Management Plan and [Project Abbreviation] Software Quality Assurance Plan for details on CER forms and processes.

Participating Organizations. This subsection should identify all groups responsible for testing. They may also participate in completing the Test Plan Walk-through Checklist in Appendix A. An example is provided:

> Test Sites. This section should provide a description of test sites. See below:
> > The planned installation level Beta test sites are listed as follows:
> > > [Installation Beta Sites]

Test Sites. This section contains the following subsections.

Software Items. This subsection should provide a complete list of all software items used in support of site testing, including versions. If the software is kept in an online repository, reference to the storage location may be cited instead of listing all items here. The following is an example:

> Software used in the testing of [Project Abbreviation]:
> > [Software Items]
> For details of client and server software specifications, see the [Project Abbreviation] Software Requirements Specifications (SRS).

Hardware and Firmware Items. This subsection should provide a complete list of all hardware and firmware items used in support of site testing. If this information is available in another project document, it may be referenced instead of listing all equipment. The following is an example:

> IBM-compatible PCs are to be used during testing, connected together in a client–server relationship through a TCP/IP-compatible network. For details of client and server hardware

specifications, see the [Project Abbreviation] Software Requirements Specifications (SRS). For integration and qualification-level testing, a representative hardware set as specified in the SRS is to be used.

Other Materials. This subsection should list all other materials required to support testing.

Proprietary Nature, Acquirer's Rights, and Licensing. Any of the issues associated with the potential proprietary nature of the software, acquirer's rights, or licensing should be addressed in this subsection. An example is provided:

> Licensing of commercial software is one bought copy for each PC it is to be used on with the exception of [group software], which also requires licensing for the number of users/client systems. Some of the information in the dataset is covered by the Privacy Act and will have to be protected in some manner.

Installation, Testing, and Control. This subsection should address environment requirements or actions required to support the installation of the application for beta testing. The following detail is provided:

> The customer will acquire the software and hardware needed as defined in the [Project Abbreviation] SRS and the SOW in order to run [Project Abbreviation].
>
> [Company Name] will install the supporting software and test it per the procedures in the [Project Abbreviation] Installation Software Test Description and as defined in the [Project Abbreviation] SDP.
>
> The CER tracking system will be used to determine eligibility for testing at any level. See the [Project Abbreviation] Software Configuration Management Plan and [Project Abbreviation] Software Quality Assurance Plan for details on CER forms and processes.

Participating Organizations. This section should identify all groups responsible for testing. They may also participate in completing the Test Plan Walk-through Checklist. An example is provided:

> The participating organizations are [Company Name], [Customer Name], and organizations at the beta test site.

The General Information subsection of the Test Identification subsection contains the following subsections.

Test Levels. This subsection should describe the different levels of testing required in support of the development effort. See Appendix A for Test Design Specification, Test Case Specification, and Test Procedure Specification. The following is provided as an example:

> Tests are to be performed at the module, integration, installation, and qualification levels prior to release for beta testing. Please refer to test design specifications #xx through xx.

Test Classes. This subsection should provide a description of all test classes. A test case specification should be associated with each of these classes. A summary of the validation method, data to be recorded, data analysis activity, and assumptions, should be provided for each case. An example text is provided:

Test class # xx. Check for correct handling of erroneous inputs.

Test Objective—Check for proper handling of erroneous inputs: characters that are not valid for this field, too many characters, not enough characters, value too large, value too small, all selections for a selection list, no selections, all mouse buttons clicked or double clicked all over the client area of the item with focus.

Validation Methods Used—Test.

Recorded Data—User action or data entered, screen/view/dialog/control with focus, resulting action

Data Analysis—Was resulting action within general fault handling defined capabilities in the [Project Abbreviation] SRS and design in [Project Abbreviation] SDD?

Assumptions and Constraints—None.

Additional examples are provided in support of test case development on the associated companion CD-ROM in *Software Test Plan.doc.*

General Test Condition. This subsection should describe the anticipated baseline test environment. Refer to the example provided below:

A sample real data set from an existing database is to be used during all tests. The sample real data set will be controlled under the configuration management system. A copy of the controlled data set is to be used in the performance of all testing.

Test Progression. This subsection should provide a description of the progression of testing. A diagram is often helpful when attempting to describe the progression of testing (Figure 4.2). The information below is provided as an example:

Qualification Testing is a qualification-level test verifying that all requirements have been met. The module and integration tests are performed as part of the Implementation phase as elements and modules are completed. All module and integration tests must be passed before performing Qualification Testing. All module tests must be passed before performing associated integration-level tests. The CER tracking system is used to determine eligibility for testing at a level. See the [Project Abbreviation] Software Configuration Management Plan and [Project Abbreviation] Software Quality Assurance Plan for details on CER forms and processes.

Planned Tests. This section should provide a detailed description of the type of testing to be employed. The following is provided as an example:

A summary of testing is provided in section 4.3. Additional information is provided in Appendix A, B, C, D, and E.

Qualification Test. This subsection should describe qualification test. An example is provided:

All of the requirement test items (refer to the Software Requirements Traceability Matrix) are to be tested as qualification-level tests. For details of the procedures and setup, see the [Project Abbreviation] Qualification Software Test Description. The resulting output of qual-

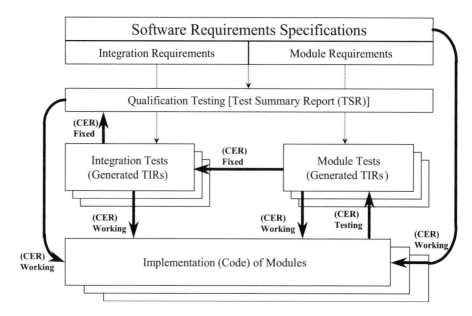

Figure 4.2. Software test progression.

ification test is the [Project Abbreviation] Software Test Report (STR) and its attached Test Problem Reports (TPR). If qualification test is passed and its results accepted by the customer, the [Project Abbreviation] software will be ready for beta release. Following Qualification Testing, the customer will review results. Signature of acceptance initiates product delivery and start of Installation beta test.

  For the qualification-level tests the following classes of tests will be used:
    Check for correct handling of erroneous inputs
    Check for maximum capacity
    User interaction behavior consistency
    Retrieving data
    Saving data
    Display screen and printing format consistency
    Check interactions between modules
    Measure time of reaction to user input
    Functional flow

Integration Test. This subsection should describe integration test. An example text is provided:

  All of the modules to be integration tested (refer to the Software Requirements Traceability Matrix) will be tested using integration-level test methodology. For details of the procedures and setup, see the [Project Abbreviation] Integration Software Test Description. The resulting outputs of this test are Internal Test Reports (ITR) or System Integration Test Reports (SITR). When these integration tests are all passed, the [Project Abbreviation] software will be ready for qualification-level testing.

  For the integration-level tests, the following classes of tests will be used:
    User interaction behavior consistency
    Display screen and printing format consistency

Check interactions between modules
Measure time of reaction to user input

Module Test. This section should describe module testing. An example text is provided:

All of the modules to be tested (refer to the Software Requirements Traceability Matrix) are to be tested using defined module-level test methodology. For details regarding test procedures and setup, see the [Project Abbreviation] Module Software Test Description. The resulting outputs of this test are Internal Test Reports (ITR) or Unit Test Reports (UTR). When all of the module tests for a module are passed, the module is ready for integration level testing.
For the module-level tests, the following classes of tests will be used:
Check for correct handling of erroneous inputs
Check for maximum capacity
User interaction behavior consistency
Retrieving data
Saving data
Display screen and printing format consistency
Measure time of reaction to user input

Installation Beta Test. This subsection should describe installation beta testing. An example text is provided:

Following qualification testing, the customer will review results. Signature of acceptance initiates product delivery and start of installation beta test. Identified tests (refer to the Software Requirements Traceability Matrix) will be tested using defined installation-level beta test procedures. Some of these tests may be documented in the Product Packaging Information in Appendix A. This test methodology will be applied at each beta test site. For details regarding test procedures and setup, see the [Project Abbreviation] Installation Software Test Description. Outputs of these tests are Test Problem Reports (TPR).
Following installation testing, the customer will review results. Signature of acceptance completes the [Project Abbreviation] Version 1.0 project.
For the installation-level tests, the following classes of tests will be used:
User interaction behavior consistency
Retrieving data
Saving data
Display screen and printing format consistency
Check interactions between modules
Measure time of reaction to user input
Functional flow

Test Schedules. This section should include test milestones as identified in the software project schedule as well as any required deliverables associated with testing. Any additional milestones may be defined here as needed. If appropriate detail is provided in the software project management plan (SRMP), then this can be referenced. The following is provided as an example:

Test schedules are defined in the [Project Abbreviation] Software Development Plan.

Risk Management. This section should identify all high-risk assumptions of the test plan. The contingency plans for each should be described and the management of risk items should be addressed.

Requirements Traceability. This section should provide information regarding the traceability of testing to the requirements and design of the software. A matrix, or database, is useful when meeting this traceability requirement. The following example is provided:

> Refer to [Project Abbreviation] Software Requirements Specification, Appendix A, for information regarding requirements traceability.

Notes. This section should provide any additional information not previously covered in the test development plan.

System Test Plan

The following information is based on IEEE Std 829, IEEE Standard for Software Test Documentation [15]; IEEE Std 1012, IEEE Standard for Software Verification and Validation [19]; IEEE Std 1028, IEEE Standard for Software Reviews [14]; and IEEE Std 1044.1, IEEE Guide to Classification of Software Anomalies [7]. A document outline is provided in Table 4-10. Additional information is provided in the document template, *System Test Plan.doc,* which is located on the companion CD-ROM.

Table 4-10. System test plan document outline

Title Page
Revision Page
Table of Contents
 1. Introduction
 2. Scope
    2.1   Identification and Purpose
    2.2   System Overview
    2.3   Definitions, Acronyms, and Abbreviations
 3. Referenced Documents
 4. System Test Objectives
 5. Types of System Testing
    5.1   Functional Testing
    5.2   Performance Testing
    5.3   Reliability Testing
    5.4   Configuration Testing
    5.5   Availability Testing
    5.6   Portability Testing
    5.7   Security and Safety Testing
    5.8   System Usability Testing
    5.9   Internationalization Testing
    5.10 Operations Manual Testing
    5.11 Load Testing
    5.12 Stress Testing
    5.13 Robustness Testing
    5.14 Contention Testing
 6. System Test Environment
    6.1   Development Test and Evaluation
    6.2   System and Software Items
    6.3   Hardware and Firmware Items

(continued)

Table 4-10 *(continued)*

The following provides section-by-section guidance in support of the creation of a system test plan. This guidance should be used to help define associated test requirements and should reflect the actual requirements of the implementing organization.

Introduction. This section should provide a brief introductory overview of the project and related testing activities described in this document.

Scope. This section contains the following subsections.

Identification and Purpose. This subsection should provide information that uniquely identifies the System effort and the associated test plan. This can also be provided in the form of a unique test plan identifier. The following text provides an example:

> This system test plan is to detail the testing planned for the [Project Name] ([Project Abbreviation]) Version [xx], Statement of Work, [date], Task Order [to number], Contract No. [Contract #], and Amendments.
>
> The goal of [Project Acronym] development is to [goal]. This system will allow [purpose].
>
> Specifics regarding the implementation of these modules are identified in the [Project Abbreviation] System Requirements Specification (SysRS) with line item descriptions in the accompanying Requirements Traceability Matrix (RTM).

System Overview. This subsection should provide a summary of all system items and system features to be tested. The need for each item and its history may be addressed here as well. References to associated project documents should be cited here. The following provides an example:

> This document describes the System Test Plan (SysTP) for the [Project Abbreviation] System. [Project Abbreviation] documentation will include this SysTP, the [Project Abbreviation] System Requirements Specification (SysRS), the System Requirements Traceability Matrix (RTM), the [Project Abbreviation] Software Development Plan (SDP), the Software Design Document (SDD), the [Project Abbreviation] User's Manual, the [Project Abbreviation] System Administrator's Manual, and the [Project Abbreviation] Data Dictionary.
>
> This SysTP describes the process to be used and the deliverables to be generated in the testing of the [Project Abbreviation]. This plan, and the items defined herein, will comply with the procedures defined in the [Project Abbreviation] Software Configuration Management (SCM) Plan and the [Project Abbreviation] Software Quality Assurance (SQA) Plan. Any nonconformance to these plans will be documented as such in this SysTP.
>
> This document is based on the [Project Abbreviation] System Test Plan template with tailoring appropriate to the processes associated with the creation of the [Project Abbreviation]. The information contained in this SysTP has been created for [Customer Name] and is to be considered to be "for official use only."
>
> Please refer to section "6. Schedule" in the SDP for information regarding documentation releases.

Definitions, Acronyms, and Abbreviations.   All relevant definitions, acronyms, and abbreviations should be included in this subsection.

Referenced Documents.   This subsection to the Definitions, Acronyms, and Abbreviations subsection should include all material referenced during the creation of the test plan.

System Test Objectives.   This subsection to the Definitions, Acronyms, and Abbreviations subsection should include the validation of the application (i.e., to determine if it fulfills its system requirements specification), the identification of defects that are not efficiently identified during unit and integration testing, and the determination of the extent to which the system is ready for launch. This subsection should provide project status measures (e.g., percentage of test scripts successfully tested). A description of some of the kinds of testing may be found in Appendix A, Software Process Work Products, Types of System Testing.

System Test Environment.   This section includes the following subsections.

Development Test and Evaluation.   This subsection provides a high level summary of all key players, facilities required, and site of test performance. The following is provided as an example:

> System tests are to be performed at [Company Name], [Site Location]. All testing will be conducted in the development center [Room Number]. The following individuals must be in attendance: [provide list of performers and observers].

System and Software Items.   This section should provide a complete list of all system and software items used in support of testing, including versions. If the system is kept in an online repository, reference to the storage location may be cited instead of listing all items here. The following is an example:

System and software used in the testing of [Project Abbreviation]:
    [System and software List]
    For details of client and server system specifications see the [Project Abbreviation] system requirements specifications (SysRS).

Hardware and Firmware Items. This subsection should provide a complete list of all hardware and firmware items used in support of testing. If this information is available in another project document, it may be referenced instead of listing all equipment. The following is an example:

> [Hardware Items] are to be used during testing, connected together in a client–server relationship through a TCP/IP-compatible network. For details of client and server hardware specifications, see the [Project Abbreviation] system requirements specifications (SysRS). For integration and qualification level testing a representative hardware set as specified in the SysRS is to be used.

Other Materials. This subsection should list all other materials required to support testing.

Proprietary Nature, Acquirer's Rights, and Licensing. Any of the issues associated with the potential proprietary nature of the software, acquirer's rights, or licensing should be addressed in this subsection. An example is provided:

> Licensing of commercial software is one purchased copy for each PC it is to be used on with the exception of [Group Software], which also requires licensing for the number of users/client systems. Some of the information in the dataset is covered by the Privacy Act and will have to be protected in some manner.

Installation, Testing, and Control. This subsection should address environment requirements or actions required to support the installation of the application for testing. The following example is provided:

> [Company Name] will acquire the software and hardware needed as per the [Project Abbreviation] software development plan (SDP) in order to run [Project Abbreviation] in accordance with contract directions and limitations.
>
>   [Company Name] will install the supporting software and test it per the procedures in the [Project Abbreviation] Installation System Test Description.
>
>   The [Project Abbreviation] Change Enhancement Request (CER) tracking system will be used to determine eligibility for testing at any level. See the [Project Abbreviation] Software Configuration Management Plan and [Project Abbreviation] Software Quality Assurance Plan for details on CER forms and processes.

Participating Organizations. This subsection should identify all groups responsible for testing. An example is provided:

> The participating organizations are [Company Name], [customer name], and organizations at the beta test site.

These organizations may participate by completing the *Test Plan Walk-through Checklist* as described in Appendix A, Software Process Work Products.

Test Site(s). This subsection should provide a description of test sites, for example:

> The planned installation-level beta test sites are listed as follows:
>   [Installation Beta Sites]

Test Indentification. This section contains the following subsections.

Test Levels. This subsection should describe the different levels of testing required in support of the development effort. See Appendix A for Test Design Specification, Test Case Specification, and Test Procedure Specification. The following is provided as an example:

> Tests are to be performed at the module, integration, installation, and qualification levels prior to release for beta testing. Please refer to test design specifications #xx through xx.

Test Classes. This subsection should provide a description of all test classes. A test case specification should be associated with each of these classes. A summary of the validation method, data to be recorded, data analysis activity, and assumptions should be provided for each case. An example text is provided in Appendix A, Example Test Classes.

General Test Conditions. This subsection should describe the anticipated baseline test environment. An example is provided below:

> A sample real dataset from an existing database is to be used during all tests. The sample real data set will be controlled under the configuration management system. A copy of the controlled data set is to be used in the performance of all testing.

Test Progression. This subsection should provide a description of the progression of testing. A diagram is often helpful when attempting to describe the progression of testing (see Figure 4.2). The information below is provided as an example:

> Qualification testing is a qualification-level test verifying that all requirements have been met. The module and integration tests are performed as part of the implementation phase as

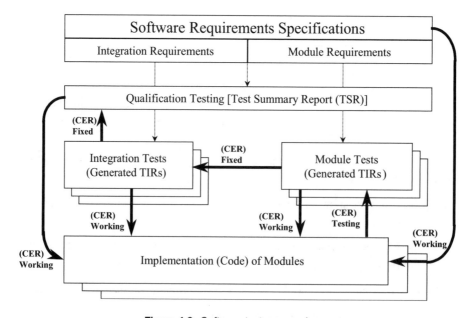

Figure 4.2. Software test progression.

elements and modules are completed. All module and integration tests must be passed before performing system testing. All module tests must be passed before performing associated integration-level tests. The CER tracking system is used to determine eligibility for testing at a level. See the [Project Abbreviation] Software Configuration Management Plan and [Project Abbreviation] Software Quality Assurance Plan for details on CER forms and processes.

Planned Testing. This subsection should provide a detailed description of the type of testing to be employed. Example text follows:

Module Test. All of the modules to be tested (refer to the Software Requirements Traceability Matrix) are to be tested using defined module-level test methodology. For details regarding test procedures and setup, see the [Project Abbreviation] Module Software Test Description. The resulting outputs of this test are internal test reports (ITR) or unit test reports (UTR). When all of the module tests are passed for a module, the module is ready for integration testing.

For the module-level tests, the following classes of tests will be used:

Check for correct handling of erroneous inputs
Check for maximum capacity
User interaction behavior consistency
Retrieving data
Saving data
Display screen and printing format consistency
Measure time of reaction to user input

Integration Test. All of the modules to be integration tested (refer to System Requirements Traceability Matrix) will be tested using integration-level test methodology. For details of the procedures and setup, see the [Project Abbreviation] Integration System Test Description. The resulting outputs of this test are internal test reports (ITR) or system integration test reports (SITR). When these integration tests are all passed, the [Project Abbreviation] System will be ready for qualification-level testing.

For the integration-level tests, the following classes of tests will be used:

User interaction behavior consistency
Display screen and printing format consistency
Check interactions between modules
Measure time of reaction to user input

System Test. All of the requirement test items (refer to the System Requirements Traceability Matrix) are to be tested as system-level tests. For details of the procedures and setup, see the [Project Abbreviation] System Test Plan Description. The resulting output of system test is the [Project Abbreviation] system test report (SysTR) and its attached test problem reports (TPR). If system test is passed and its results accepted by the [Stakeholder], the [Project Abbreviation] System will be ready for qualification test. Following qualification testing, the customer will review results. Signature of acceptance initiates product delivery and start of installation beta test.

Qualification Test. All of the requirement test items (refer to the System Requirements Traceability Matrix) are to be tested as qualification-level tests. For details of the procedures and setup, see the [Project Abbreviation] Qualification System Test Description. The resulting output of qualification test is the [Project Abbreviation] system test report (SysTR) and its attached test problem reports (TPR). If qualification test is passed and its results accepted by the customer, the [Project Abbreviation] system will be ready for beta release. Following qualification testing, the customer will review results. Signature of acceptance initiates product delivery and start of installation beta test.

For the qualification-level tests the following classes of tests will be used:
Check for correct handling of erroneous inputs
Check for maximum capacity
User interaction behavior consistency
Retrieving data
Saving data
Display screen and printing format consistency
Check interactions between modules
Measure time of reaction to user input
Functional Flow

Installation Beta Test. Following qualification testing, the customer will review results. Signature of acceptance initiates product delivery and start of installation beta test. Identified tests (refer to the System Requirements Traceability Matrix) will be tested using defined installation level beta test procedures. Some of these tests may be documented in the Product Packaging Information in Appendix A. This test methodology will be applied at each beta test site. For details regarding test procedures and setup, see the [Project Abbreviation] Installation System Test Description. Outputs of these tests are test problem reports (TPR). Following installation testing, the customer will review results. Signature of acceptance completes the [Project Abbreviation] Version 1.0 project.

For the installation-level tests, the following classes of tests will be used:
User interaction behavior consistency
Retrieving data
Saving data
Display screen and printing format consistency
Check interactions between modules
Measure time of reaction to user input
Functional flow

Test Schedules. This section should include test milestones as identified in the system project schedule as well as any required deliverables associated with testing. Any additional milestones may be defined here as needed. If appropriate detail is provided in the Software Project Management Plan (SRMP), then this can be referenced. The following is provided as an example:

Test schedules are defined in the [PROJECT ABBREVIATION] Software Development Plan.

Risk Management. This section should identify all high-risk assumptions of the test plan. The contingency plans for each should be described and the management of risk items should be addressed.

Requirements Traceability. This section should provide information regarding the traceability of testing to the requirements and design of the system. A matrix, or database, is useful when meeting this traceability requirement. The following example is provided:

Refer to [Project Abbreviation] System Requirements Specification, Appendix A for information regarding requirements traceability.

Notes. This section should provide any additional information not previously covered in the test development plan.

System Integration Test Plan

The following information is based on IEEE Std 829, IEEE Standard for Software Test Documentation [15], and IEEE Std 12207.0, Standard for Information Technology—Life cycle processes [40]. Additional information is provided in the document template, *System Integration Test Plan.doc,* which is located on the companion CD-ROM. Table 4-11 describes an outline of suggested document content.

The following provides section-by-section guidance in support of the creation of a System Integration Test Plan. This guidance should be used to help define associated test-

Table 4-11. System integration test plan document outline

Title Page
Revision Page
Table of Contents
1. Introduction
2. Scope
    2.1 Identification and Purpose
    2.2 System Overview
    2.3 Definitions, Acronyms, and Abbreviations
3. Referenced Documents
4. System Integration Test Objectives
5. System Integration Testing Kinds and Approaches
    5.1 Software Integration Testing
    5.2 COTS Integration Testing
    5.3 Database Integration Testing
    5.4 Hardware Integration Testing
    5.5 Prototype Usability Testing
    5.6 Approaches
6. System Integration Test Environment
    6.1 Development Test and Evaluation
    6.2 System and Software Items
    6.3 Hardware and Firmware
    6.4 Other Materials
    6.5 Proprietary Nature, Acquirer's Rights, and Licensing
    6.6 Participating Organizations
    6.7 Test Sites
7. Test Identification
    7.1 General Information
    7.2 Test Levels
    7.3 Test Classes
    7.4 General Test Conditions
    7.5 Test Progression
    7.6 Planned Testing
    7.7 Integration Test
8. Test Schedules
9. Risk Management
10. Requirements Traceability
11. Notes
Appendix A. System Test Requirements Matrix
Appendix B. System Integration Test Description

ing requirements and should reflect the actual requirements of the implementing organization.

Introduction. This section should provide a brief introductory overview of the project and related testing activities described in this document.

Identification and Purpose. This subsection should provide information that uniquely identifies the system effort and the associated test plan. This can also be provided in the form of a unique test plan identifier. The following text provides an example:

> This system integration test plan is to detail the testing planned for the [Project Name] ([Project Abbreviation]) Version [xx], Statement of Work, [date], Task Order [to number], Contract No. [Contract #], and Amendments.
>
> The goal of [Project Acronym] development is to [goal]. This system will allow [purpose].
>
> Specifics regarding the implementation of these modules are identified in the [Project Abbreviation] System Requirements Specification (SysRS) with line item descriptions in the accompanying Requirements Traceability Matrix (RTM) and references to the Interface Control Documentation (ICD).

System Overview. This subsection should provide a summary of all system items and system features to be tested. The need for each item and its history may be addressed here as well. References to associated project documents should be cited here. The following provides an example:

> This document describes the system integration test plan (SysITP) for the [v] system. [Project Abbreviation] documentation will include this SysITP, the [Project Abbreviation] System Requirements Specification (SysRS), the System Requirements Traceability Matrix (RTM), the Interface Control Documentation (ICD), the [Project Abbreviation] Software Development Plan (SDP), the Software Design Document (SDD), the [Project Abbreviation] User's Manual, the [Project Abbreviation] System Administrator's Manual, and the [Project Abbreviation] Data Dictionary.
>
> This SysITP describes the process to be used and the deliverables to be generated in the testing of the [Project Abbreviation]. This plan, and the items defined herein, will comply with the procedures defined in the [Project Abbreviation] Software Configuration Management (SCM) Plan and the [Project Abbreviation] Software Quality Assurance (SQA) Plan. Any nonconformance to these plans will be documented as such in this SysITP.
>
> This document is based on the [Project Abbreviation] system integration test plan template with tailoring appropriate to the processes associated with the creation of the [Project Abbreviation]. The information contained in this SysITP has been created for [Customer Name] and is to be considered "for official use only."
>
> Please refer to section "6. Schedule" in the SDP for information regarding documentation releases.

Definitions, Acronyms, and Abbreviations. All relevant definitions, acronyms, and abbreviations should be included in this subsection.

Referenced Documents. This section should include all material referenced during the creation of the test plan.

System Integration Test Objectives. Test objectives should include the verification of the product integration (i.e., to determine if it fulfills its system requirements specification and architectural design), the identification of defects that are not efficiently identified during unit testing, and the determination of the extent to which the subsystems/system is ready for system test. This section should provide project status measures (e.g., percentage of test scripts successfully tested).

System Integration Testing Kinds and Approaches. This section contains the following subsections.

Software Integration Testing. Incremental integration testing of two or more integrated software components on a single platform or multiple platforms is conducted to produce failures caused by interface defects. This subsection should provide references to the SysRS, design documentation, and interface control document to show a complete list of subsystem or system modules and the resulting test strategy.

COTS Integration Testing. Incremental integration testing of multiple commercial-off-the-shelf (COTS) software components should be conducted to determine if they are not interoperable (i.e., if they contain any interface defects). This subsection should provide references to the SysRS, design documentation, and interface control document to show a complete list of subsystem or system modules and the resulting test strategy.

Database Integration Testing. Incremental integration testing of two or more integrated software components is conducted to determine if the application software components interface properly with the database(s). This subsection should provide references to the SysRS, design documentation, and interface control document to show a complete list of subsystem or system modules and resulting test strategy.

Hardware Integration Testing. Incremental integration testing of two or more integrated hardware components in a single environment is conducted to produce failures caused by interface defects. This section should provide references to the SysRS, design documentation, and interface control document to show a complete list of subsystem or system modules and the resulting test strategy.

Prototype Usability Testing. Incremental integration testing of a user interface prototype against its usability requirements is conducted to determine if it contains any usability defects. This section should provide references to the SysRS, design documentation, and user manual to show a complete list of subsystem or system modules and the resulting test strategy.

Approaches. Top-down testing consists of integrating high-level components of a system before their design and implementation has been completed. Bottom-up testing consists of integrating low-level components before the high-level components have been developed. Object-oriented testing consists of use-case or scenario-based testing, thread testing, and/or object interaction testing. Interface testing consists of integrating modules or subsystems to create large systems.

System Integration Test Environment. This section contains the following subsections.

Development Test and Evaluation. Provide a high-level summary of all key players, facilities required, and site of test performance. The following is provided as an example:

System integration tests are to be performed at [company name], [site location]. All integration testing will be conducted in the development center [room number]. The following individuals must be in attendance: [provide list of performers and observers].

System and Software Items. This subsection should provide a complete list of all system and software items used in support of integration testing, including versions. If the system is kept in an online repository, reference to the storage location may be cited instead of listing all items here. The following is an example:

System and software used in the integration testing of [Project Abbreviation]:
[System and software List]
For details of client and server system specifications see the [Project Abbreviation] system requirements specifications (SysRS).

Hardware and Firmware Items. This subsection should provide a complete list of all hardware and firmware items used in support of integration testing. If this information is available in another project document, it may be referenced instead of listing all equipment. The following is an example:

[Hardware Items] are to be used during integration testing, connected together in a client-server relationship through a TCP/IP-compatible network. For details of client and server hardware specifications, see the [Project Abbreviation] System Requirements Specifications (SysRS). For integration testing, a representative hardware set as specified in the SysRS is to be used.

Other Materials. This subsection should list all other materials required to support integration testing.

Proprietary Nature, Acquirer's Rights, and Licensing. Any of the issues associated with the potential proprietary nature of the software, acquirer's rights, or licensing should be addressed in this subsection. An example is provided:

Licensing of commercial software is one purchased copy for each PC it is to be used on with the exception of [Group Software], which also requires licensing for the number of users/client systems. Some of the information in the dataset is covered by the Privacy Act and will have to be protected in some manner.

Participating Organizations. This subsection should identify all groups responsible for integration testing. An example is provided:

The participating organizations are [Company Name], [Customer Name], and organizations at the beta test site.

They may participate in completing the Test Plan Walk-through Checklist. An example is identified in Appendix A of this text, Software Process Work Products.

Test Site(s). This subsection should provide a description of test sites. See below:

The planned integration test sites are listed as follows:
[Integration Sites]

Test Identification. This section includes the following subsections.

Test Levels. This subsection should describe the different levels of testing required in support of the development effort. See Appendix A, *Test Design Specification, Test Case Specification,* and *Test Procedure Specification.* The following is provided as an example:

> Tests are to be performed at the integration level prior to release for system testing. Please refer to test design specifications #xx through xx.

Test Classes. This subsection should provide a description of all test classes. A test case specification should be associated with each of these classes. A summary of the validation method, data to be recorded, data analysis activity, and assumptions, should be provided for each case.

General Test Conditions. This subsection should describe the anticipated baseline test environment. Refer to the example provided below:

> A sample real dataset from an existing database is to be used during all tests. The sample real dataset will be controlled under the configuration management system. A copy of the controlled dataset is to be used in the performance of all testing.

Test Progression. This subsection should provide a description of the progression of testing. A diagram is often helpful when attempting to describe the progression of testing. The information below is provided as an example:

> The system integration testing is testing that verifies that all subsystems or modules work together. The module tests are performed prior or during the implementation phase as elements and modules are completed. All module and integration tests must be passed before performing system testing. The Change Enhancement Request (CER) tracking system is used to determine eligibility for testing at a level. See the [Project Abbreviation] Software Configuration Management Plan and [Project Abbreviation] Software Quality Assurance Plan for details on CER forms and processes.

Planned Testing. This subsection should provide a detailed description of the type of testing to be employed. The following is provided as an example:

> A summary of testing is provided in section xxx. Additional information is provided in Appendices A and B.

Integration Test. This subsection should provide a detailed description of the type of integration testing to be employed, for example,

> All of the modules to be integration tested (refer to the System Requirements Traceability Matrix) will be tested using integration-level test methodology. For details of the procedures and setup, see the [Project Abbreviation] System Integration Test Description. The resulting outputs of this test are Internal Test Reports (ITR) or System Integration Test Reports (SITR). When these integration tests are all passed, the [Project Abbreviation] system will be ready for system-level testing.

For the integration-level tests the following classes of tests will be used:
  User interaction behavior consistency
  Display screen and printing format consistency
  Check interactions between modules
  Measure time of reaction to user input

Test Schedules. This section should include test milestones as identified in the system project schedule as well as any required deliverables associated with testing. Any additional milestones may be defined here as needed. If appropriate detail is provided in the Software Project Management Plan (SRMP), then this can be referenced. The following is provided as an example:

  Test schedules are defined in the [Project Abbreviation] Software Development Plan.

Risk Management. This section should identify all high-risk assumptions of the test plan. The contingency plans for each should be described and the management of risk items should be addressed.

Requirements Traceability. This section should provide information regarding the traceability of testing to the requirements and design of the System. A matrix, or database, is useful when meeting this traceability requirement. The following is provided as an example:

  Refer to [Project Abbreviation] System Requirements Specification, Appendix A for information regarding requirements traceability.

Notes. This section should provide any additional information not previously covered in the test development plan.

Appendix A. System Test Requirements Matrix. This is redundant. Reference should only be made to SysRS Appendix A and test information attached as another column.

Appendix B. System Integration Test Description. This should contain a detailed description of each test.

JOINT REVIEW

The joint review process evaluates the status and products of an activity of a project as appropriate. Joint reviews are at both project management and technical levels and are held throughout the life of the contract. Joint reviews may include the customer and/or the supplier. Examples are project management reviews and technical reviews. Throughout the development process, the technical reviews complete the activities of:

- Software requirements analysis
- Software architectural design
- Software detailed design
- Software integration
- Software acceptance support

Table 4-12 provides a list of the Joint Review process objectives.

Table 4-12. Joint review process objectives

---
a) Evaluate the status and products of an activity of a process through joint review activities between the parties to a contract;
b) Establish mechanisms to ensure that action items raised are recorded for action.

Technical Reviews

The information provided here is based upon the recommendations provided in IEEE Std 1028, IEEE Standard for Software Reviews. Technical reviews are an effective way to detect and identify problems early in the software development process. The criteria for technical reviews are presented here to help organizations define their review practices.

Introduction. Technical reviews are an effective way to evaluate a software product by a team of qualified personnel to identify discrepancies from specifications and standards. Although technical reviews identify anomalies, they may also provide the recommendation and examination of various alternatives. The examination need not address all aspects of the product and may only focus on a selected aspect of a software product.

Software requirements specifications, design descriptions, test and user documentation, installation and maintenance procedures, and build processes are all examples of items subject to technical reviews.

Responsibilities. The roles in support of a technical review are:

Decision maker. The decision maker is the individual requesting the review and determines whether objectives have been met.

Review leader. The review leader is responsible for the review and must perform all administrative tasks (including summary reporting), ensure that the review is conducted in an orderly manner, and ensure that the review meets its objectives.

Recorder. The recorder is responsible for the documentation of all anomalies, action items, decisions, and recommendations made by the review team.

Technical staff. The technical staff should actively participate in the review and evaluation of the software product.

The following roles are optional and may also be established for the technical review:

Management staff. Management staff may participate in the technical review for the purpose of identifying issues that will require management resolution.

Customer or user representative. The role of the customer or user representative should be determined by the review leader prior to the review.

Input. Input to the technical review should include a statement of objectives, the software product being examined, existing anomalies and review reports, review procedures, and any standard against which the product is to be examined. Anomaly categories should be defined and available during the technical review. For additional information in support of the categorization of software product anomalies, refer to IEEE Std 1044 [7].

Authorization. All technical reviews should be defined in the SPMP. The plan should describe the review schedule and all allocation of resources. In addition to those technical reviews required by the SPMP, other technical reviews may be scheduled.

Preconditions. A technical review should be conducted only when the objectives have been established and the required review inputs are available.

Management Preparation. Managers are responsible for ensuring that all reviews are planned, that all team members are trained and knowledgeable, and that adequate resources are provided. They are also responsible for ensuring that all review procedures are followed.

Planning the Review. The review leader is responsible for the identification of the review team and their assignment of responsibility. The leader should schedule and announce the meeting, prepare participants for the review by providing them the required material, and collect all comments.

Overview of Review Procedures. The team should be presented with an overview of the review procedures. This overview may occur as a part of the review meeting or as a separate meeting.

Overview of the Software Product. The team should receive an overview of the software product. This overview may occur either as a part of the review meeting or as a separate meeting.

Preparation. All team members are responsible for reviewing the product prior to the review meeting. All anomalies identified during this prereview process should be presented to the review leader. Prior to the review meeting, the leader should classify all anomalies and forward these to the author of the software product for disposition.

The review leader is also responsible for the collection of all individual preparation times to determine the total preparation time associated with the review.

Examination. The review meeting should have a defined agenda that should be based upon the premeeting anomaly summary. Based upon the information presented, the team should determine whether the product is suitable for its intended use, whether it conforms to appropriate standards, and whether it is ready for the next project activity. All anomalies should be identified and documented.

Rework/Follow-up. The review leader is responsible for verifying that the action items assigned in the meeting are closed.

Exit Criteria. A technical review is considered complete when all follow-up activities have been completed and the review report has been published.

Output. The output from the technical review should consist of the project being reviewed, a list of the review team members, a description of the review objectives and a list of resolved and unresolved software product anomalies. The output should also include a list of management issues, all action items and their status, and any recommendations made by the review team.

Management Reviews

The information provided here is based upon the recommendations provided in IEEE Std 1028, IEEE Standard for Software Reviews. Management reviews are an effective way to detect and identify problems early in the software development process. The criteria for management reviews are presented here to help organizations define their review practices. It is important to remember that management reviews are not only about the specific review of products, but also may cover aspects of the software process.

Introduction. The purpose of a management review is to monitor project progress and to determine the status according to documented plans and schedules. Reviews can also be used to evaluate the effectiveness of the management approaches. The effective use of management reviews can support decisions about corrective action, resource allocation, or changes in scope.

Management reviews may not address all aspects of a given project during a single review but may require several review cycles to completely evaluate a software product. Examples of software products subject to management review include:

Anomaly reports
Audit reports
Backup and recovery plans
Contingency plans
Customer complaints
Disaster plans
Hardware performance plans
Installation plans
Maintenance plans
Procurement and contracting methods
Progress reports
Risk management plans
Software configuration management plans
Software project management plans
Software quality assurance plans
Software safety plans
Software verification and validation plans
Technical review reports
Software product analyses
Verification and validation reports

Responsibilities. The management personnel associated with a given project should carry out management reviews. The individuals qualified to evaluate the software product should perform all management reviews.

Decision Maker. The review is conducted for the decision maker and it is up to this individual to determine whether the objectives of the review have been met.

Review Leader. The review leader is responsible for all administrative tasks required in support of the review. The review leader should ensure that all planning and preparation have been completed, that all objectives are established and met, and that the review is conducted and review outputs are published.

Recorder. The recorder is responsible for the documentation of all anomalies, action items, decisions, and recommendations made by the review team.

Management Staff. The management staff is responsible for carrying out the review. They are responsible for active participation in the review.

Technical Staff. The technical staff are responsible for providing the information required in support of the review.

Customer or User Representative. The role of the customer or user representative should be determined by the review leader prior to the review.

Input. Input to the management review should include the following:

   A statement of review objectives
   The software product being evaluated
   The software project management plan
   Project status
   A current anomalies list
   Documented review procedures
   Status of resources, including finance, as appropriate
   Relevant review reports
   Any associated standards
   Anomaly categories

Authorization. The requirement for conducting management reviews should initially be established in the project planning documents. The completion of a software product or completion of a scheduled activity may initiate a management review. In addition to those management reviews required by a specific plan, other management reviews may be announced and held.

Preconditions. A management review should be conducted only when the objectives for the review have been established and all required inputs are available.

Management Preparation. Managers should ensure that the review is performed as planned and that appropriate time and resources are allocated in support of the review process. They should ensure that all participants are technically qualified and have received training and orientation in support of the review. They are responsible for ensuring that all reviews are carried out as planned and that any resulting action items are completed.

Planning the Review. The review leader is responsible for identifying the review team and the assignment of specific responsibilities. They are responsible for scheduling the meeting and for the distribution of all materials required in support of review preparation activities. Prereview comments should be collected by the review leader, classified, and presented to the author prior to the review.

Overview of Review Procedures. A qualified person should present an overview of the review procedures for the review team if requested by the review leader.

Preparation. Each review team member should review the software product and any other review inputs prior to the review meeting. All anomalies detected during this examination should be documented and sent to the review leader.

Examination. The management review should consist of one or more meetings of the review team. The meetings should review the objectives of the review, evaluate the software product and product status, review all items identified prior to the review meeting, and generate a list of action items, including associated risk. The meeting should be documented. This documentation should include any risk issues that are critical to the success of the project, provide a confirmation of the software requirements, list action items, and identify other issues that should be addressed.

Rework/Follow-up. The review leader is responsible for ensuring that all action items assigned in the meeting are closed.

Exit Criteria. The management review is complete when the activities listed when all required outputs and follow-up activities are finished.

Output. The output from the management review should identify:

The project being reviewed
The review team members
Review objectives
Software product reviewed
Specific inputs to the review
All action items
A list of all anomalies

AUDIT

The audit process determines the degree of organizational and project compliance with the processes, requirements, plans, and contract. This process is employed by one party (auditing party) who audits the software products or activities of another party (audited party). The audit produces a list of detected issues or problems, which are recorded and entered into the Problem Resolution Process. Table 4-13 provides a list of the audit process objectives.

Table 4-13. Audit process objectives

a) Determine compliance with requirements, plans, and contract, as appropriate.
b) Arrange the conduct of audits of work products or process performance by a qualified independent party, as specified in the plans.
c) Conduct follow-up audits to assess corrective action(s), closure, and root-cause actions.

Audits

The information provided here is based upon the recommendations provided in IEEE Std 1028, IEEE Standard for Software Reviews. Audits an effective way to detect and identify problems in the software development process and to ensure that practices and procedures are being implemented as expected. Audits can also be performed by comparing what people are doing against established plans and procedures. The criteria for audits are presented here to help organizations define their review practices.

Introduction. A software audit provides an independent evaluation of conformance. Examples of software products subject to audit include the following:

    Backup and recovery plans
    Contingency plans
    Contracts
    Customer complaints
    Disaster plans
    Hardware performance plans
    Installation plans and procedures
    Maintenance plans
    Management review reports
    Operations and user manuals
    Procurement and contracting methods
    Reports and data
    Risk management plans
    Software configuration management plans
    Software design descriptions
    Source code
    Unit development folders
    Software project management plans
    Software quality assurance plans
    Software requirements specifications
    Software safety plans
    Software test documentation
    Software user documentation
    Software verification and validation plans

Responsibilities. The following roles should be established for an audit:

Lead Auditor. The lead auditor is responsible for all administrative tasks, assembly of the audit team and their management, and for ensuring that the audit meets its objectives. The lead auditor should prepare a plan for the audit and summary activities with an audit report.

    The lead auditor should be free from bias and influence that could reduce any ability to make independent, objective evaluations.

Recorder. The recorder should document all anomalies, action items, decisions, and recommendations made by the audit team.

Auditor. The auditors should examine all products as described in the audit plan. They should record their observations and recommend corrective actions. All auditors should be free from bias and influences that could reduce their ability to make independent, objective evaluations.

Initiator. The initiator determines the need, focus, and purpose of the audit. The initiator determines the members of the audit team and reviews the audit plan and audit report.

Audited Organization. The audited organization should provide a liaison to the auditors and is responsible for providing all information requested by the auditors. When the audit is completed, the audited organization should implement corrective actions and recommendations.

Input. Inputs to the audit should be listed in the audit plan and should include the purpose and scope of the audit, a list of products to be audited, and audit evaluation criteria.

Entry Criteria
Authorization. An initiator decides upon the need for an audit. A project milestone or a nonroutine event, such as the suspicion or discovery of a major nonconformance, may drive this decision. The initiator selects an auditing team and provides the auditors with all supporting information. The lead auditor produces an audit plan and the auditors prepare for the audit.

Preconditions. An audit should only be conducted if the audit has proper authorization, the audit objectives have been established, and all audit items are available.

Procedures
Management Preparation. Managers are responsible for ensuring that the appropriate time and resources have been planned in support of the audit activities. Managers should also ensure that adequate training and orientation on audit procedures is provided.

Planning the Audit. The audit plan should describe the purpose and scope of the audit, the audited organization, including location and management, and a list of software products to be audited. The plan should describe all evaluation criteria, auditor's responsibilities, examination activities, resource requirements and schedule, and required reporting.

The initiator should approve the audit plan but the plan should be flexible enough to allow for changes based on information gathered during the audit, subject to approval by the initiator.

Opening Meeting. An opening meeting between the audit team and audited organization should be scheduled to occur at the beginning of the examination activity of the audit. This meeting should cover the purpose and scope of the audit, all software products being audited, all audit procedures and outputs, audit requirements, and audit schedule.

Preparation. The initiator should notify the audited organization's management in writing before the audit is performed, except for unannounced audits. The purpose of notification is to ensure that the people and material to be examined in the audit are available.

Auditors should prepare for the audit by reviewing the audit plan and any information available about the audited organization and the products to be audited. The lead auditor should prepare for the team orientation and any necessary training. The lead auditor should also ensure that facilities are available for the audit interviews and that all materials are available.

Examination. The examination should consist of evidence collection and analysis with respect to the audit criteria as describe in the audit plan. A closing meeting between the auditors and audited organization should be conducted, and then an audit report should be prepared.

Evidence Collection. The auditors should collect evidence of conformance and nonconformance by interviewing audited organization staff, examining documents, and witnessing processes. The auditors should attempt all the examination activities defined in the audit plan and undertake additional activities if they consider them to be required to determine conformance. Auditors should document all observations of conformance or nonconformance. These observations should be categorized as major or minor. An observation should be classified as major if the nonconformity will likely have a significant effect on product quality, project cost, or project schedule. All observations should be discussed with the audited organization before the closing audit meeting.

Closing Meeting. The lead auditor should convene a closing meeting with the audited organization's management. The closing meeting should review the progress of the audit, all audit observations, and preliminary conclusions and recommendations. Agreements should be reached during the closing audit meeting and must be completed before the audit report is finalized.

Reporting. The lead auditor is responsible for the preparation of the audit report. The audit report should be prepared as quickly as possible following the audit. Any communication between auditors and the audited organization made between the closing meeting and the issue of the report should pass through the lead auditor. The lead auditor should send the audit report to the initiator who will distribute the audit report within the audited organization.

Follow-up. The initiator and audited organization should determine what corrective action may be required and the type of corrective action to be performed.

Exit Criteria. An audit is complete when the audit report has been delivered to the initiator and all audit recommendations have been performed.

Output. The output of the audit is the audit report. The audit report should describe the purpose and scope of the audit; the audited organization, including participants; identification of all products included in the audit, and recommendations. All criteria (e.g., standards and procedures) in support of the audit should be identified in the plan. A summary of all audit activities should be provided. All classified observations should be included, as well as a summary and interpretation of the findings. A schedule and list of all follow-up activities should be described.

Software Measurement and Measures Plan

The importance of software process measures and measurements is critical for both the software engineering foundation and the implementation and success of Lean Six Sigma.

In every industry there are significant differences between the leaders and the laggards in terms of market shares, technological sophistication, and quality and productivity levels. In the software industry one of the most significant differences is that the leaders know their quality and productivity levels because they measure them. The laggards do not measure. Therefore the laggards do not have a due as to how good or bad they are.

Companies that do not measure tend to waste scarce investment dollars on approaches that consume time and energy but accomplish very little. Surprisingly, investment in good quality and productivity measurement programs has one of the best returns on investment of any known software technology. [187]

The following provides a suggested format for a project-level measurement and measures* plan. This plan should contain a description of all measurement and analysis used in support of an identified software effort.

Additional information in support of related work products is provided in Appendix A, Software Process Work Products. These include a list of measures for reliable software and the measurement information model as described in ISO/IEC 15939. Table 4-14 provides a proposed software measurement and measures plan document outline in support of Lean Six Sigma requirements.

Table 4-14. Software measurement and measurements plan document outline

Title Page
Revision Page
Table of Contents
1. Introduction
    1.1 Purpose
    1.2 Scope
    1.3 Definitions, Acronyms, and Abbreviations
    1.4 References
2. Measurements Process Management
    2.1 Responsibilities
    2.2 Life Cycle Reporting
    2.3 General Information
3. Measurement and Measures

Software Measurement and Measures Plan Document Guidance

The following provides section-by-section guidance in support of the creation of a software measurement and measures plan (MMP). The MMP should be considered to be a living document. The MMP should change to reflect any process improvement or changes in procedures relating to software measurement and measures activities. This guidance should be used to help define a measurement and measures process and should reflect the actual processes and procedures of the implementing organization. Additional information is provided in the document template, *Software Measurement and Measures Plan.doc,* which is located on the companion CD-ROM.

---

*As applied here, "measure" is a problematic term because it has no counterpart in generally accepted metrology. Generally, it is used in software engineering to mean a measurement coupled with a judgmental threshold of acceptability.

Introduction. The introduction is not required, but can be used to state the goals for measurement activities. The following information is provided as an example introduction:

> The goal of the [Project Name] ([Project Abbreviation]) software development effort is to [Project Goal]. This software will provide a method for [Customer Name] to efficiently meet their customer's requirements.
>
> This software measurement and measures (SMM) plan has been developed to ensure that [Project Abbreviation] software products conform to established technical and contractual requirements.

Purpose. This subsection should describe the objectives of measurement and analysis activities and how these activities support the software process. An overview of all measures, data collection and storage mechanisms, analysis techniques, and reporting and feedback mechanisms should be provided. An example is provided:

> The purpose of this SMM plan is to establish, document, and maintain consistent methods for measuring the quality of all software products developed and maintained by the [Project Abbreviation] development team. This plan defines the minimum standardized set for information gathering over the software life cycle. These measures serve as software and system measures and indicators that critical technical characteristics and operational issues have been achieved.

Scope. This subsection should provide a brief description of scope, including the identification of all associated projects and a description of the measure methodology employed. An example is provided below:

> The measures to be utilized by [Project Abbreviation] are described in detail. These measures deemed to be the most feasible and cost-effective to initially implement are described in Table X-X below.

Table X-X. Example Program Measures

Phase	Common issues	Example core metric/ Recommended example measures	Life cycle phase(s)
Requirements	Product Stability	Requirements stability/Numbers of changes in requirements	Tracked throughout project measured against baseline
Requirements	Resources and Cost	Software cost estimation/WBS and cost-based	Tracked throughout project measured against baseline
Requirements	Schedule and Progress	Schedule; development progress/Milestone; Gantt chart	Tracked throughout project measured against baseline
Design	Product Size	Software size; Complexity/SLOC, the number of files, the skill levels of programmers, the size and complexity of the software	Track actual versus planned throughout project
Design	Technology Effectiveness	Requirements traceability/Database or table showing traceability of software requirements	Track throughout life cycle
Implementation	All	Overall project risk, risk tracking/reports of risk	Track throughout life cycle

Table X-X. *Continued*

Phase	Common issues	Example core metric/ Recommended example measures	Life cycle phase(s)
Test	Product quality	Breadth of testing/Completeness of testing; number of trouble reports opened, closed, etc. during test period	Dependent somewhat upon the complexity of the project. Issues associated with breadth of testing should be addressed soon after the definition of requirements. These should be tracked during test and reported to project completion.
Test	Product quality	Fault density/Number of defects discovered in A/size of A (where A is the software product)	These should be tracked during test and reported to project completion.
Deployment	Customer satisfaction	Customer feedback; Support/performance rating, requests for support.	This activity is usually completed at the end of the project but is very effective if performance is milestone driven.

The measures employed by {project abbreviation}deal with the common issues associated with product quality as shown in Table X-X.

Additional measures were evaluated for [Project Abbreviation] Program implementation. Those measures reviewed and deemed "not applicable" are noted as NA.

The specific products developed and maintained by the [Project Abbreviation] development team may be found in the [Project Abbreviation] software configuration management (SCM) Plan.

Definitions, Acronyms, and Abbreviations. This subsection should include all definitions and acronyms used during the development of this plan.

References. This subsection should include a list of all references used during the development of this plan.

Measures Process Management. This section should describe the process for the definition, selection, collection, and reporting of measurement and measures. Using a diagram to help describe the process flow is often helpful. The following provides an example of information typical to this section:

The [Project Abbreviation] Program Manager, the [Project Abbreviation] Project Manager, and the [Company Name] Engineering Process Group (EPG) will work together to develop, maintain, and implement an effective software measurements and measures program. The results of all measures should be reported at Program Management Reviews (PMRs), Configuration Control Board (CCB) reviews, and Pre-Release/validation reviews. Figure X-X (see next page) provides an overview of the [Project Abbreviation] measures collection and reporting process.

Responsibilities. This subsection should describe all responsibilities related to the measurement and measures process. Examples follow.

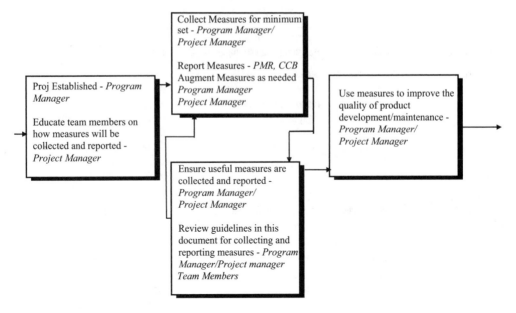

Figure X-X. [Project Abbrev.] Software quality measurement and measures process.

Program Manager. The [Project Abbreviation] Program Manager is ultimately responsible for ensuring that software measurement and measures (SMM) procedures are developed, implemented and maintained. In this capacity, the Program Manager will work with the [Company Name] EPG and applicable [Project Abbreviation] CCBs in establishing standards and implementing EPG-recommended measures.

The Program Manager is ultimately responsible for ensuring that the [Project Abbreviation] Project Manager and the associated [Project Abbreviation] team is trained in SMM policies and procedures that apply to their function on the team. This will consist of initial training for new members and refresher training, as required, due to process or personnel changes.

Project Manager. The [Project Abbreviation] Project Manager shall provide direct oversight of SMM activities associated with project software development, modification, or maintenance tasks assigned within project functional areas. (Refer to Figure X-X.)

The Project Manager is responsible for ensuring that project SMM are collected and reported to the Program Manager. (Refer to Figure X-X.)

Life Cycle Reporting. This subsection should describe all minimum reporting requirements. The following provides an example:

Figure Y-Y shows the applicability of the minimum measure set over the software life cycle. These measures can provide valuable insight into a program, especially with regard to demonstrated results and readiness for test. The results of all measures should be reported at Program Reviews, Staff (Project) Reviews, and all Validation/Test Reviews. Figure Y-Y (see next page) also identifies the measures that should be reported at each major decision milestone. Any other measure that indicates the potential for serious problems should also be reported at indicated milestones.

The applicable time periods for data collection and analysis are provided with each mea-

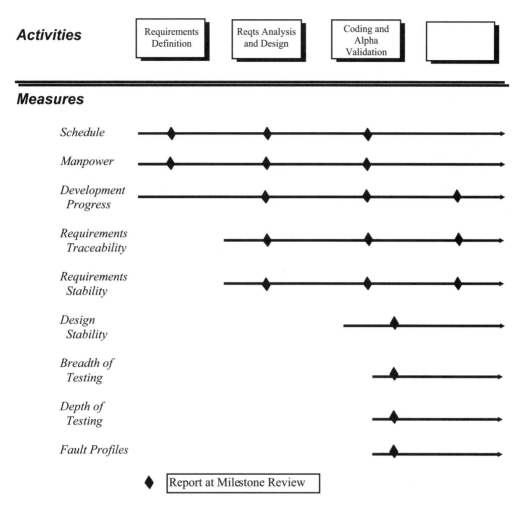

Figure Y-Y. Measures and the Project Life Cycle

sure description. For most measures, data continues to be collected after the system is fielded. The measures collected will be used as program maturity status indicators. They will be used to portray trends over time, rather than single point values. When a measures base is established, trends will be compared with data from similar systems.

General Information. This subsection should provide any additional relevant information not previously provided. This information may include a description of tools, procedures, and/or data storage provision not previously addressed. The following is an example:

The graphical displays shown in this plan (see Figure Z-Z on next page) are provided for illustrative purposes. There may be other ways of processing and displaying the data that are more appropriate for a specific [Project Abbreviation] event. Specific project reports will be archived by [reporting date] by the Program Manager or designee.

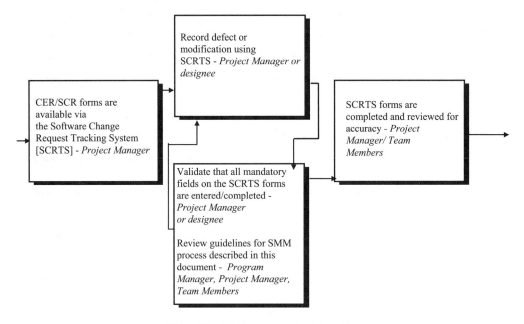

Figure Z-Z. Collecting measures for each activity.

Measurement and Measures. This section (the rest of the plan) should provide a detailed description of all measurement collection and measures calculation. This should also include data requirements, the frequency and type of reporting, interpretation recommendations, and examples when possible. It is often helpful to break this down by categories (e.g., management, requirements, quality). Detailed information should be provided in support of all measures defined within a given category. Example measures are provided in Appendix A, Example Measurements, of this book.

PROBLEM RESOLUTION

The problem resolution process analyzes and resolves the problems (including nonconformances), whatever their nature or source, that are discovered during the execution of development, operation, maintenance, or other processes. Normally, problems found in products under author control are resolved by the author, not in this process. Table 4-15 provides a list of the problem resolution process objectives.

Table 4-15. Validation process objectives

a) Provide a timely, responsive, and documented means to ensure that all discovered problems are analyzed and resolved.

b) Provide a mechanism for recognizing and acting on trends in problems identified.

Risk Management Plan

Risk management focuses on the assessment and control (mitigation) of associated project risk. Risk assessment essentially involves the identification of all of the potential dangers, and their severity, that will affect the project and the evaluation of the probability of occurrence and potential loss associated with each item listed. The control of risk requires risk mitigation through the development of techniques and strategies for dealing with the highest-ordered risks. Risk control also involves the evaluation of the effectiveness of the techniques and strategies employed throughout the project life cycle. Table 4-16 provides a suggested outline for a risk management plan.

The following provides section-by-section guidance in support of the creation of a risk management plan. Additional information is provided in the document template, *Risk Management Plan.doc,* which is located on the companion CD-ROM. This risk management plan template is designed to facilitate the development of a planned approach to risk management activities. This document template used IEEE Std 1540-2001, IEEE Standard for Software Life Cycle Processes—Risk Management [38] and IEEE Std 12207.0 and 12207.1, Standards for Information Technology—Software Life Cycle Processes [40] as primary reference material. Additional information in support of risk-management are Risk Taxonomy, Risk Taxonomy Questionnaire, Risk Action Request, Risk Mitigation Plan, and Risk Matrix Sample in Appendix A, Software Work Products.

Introduction. This section should provide the reader with a basic description of the project, including associated contract information and any standards used in the creation of the document. A description of the document approval authority and process should be provided.

Purpose. This section should provide a description of the motivation for plan implemen-

Table 4-16. Risk management plan document outline

Title Page
Revision Page
Table of Contents
  1. Introduction
  2. Purpose and Scope
  3. Definitions and Acronyms
  4. References
  5. Risk Management Overview
  6. Risk Management Policies
  7. Organization
  8. Responsibilities
  9. Orientation and Training
10. Costs and Schedules
11. Process Description
12. Process Evaluation
13. Communication
14. Plan Management

tation and the expectations for anticipated benefit. Define the perceived benefit from plan execution.

Scope. This section should describe the specific products and the portion of the software life cycle covered by the risk management plan. An overview of the major section of the risk management plan should also be provided.

Definitions and Acronyms. This section should provide a complete list of all acronyms and all relevant key terms used in the document that are critical to reader comprehension.

References. List all references used in the preparation of the document in this section.

Risk Management Overview. This section should describe the project's organization structure, its tasks, and its roles and responsibilities. A reference to any associated software project management plan should be provided here.

Risk Management Policies. This section should identify all organizational risk management policies. Refer to IEEE Std 1540 for detailed information supporting risk management requirements.

Organization. This section should depict the organizational structure that assesses and controls any associated project risk. This should include a description of each major element of the organization together with the roles and delegated responsibilities. The amount of organizational freedom and objectivity to evaluate and monitor the quality of the software, and to verify problem resolutions, should be clearly described and documented. In addition, the organization responsible for preparing and maintaining the risk management plan should be identified.

Responsibilities. This section should identify the specific organizational element that is responsible for performing risk management activities. Individuals who typically participate in risk management activities and should be described are: program manager, project manager, risk management manager, software team members, quality assurance manager, and configuration management manager.

Orientation and Training. This section should describe the orientation or training activities required in support of risk management activities. If this information is supplied in a related training plan, a reference to this plan may be provided here.

Costs and Schedules. Describe all associated cost and schedule impacts in this section. If this information is supplied in a related software project management plan, a reference to this plan may be provided here.

Process Description. Describe the risk management processes employed by the project in this section. If an organizational risk management process exists, then a reference to these processes may be provided here. If any adaptations are required during the adoption of organizational processes at the project level, then these adaptations must be described in this section. A diagram is often useful when communicating process descriptions.

All techniques and tools used during the risk management process should be described. Risk taxonomies, forms, and databases are often used in support of project-level risk management activities. If these items are used, a description should be provided.

Process Evaluation. Describe how measurement information will be collected in support of the evaluation of the risk management process. It this information is provided in a related software measurement and measurements plan, and then a reference to this plan may be provided here.

Communication. This section should describe how all associated project risk information will be coordinated and communicated among all project stakeholders. It is important to address this when discussing risk elevation procedures.

Plan Management. The management of the risk management plan, including revisions, approval, and associated configuration management requirements, should be addressed in this section.

Probability/Impact Risk Rating Matrix

IEEE Std 1490, IEEE Guide—Adoption of the PMI Standard, A Guide to the Project Management Body of Knowledge (PMBOK) [35], describes risk management as one of the project management knowledge areas. The PMBOK dedicates an entire chapter to the definition and analysis of and appropriate response to associated project risk. The phrase used by the PMBOK to describe the evaluation of project risk and associated likely outcomes is "risk quantification." Several methods for risk assessment are presented in the PMBOK; these methods range from statistical summaries to expert judgment. The Probability/Impact Risk Rating Matrix, Figure 4-3, is provided as an example of a way to leverage expert judgment and rate identified project risk.

	Consequences				
	Insignificant	Minor	Moderate	Major	Catastrophic
Likelihood	1	2	3	4	5
A (Almost certain)	H	H	E	E	E
B (Likely)	M	H	H	E	E
C (Possible)	L	M	H	E	E
D (Unlikely)	L	L	M	H	E
E (Rare)	L	L	M	H	H

E	**Extreme Risk – Immediate action; senior management involved**
H	**High Risk – Management responsibility should be specified**
M	**Moderate Risk – Manage by specific monitoring or response**
L	**Low Risk – Manage by routine process**

Figure 4-3. Probability/Impact Risk Rating Matrix.

5

12207 ORGANIZATIONAL PROCESSES

IEEE Std 12207 describes four organizational life cycle processes: the management process, infrastructure process, improvement process, and training process. These processes work at the organizational level for all projects and should be in place prior to performing the five primary life cycle processes of acquisition, supply, development, maintenance, and/or operation, and the eight supporting processes. The organization employing and performing a primary and supporting process, needs to

- Manage it at the project level following the management process
- Establish an infrastructure under it following the infrastructure process
- Tailor it for the project following the tailoring process
- Manage it at the organizational level following the improvement process and the training process
- Assure its quality through joint reviews, audits, verification, and validation

Organizational life cycle processes evaluate whether a new, changed, or outsourced process could be supported. Figure 5-1 provides an overview of these processes.

MANAGEMENT

The management process contains the generic activities and tasks managing their respective processes. The manager is responsible for product management, project management, and task management of the applicable processes, such as the acquisition, supply, development, operation, maintenance, or supporting processes. This process consists of the following activities, starting with initiation and scope definition, leading to planning, followed by execution and control, including review and evaluation, and ending with closure. Table 5-1 provides a list of the management process objectives.

Practical Support for Lean Six Sigma Software Process Definition. By S. Land, D. Smith, and J. Walz **133**
Copyright © 2008 IEEE Computer Society

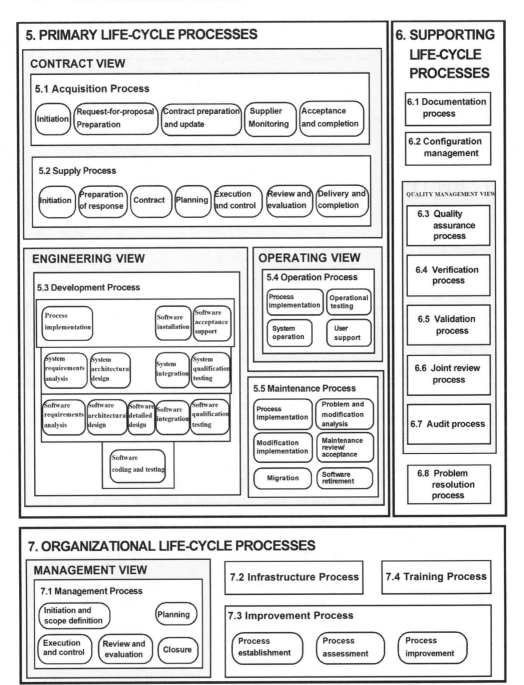

Figure 5-1. IEEE 12207 Organizational Life Cycle Processes [40].

Table 5-1. Management process objectives

a) Define the project work scope.
b) Identify, size, estimate, plan, track, and measure the tasks and resources necessary to complete the project.
c) Identify and manage interfaces between elements in the project and with other projects and organizational units.
d) Take corrective action when project targets are not achieved.
e) Establish quality goals, based on the customer's quality requirements, for various checkpoints within the project's software life cycle.
f) Establish product performance goals, based on the customer's requirements, for various checkpoints within the project's software life cycle.
g) Define and use measures that reflect the results of project activities or tasks, at checkpoints within the project's life cycle, to assess whether the technical, quality, and product performance goals have been achieved.
h) Establish criteria, measures, and procedures for identifying software engineering practices and integrate improved practices into the appropriate software life cycle processes and methods.
i) Perform the identified quality activities and confirm their performance.
j) Take corrective action when technical, quality, and product performance goals are not achieved.
k) Determine the scope of risk management to be performed for the project.
l) Identify risks to the project as they develop.
m) Analyze risks and determine the priority in which to apply resources to mitigate those risks.
n) Define, implement, and assess appropriate risk mitigation strategies.
o) Define, apply, and assess risk measures to reflect the change in the risk state and the progress of the mitigation activities.
p) Establish an environment that supports effective interaction between individuals and groups.
q) Take corrective action when expected progress is not achieved.

Software Requirements Management Plan

IEEE Standard 830, IEEE Recommended Practice for Software Requirements Specification [6], provides detailed guidance in support of the development of a software requirements management plan. This standard provides detail on what is required to effectively manage software requirements and how to document these requirements in a management plan.

IEEE Std 12207.0, Standard for Information Technology—Software Life Cycle Processes [40], describes seventeen processes spanning the entire life cycle of a software product or service. Even if an organization's processes were defined using other sources, the standard is useful in characterizing the essential characteristics of these software processes and should be considered prior to the implementation of process improvement activities. Referencing this standard, and reviewing what is required for each of these primary process areas, can provide valuable additional guidance in support of the activities associated with requirements management and analysis. It is important that IEEE 12207 be considered prior to the implementation of process improvement activities associated with requirements management and requirements analysis.

The implementation of IEEE Std 830, Recommended Practice for Software Requirements Specifications, assures the development of a software requirements specification but does not support the requirement of the institutionalization of software requirements management. At a minimum, IEEE Std 830 must be used in conjunction with IEEE Std

1058, Standard for Software Project Management Plans, and IEEE Std 12207.0, Standard for Information Technology—Software Life Cycle Processes, in support of this goal. Incorporating requirements management into the project planning, organizational processes, and management helps to ensure that it is integrated into the life cycle of the project.

In order for the IEEE Std 830, IEEE Recommended Practice for Software Requirements Specifications [6], to effectively support Lean Six Sigma improvement activities, information needs to be included in a software requirements management plan (SRMP) that will demonstrate that the resources, people, tools, funds, and time have been considered for the management of the requirements that are defined.

This information should relate directly to the software project management plan (SPMP). For example, if one person is required to perform the duties of the requirements manager, this should be identified in the SPMP. This section could reference the SPMP, if the information is contained in that document. However, the present IEEE Standard for SPMPs does not require the identification of requirements management issues such as resources and funding, reporting procedures, and training. Table 5-2 provides a suggested software requirement specification document outline.

Table 5-2. Software requirements management plan document outline

Title Page
Revision Page
Table of Contents
  1.0 Introduction
  2.0 Purpose
        2.1 Scope
        2.2 Definitions
        2.3 Goals
  3.0 References
        3.1 Key Acronyms
        3.2 Key Terms
        3.3 Key References
  4.0 Management
        4.1 Organization
        4.2 Tasks
        4.3 Responsibilities
               4.3.1 Management
               4.3.2 Program Manager
               4.3.3 Project Lead
               4.3.4 Team Members
               4.3.5 Customer
  5.0 Software Requirements Management Overview
        5.1 Software Requirements Modeling Techniques
               5.1.1 Functional Analysis
               5.1.2 Object-Oriented Analysis
               5.1.3 Dynamic Analysis
        5.2 Software Requirements Management Process
               5.2.1 Requirements Elicitation
               5.2.2 Requirements Analysis
               5.2.3 Requirements Specification
        5.3 Characteristics of a Good SRS

(continued)

Table 5-2.   *(continued)*

Software Requirements Management Plan Document Guidance

The following provides section-by-section guidance in support of the creation of a software requirements management plan. The development of an organizational SRMP not only helps baseline the process of software requirements definition and management, but the SRMP can also provide valuable training and guidance to project performers. This guidance should be used to help define a management process and should reflect the actual processes and procedures of the implementing organization. Additional information is provided in the document template, *Software Requirements Management Plan.doc,* which is located on the companion CD-ROM. Additional information in support of requirements traceability is provided in Appendix A, Software Process Work Products.

Purpose. This section should describe the purpose of the software requirements management plan and why effective requirements management is critical to the success of any software development or maintenance effort.

Software requirements management activities are concerned with two major areas:

1. Ensuring that all software requirements are clearly defined
2. Preventing the requirements process from becoming burdened by design, verification, or management details that would detract from specifying requirements

Scope. This subsection should address the scope of the plan and its applicability.

Definitions. This subsection should list all relevant defined terminology.

Goals. This subsection should identify all organizational goals associated with software requirements management.

References. A list of key acronyms, definitions, and references should be provided in this section of the document.

Management. This section contains the following subsections.

Organization. This subsection should describe the individuals responsible for software requirements management. The following is provided as an example:

> The [Project Abbreviation] program manager will ensure completion of software requirements management by coordinating the [Project Abbreviation] team development of an SRS. The program Manager will ensure that the training and time is provided to the project leader to have the SRS properly developed. The SRS will be developed to ensure that requirements are properly captured.

Tasks. This subsection should describe the high-level tasks associated with software requirements management and the specification of requirements, the following is provided as an example:

> The tasks to be performed in the development of an SRS are:
> 1. Specifying software requirements as completely and thoroughly as possible
> 2. Defining software product functions
> 3. Defining design constraints
> 4. Creating the software requirements traceability matrix
> 5. Defining assumptions and dependencies

Responsibilities. This subsection should describe the responsibilities of those individuals at all levels within the organization. It should address the roles of senior management, program, project lead, team members, and the customer.

Software Requirements Management Overview. This section should provide an overview of the software requirements management process. The following is provided as an example:

> Software requirements management is the set of processes (activities and tasks) and techniques used to define, analyze, and capture the customer's needs for a software-based system. This process starts with the customer's recognition of a problem and continues through the development and evolution of a software product that satisfies the customer's requirements. At that point, the software solution is terminated. For requirements management to work effectively, the requirements must be documented and agreed to by both the customer and [Company Name]. The formal document that contains this agreement is an SRS. The focus of the requirements management effort is then to create, approve, and maintain an SRS that meets and continues to meet the customer's needs.

Software Requirements Modeling Techniques. This subsection should provide a description of the software requirements modeling techniques.

Software Requirements Management Process. This subsection should provide a detailed description of the software requirements management process. The following information is provided as an example:

> The process of creating and approving an SRS can consist of three activities: requirements elicitation, requirements analysis, and requirements specification. Some problems are so clearly stated that only requirements specification is needed. Other problems require more re-

finement and, therefore, also require elicitation and analysis. This description will cover all three activities, assuming a worst-case scenario.

Projects following this plan will assess the state of their requirements and enter the process at the appropriate point. (Include a figure that represents the requirements management process, with all key players identified at appropriate entry points.)

Requirements Elicitation. Requirements elicitation is the process of extracting information about the product from the customer. Requirements elicitation is highly user interactive. A series of activities may be used to help both the customer and the developer create a clear picture of what the customer requires. These activities are discussed below. They include interviews, scenarios, ambiguity resolution, context analysis, rapid prototyping and clarification, and prioritization. This particular ordering of techniques is most useful for moving from the most obscure request to the most definitive. Other orderings also will work well and should be adapted to the project fit.

Interviews. Interviews are the most natural form of gaining an understanding of the customer's request. However, interviews are often performed with little advanced preparation. This is counterproductive. The customer becomes frustrated at being asked unimportant questions; the developer is forced to make assumptions about what is not understood. One way to reduce the failure rate of interviews is to focus the interview using the following strategy:

1. Lose your preconceptions regarding the request. Assuming that you know what must be accomplished prevents you from listening carefully to the customer and may give the customer the perception that the interview is just a formality.

2. Prepare for the interview. Find out what you can about the customer, the   people to be interviewed, and the request in general. This gives you a background for understanding the problem. Clarify those areas of uncertainty with the customer, this may help to better formulate requirements.

3. Ask questions that will help you understand the request from a design perspective. These questions lead to clarification of for whom the project is being built and why it is being built. This is similar to the current commercial that says, "How did your broker know you wanted to retire in ten years?" The answer is, "He asked!" These questions might include:
   a) Who is the client? This addresses for whom the system is really being built.
   b) What is a highly successful solution really worth to this client? This addresses the relative value of the solution.
   c) What is the real reason for this request? This might lead to alternative approaches that might be more satisfactory to the customer.
   d) Who should be on the development team? This may identify particular skills or knowledge that the customer recognizes as necessary.
   e) What are the trade-offs between time and value? This gives a better perspective on urgency.

4. Ask questions that will help you understand the product better. These might include:
   a) What problem does this product solve?
   b) What problem could this product create? It is a good idea to think about this early. This might also help to bound the problem area.

c) What environment is this product likely to encounter? Again, this helps to bound the request, and it may help identify some of the restrictions on the solution.

d) What kind of precision is required or desired in the product?

Scenarios. A scenario is a form of prototyping used to describe a customer request. In a scenario, examples of how a solution might react are examined. Again, this is a very natural process, but one that needs to be prepared to work effectively. Scenarios are built by determining how implementation might satisfy the request. Assumptions, problems, and the customers' reactions to the scenarios are used for further evaluation.

Once the interviews and scenarios are progressing, a work statement can be formed. A work statement is a textual description of the customer's interpretation of the request. This may range from very formal to very abstract. The purpose of interviews and scenarios is to clarify the work statement.

Ambiguity Resolution. Ambiguity resolution is an attempt to identify all the possible misinterpretable aspects of a work statement. Examples of ambiguous areas are: terms, the definitions that are often interpreted differently; acronyms that are not clearly defined; and misleading statements of what is required. Question everything; nothing should be assumed while reading the work statement.

The output of this step is a redefined work statement, possibly with a list of definitions. The list of definitions should be the starting point for a future data dictionary.

Context Analysis. Context analysis is an early modeling of the system to identify key objects as named links. This helps to organize pieces of the request and shows connectivity to highlight communication.

Clarification and Prioritization. The final step in requirements elicitation is to create a list of the desired functions, attributes, constraints, preferences, and expectations. This moves the work statement toward a more formal software requirements specification.

1. Functions are the "what" of the product. They describe what the product must do or accomplish to satisfy the customer's request. Functions should be classified as *evident, hidden,* and *frill.* Once this classification is complete, the developer has set the stage for what must be demonstrated to the customer (*evident* functions) and what must be developed to satisfy the customer (*hidden* functions). The *frills* are kept on a standby list to be obtained, if possible.

2. Attributes are characteristics of the product. They describe the way the product should satisfy customer needs. Attributes and attribute lists (linear descriptions of how the attribute is recognized) can be classified as *must, want,* and *ignore.* Attributes and attribute lists are attached to the functions described above. To satisfy the customer all *must* attributes have to be met, and as many *want* attributes as possible should be met. *Ignore* attributes are ignored.

3. Constraints are mandatory conditions placed on an attribute. These may come from regulations, policies, or customer experience. (Developers should always attempt to recognize and validate all constraints. Constraints limit the options available and they should be investigated before they limit the system.)

4. Preferences are desirable, but optional, conditions placed on an attribute. Preferences help to satisfy the customer, so they should be achieved whenever possible.

5. Expectations are the hardest items to capture. These are what the customer expects from the product; they are seldom clearly defined. Every attempt should be made to determine what the customer expects the system to do, and these expectations should be met if possible. If the customer's expectations are exorbitant, then these issues should be addressed at the project onset to avoid later disappointment.

The output of the requirements elicitation process is a well-formed, redefined work statement clearly stating the customer's request. This work statement provides the basis for the requirements analysis.

Requirements Analysis. Requirements analysis involves taking the work statement and creating a model of the customer's request. The type of request should determine which technique to select.

Requirements Specification. The specification, called the SRS, is the formal definition of the statement of the customer's problem. The requirements specification combines the knowledge gained during requirements elicitation (attributes, constraints, and preferences) with the model produced during requirements analysis.

Characteristics of a Good SRS. This subsection should describe the characteristics of a good SRS. The following are offered as criteria.

Correct. "An SRS is correct if and only if every requirement stated therein represents something required of the system to be built" [137].

Nonambiguous. "An SRS is nonambiguous if and only if every requirement stated therein has only one interpretation" [137].

Complete. "An SRS is complete if it possesses the following four qualities. (1) Everything that the software is supposed to do is included in the SRS. (2) All definitions of software responses to all realizable classes of input data, in all realizable classes of situations. (3) All pages are numbered; all figures and tables are numbered, named, and referenced; all terms and units of measure are provided; and all referenced material and sections are present. (4) No sections are marked 'To Be Determined (TBD)'" [137].

Verifiable. An SRS is verifiable only if every requirement stated therein is verifiable. A requirement is verifiable if there exists some finite cost-effective process with which a person, or machine, can check that the actual as-built software product meets the requirement.

Consistent. An SRS is consistent if and only if no subset of individual requirements stated therein conflict. This may be manifested in a number of ways: (1) conflicting terms, (2) conflicting characteristics, and (3) temporal inconsistency.*

Modifiable. "An SRS is modifiable if its structure and style are such that any necessary changes to the requirements can be made easily, completely, and consistently" [137].

Traceable. "An SRS is traceable if the origin of each of its requirements is clear and if it facilitates the referencing of each requirement in future development or enhancement documentation" [137].

Usable During Operation and Maintenance. This is a combination of being modifiable as described above and being understandable to users of the SRS. The SRS should last as long as the system built from it and always accurately reflect the state of the system. SRS requirements that do not meet the above tests should be considered a risk or a reportable issue for the Project Manager's Open Issues List.

---

*Temporal inconsistency is when two or more parts of the SRS require the product to obey contradictory timing constraints.

Standards and Practices. This section should provide references to applicable corporate policies and procedures and external standards if appropriate. The following is provided as an example:

> [Project Abbreviation] Projects will document and maintain the current status of all customer-approved SRSs. At this time, the standard for software requirements management is limited to the need to establish and follow the format of Appendix A to create specifications for each new project. This format will be followed unless the program is required, by contract, to use another format. As [Company Name]'s maturity within the Capability Maturity Model increases, more detailed standards and practices will evolve. These standards will be based on measurements to ensure consistent capability to complete the requirements.
>
> In developing the software requirements specification, the program manager will incorporate all existing [Company Name] standards and policies for software, including, but not limited to, software configuration management and software quality assurance.

Software Measurement. This section should describe all measurement activities associated with an organizations' management of software requirements. The following is provided as an example:

> Software measurements will be applied to the process of software requirements management on a continual basis in order to ensure that customer requirements are met and to improve the SRM process. Measurement from a [Project Abbreviation] perspective means measuring progress toward the [Project Abbreviation] goal of accurately identifying project software requirements. (Reference project level software measurement plan if appropriate.)
>
> In order to improve, the results of projects must be gathered, including deviations from the SRSs. Recording and learning from the lessons of early projects will improve the software requirements management process and serve to move [Project Abbreviation] toward the goal of effective requirements management. Both corporate and project configuration management will help review SRSs and identify deviations.

Verification and Validation. This section should describe the minimum acceptable verification and validation of software requirements. Verification and validation are two important aspects of any project software life cycle. They provide the checks and balances to the development process that ensure that the system is being built (validation). According to IEEE/EIA 12207.1, Verification is "confirmation by examination and provision of objective evidence that specified requirements have been fulfilled," and validation is "confirmation by examination and provision of objective evidence that the particular requirements for a specific intended use are fullfilled" [40]. In other words, verification is "are we building the right product?" and validation is "are we building the product right?"

Verification and validation differ in scope and purpose. Verification is concerned with comparing the results of each activity of the software life cycle with the specification from the previous activity. The scope of verification therefore, is limited to two activities (current and prior). For this reason, the context of verification changes depends upon the pair of activities under review.

Validation takes a birds-eye view of requirements, and the implementation of those requirements. Each functional and nonfunctional requirement listed in the SRS will be testable by one of four techniques. The SRS will reference a requirements traceability matrix (RTM), that lists all requirements and the method used to validate that requirement. Four possible methods of validation are:

1. **Demonstration.** The end product is executed for the customer to demonstrate the desired properties have been met.
2. **Test.** The end product is executed through predefined, scripted tests for the customer to show functionality and demonstrate correctness.
3. **Analysis.** Results of the product execution are examined to ensure program correctness.
4. **Inspection.** The product itself is inspected to demonstrate that it is correct.

Software Configuration Management. This section should provide a cross-reference to the supporting software configuration management plan and associated activities. The following is provided as an example:

> The software project management plan will present a schedule for developing the SRS. The schedule will indicate the expected time for baselining the completed SRS. The baselined SRS will be placed under configuration management defined in the program configuration management process.

Developing a Software Requirements Specification. This section should provide detailed guidance in support of the development of a SRS. The following is provided as guidance:

> Each program manager should ensure that all new projects develop an SRS and submit it through the Corporate SQA Manager to [Company Name] Management for approval. Within each of the sections listed in Appendix A, a general explanation of the section contents is given. Appendix A should be considered a boilerplate for developing your SRS. If a section does not pertain to your project, the following should appear below the section heading, "This section is not applicable to this plan," together with appropriate reasons for the exclusion. Additional sections and appendices may be added as needed. Some of the required material may appear in other project documents. If so, then references to these documents should be made in the body of the software requirements specification. In any case, the contents of each section of the plan should be specified either directly or by reference to another document.

Appendix A. Project Software Requirements Specification. This appendix should contain a document template for project use.

Appendix B. Template for Requirements Traceability Matrix. This appendix should contain a traceability matrix template for project use.

Software Project Management Plan

IEEE Std 1058, IEEE Standard for Software Project Management Plans [13], and IEEE/EIA 12207.0, IEEE Standard for Life Cycle Processes [40], are effective instruments in support of software project planning. However, information regarding the measurements required in support of software project planning activities needs to be added and stated explicitly. Table 5-3 provides a suggested software project management plan document outline.

Table 5-3. Software project management plan document outline

Title Page
Revision Page
Table of Contents
1. Introduction
    1.1 Project Overview
    1.2 Project Deliverables
    1.3 Document Overview
    1.4 Acronyms and Definitions
2. References
3. Project Organization
    3.1 Organizational Policies
    3.2 Process Model
    3.3 Organizational Structure
    3.4 Organizational Boundaries and Interfaces
    3.5 Project Responsibilities
4. Managerial Process
    4.1 Management Objectives and Priorities
    4.2 Assumptions, Dependencies, and Constraints
    4.3 Risk Management
    4.4 Monitoring and Controlling Mechanisms
    4.5 Staffing Plan
5. Technical Process
    5.1 Tools, Techniques, and Methods
    5.2 Software Documentation
    5.3 Project Support Functions
6. Work Packages
    6.1 Work Packages
    6.2 Dependencies
    6.3 Resource Requirements
    6.4 Budget and Resource Allocation
    6.5 Schedule
7. Additional Components

Software Project Management Plan Document Guidance

The following provides section-by-section guidance in support of the creation of a SPMP. The SPMP should be considered to be a living document. The SPMP should change, in particular any associated schedules, and reflect any required change during the life cycle of a project. This guidance should be used to help define a management process and should reflect the actual processes and procedures of the implementing organization. Additional information is provided in the document template, *Software Project Management Plan.doc* that is located on the companion CD-ROM. Additional information is provided in Appendix A, Software Process Work Products, that describes the work breakdown structure, workflow diagram, and stakeholder involvement matrix work products.

Introduction. This section contains the following four subsections.

Project Overview. This subsection should briefly state the purpose, scope, and objectives of the system and the software to which this document applies. It should describe the general nature of the system and software; summarize the history of system development, operation, and maintenance; and identify the project sponsor, acquirer, user, developer, and support agencies; and identify current and planned operating sites. The project overview should also describe the relationship of this project to other projects, as appropriate, addressing any assumptions and constraints. This section should also provide a brief schedule and budget summary. This overview should not be construed as an official statement of product requirements. Reference to the official statement of product requirements should be provided in this subsection.

Project Deliverables. This subsection should list all of the items to be delivered to the customer, the delivery dates, delivery locations, and quantities required to satisfy the terms of the project agreement. This list of project deliverables should not be construed as an official statement of project requirements.

Document Overview. This subsection should summarize the purpose and contents of this document and describe any security or privacy considerations that should be considered associated with its use. This subsection should also specify the plans for producing both scheduled and unscheduled updates to the SPMP. Methods of disseminating the updates should be specified. This subsection should also specify the mechanisms used to place the initial version of the SPMP under change control and to control subsequent changes to the SPMP.

Acronyms and Definitions. This subsection should identify acronyms and definitions used within the project SPMP. The project SPMP should only list acronyms and definitions used within the SPMP.

References. This section should identify the specific references used within the project SPMP. The project SPMP should only contain references used within the SPMP.

Project Organization. This section contains the following five subsections.

Organizational Policies. This subsection of the SPMP should identify all organizational policies relative to the software project.

Process Model. This subsection of the SPMP should specify the (life cycle) software development process model for the project, describe the project organizational structure, identify organizational boundaries and interfaces, and define individual or stakeholder responsibilities for the various software development elements.

Organizational Structure. This subsection should describe the makeup of the team to be used for the project. All project roles and stakeholders should be identified, and a description of the internal management structure of the project should be given. Diagrams may be used to depict the lines of authority, responsibility, and communication within the project.

Organizational Boundaries and Interfaces. This subsection should describe the limits of the project, including any interfaces with other projects or programs, the application of the program's SCM and SQA (including any divergence from those plans), and the interface with the project's customer. This section should describe the administrative and managerial boundaries between the project and each of the following entities: the parent organization, the customer organization, subcontracted organizations, or any other organizational entities that interact with the project. In addition, the administrative and managerial interfaces of the project support functions, such as configuration management, quality assurance, and verification, should be specified in this subsection.

Project Responsibilities. This subsection should describe the project's approach through a description of the tasks required to complete the project (e.g., requirements → design → implementation → test) and any efforts (update documentation, etc.) required to successfully complete the project. It should state the nature of each major project function and activity, and identify the individuals or stakeholders who are responsible for those functions and activities.

Managerial Process. This section should specify management objectives and priorities; project assumptions, dependencies, and constraints; risk management techniques; monitoring and controlling mechanisms to be used; and the staffing plan.

Management Objectives and Priorities. This subsection should describe the philosophy, goals, and priorities for management activities during the project. Topics to be specified may include, but are not limited to, the frequency and mechanisms of reporting to be used; the relative priorities among requirements, schedule, and budget for this project; risk management procedures to be followed; and a statement of intent to acquire, modify, or use existing software.

Assumptions, Dependencies, and Constraints. This subsection should state the assumptions on which the project is based, the external events the project is dependent upon, and the constraints under which the project is to be conducted.

Risk Management. This subsection should identify the risks for the project. Completed risk management forms should be maintained and tracked by the project leader with associated project information. These forms should be reviewed at weekly staff meetings. Risk factors that should be considered include contractual risks, technological risks, risks due to size and complexity of the project, risks in personnel acquisition and retention, and risks in achieving customer acceptance of the product.

Monitoring and Controlling Mechanisms. This subsection should define the reporting mechanisms, report formats, information flows, review and audit mechanisms, and other tools and techniques to be used in monitoring and controlling adherence to the SPMP. A typical set of software reviews is listed in Appendix A. Project monitoring should occur at the level of work packages. The relationship of monitoring and controlling mechanisms to the project support functions should be delineated in this subsection. This subsection should also describe the approach to be followed for providing the acquirer or its authorized representative access to developer and subcontractor facilities for review of software products and activities.

Staffing Plan. This subsection should specify the numbers and types of personnel required to conduct the project. Required skill levels, start times, duration of need, and methods for obtaining, training, retaining, and phasing out of personnel should be specified.

Technical Process. This section should specify the technical methods, tools, and techniques to be used on the project. In addition, the plan for software documentation should be specified, and plans for project support functions such as quality assurance, configuration management, and verification and validation may be specified.

Tools, Techniques, and Methods. This subsection of the SPMP should specify the computing system(s), development methodology(s), team structures(s), programming language(s), and other notations, tools, techniques, and methods to be used to specify, design, build, test, integrate, document, deliver, and modify or maintain or both (as appropriate) the project deliverables.

This subsection should also describe any tools (compilers, CASE tools, and project management tools), any techniques (review, walk-through, inspection, prototyping) and the methods (object-oriented design, rapid prototyping) to be used during the project.

Software Documentation. This subsection should contain, either directly or by reference, the documentation plan for the software project. The documentation plan should specify the documentation requirements, and the milestones, baselines, reviews, and sign-offs for software documentation. The documentation plan may also contain a style guide, naming conventions and documentation formats. The documentation plan should provide a summary of the schedule and resource requirements for the documentation effort. IEEE Std for Software Test Documentation (IEEE Std 829) [15] provides a standard for software test documentation.

Project Support Functions. This subsection should contain, either directly or by reference, plans for the supporting functions for the software project. These functions may include, but are not limited to, configuration management, software quality assurance, and verification and validation.

Work Packages. This section of the SPMP should specify the work packages, identify the dependency relationships among them, state the resource requirements, provide the allocation of budget and resources to work packages, and establish a project schedule. The work packages for the activities and tasks that must be completed in order to satisfy the project agreement must be described in this section. Each work package should be uniquely identified; identification may be based on a numbering scheme or descriptive titles. A diagram depicting the breakdown of activities into subactivities and tasks may be sued to depict hierarchical relationships among work packages.

Dependencies. This subsection should specify the ordering relations among work packages to account for interdependencies among them and dependencies on external events. Techniques such as dependency lists, activity networks, and the critical path may be used to depict dependencies.

Resource Requirements. This subsection should provide, as a function of time, estimates of the total resources required to complete the project. Numbers and types of personnel, computer time, support software, computer hardware, office and laboratory facilities, travel, and maintenance requirements for the project resources are typical resources that should be specified.

Budget and Resource Allocation. This subsection should specify the allocation of budget and resources to the various project functions, activities, and tasks. Defined resources should be tracked.

Schedule. This subsection should be used to capture the project's schedule, including all milestones and critical paths. Options include Gantt charts (*Milestones Etc.*™ or *Microsoft Project*™), Pert charts, or simple time lines.

Additional Components. This section should address additional items of importance on any particular project. These may include subcontractor management plans, security plans, independent verification and validation plans, training plans, hardware procurement plans, facilities plans, installation plans, data conversion plans, system transition plans, or product maintenance plans.

Stakeholder Involvement

IEEE Std 1058 addresses the identification of all key participants (project stakeholders) associated with a development effort. This standard requires the definition of the organizational structure, all customer relationships, roles, and responsibilities, resources, and how involvement is monitored. See Stakeholder Involvement Matrix in Appendix A, Software Process Work Products.

Work Breakdown Structure (WBS)

A work breakdown structure (WBS) defines and breaks down all work associated with a project into manageable parts. It describes all activities that have to occur to accomplish the project. The WBS can serve as the foundation for the integration of project component schedules, budget, and resource requirements. IEEE Std 1490-2003, IEEE Guide—Adoption of the PMI Standard, A Guide to the Project Management Body of Knowledge [35], recommends the use of the structure shown in Figure 5-2. Additional information in support of associated IPM work products is provided in Appendix A, Software Process Work Products.

A work breakdown structure (WBS) defines and breaks down the work associated with a project into manageable parts. It describes all activities that have to occur to accomplish the project. The WBS serves as the foundation for the development of project schedules, budget, and resource requirements.

A WBS may be structured by project activities or components, functional areas or types of work, or types of resources, and is organized by its smallest component—a work package. A work package is defined as a deliverable or product at the lowest level of the WBS. Work packages may also be further subdivided into activities or tasks. IEEE Std 1490-2003, IEEE Guide—Adoption of the PMI Standard, A Guide to the Project Management Body of Knowledge, recommends the use of nouns to represent the "things" in a WBS. Figure 5-3 provides an example of a sample WBS organized by activity.

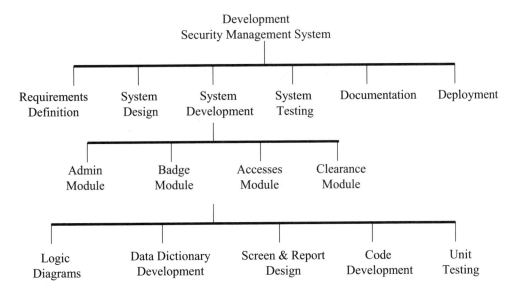

Figure 5-2. Example work breakdown structure.

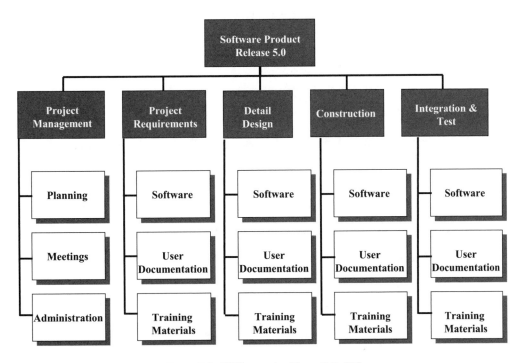

Figure 5-3. WBS organized by activity [35].

Work Breakdown Structure (WBS) for Postdevelopment Stage

The example provided in Table 5-4 provides suggested content for a WBS in support of the post development stage of a software development life cycle. This WBS content is illustrative only and should be customized to meet specific project requirements.

Table 5-4. WBS content for postdevelopment activity

Product Installation
  Distribute Product (Software)
      Package and distribute product
      Distribute installation information
      Conduct integration test for product
      Conduct regression test for product
      Conduct user acceptance test for software
      Conduct reviews for product
      Perform configuration control for product
      Implement documentation for product
  Install Product (Software)
      Install product (packaged software) and database data
      Install any related hardware for the product
      Document installation problems
  Accept Product (Software) in Operations Environment
      Compare installed product (software) to acceptance criteria
      Conduct reviews for installed product (software)
      Perform configuration control for installed product (software)
Operations and Support
      Utilize installed software system
      Monitor performance
      Identify anomalies
      Produce operations log
      Conduct reviews for operations logs
      Perform configuration control for operations logs
Provide Technical Assistance and Consulting
    Provide Response to Technical Questions or Problems
    Log Problems
Maintain Support Request Logs
    Record Support Requests
    Record Anomalies
    Conduct Reviews for Support Request Logs
Maintenance
    Identify Product (Software) Improvements Needs
        Identify product improvements
        Develop corrective/perfective strategies
        Produce product (software) improvement recommendations
    Implement Problem Reporting Method
        Analyze reported problems
        Produce report log
        Produce enhancement problem reported information
        Produce corrective problem reported information
        Perform configuration control for reported information
    Reapply Software Life Cycle methodology

INFRASTRUCTURE

The infrastructure process is a process to establish and maintain the infrastructure needed for any other process. The infrastructure may include hardware, software, tools, techniques, standards, and facilities for development, operation, or maintenance. Table 5-5 describes the infrastructure process objectives.

Table 5-5. Infrastructure process objectives

a) Establishing and maintaining a well-defined software engineering environment, consistent with, and supportive of, the set of standard processes and organizational methods and techniques.
b) Tailoring the software engineering environment to the needs of the project and the project team.
c) Developing a software engineering environment that supports project team members regardless of the performance location of process activities.
d) Implementing a defined and deployed strategy for reuse.

Organization's Set of Standard Processes

The information provided here is based upon IEEE Std1074, IEEE Standard for Developing Software Life Cycle Processes [11]; IEEE/EIA Std 12207.0, Industry Implementation of International Standard ISO/IEC 12207 (ISO/IEC 12207), Standard for Information Technology–Software life cycle processes [40]. Table 5-6 describes the three categories of life cycle processes that are found in IEEE Std. 12207.0.

Table 5-6. IEEE Std 12207 life cycle processes [40]

Primary life cycle processes	Acquisition process
	Supply process
	Development process
	Operation process
	Maintenance process
Supporting life cycle processes	Documentation process
	Configuration management process
	Quality assurance process
	Verification process
	Validation process
	Joint review process
	Audit process
	Problem resolution process
Organizational life cycle processes	Management process
	Infrastructure process
	Improvement process
	Training process

This three-dimensional view of standard processes is supported by Figure 5-4. This figure is from IEEE Std 12207 and also provides an illustration of representative viewpoints.

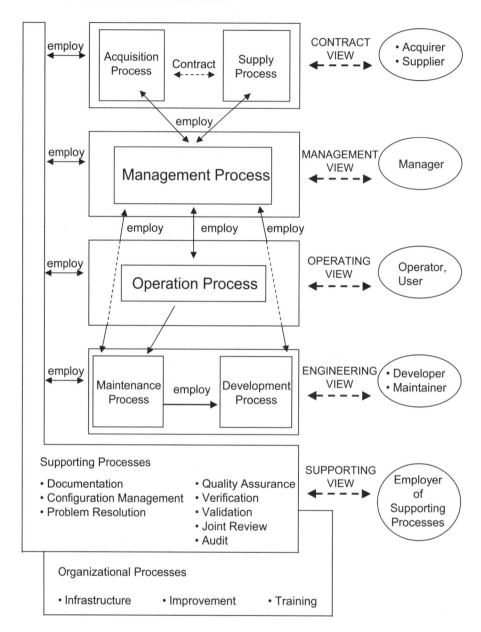

Figure 5-4. Life cycle processes and views [40].

IMPROVEMENT

The improvement process is a process for establishing, assessing, measuring, controlling, and improving a software life cycle process. The improvement process uses software life cycle data as it provides a history of what happened during development and maintenance. Table 5-7 describes the improvement process objectives.

Table 5-7. Improvement process objectives

---

a) Establish a well-defined and maintained standard set of processes, along with a description of the applicability of each process.

b) Identify the detailed tasks, activities, and associated work products for each standard process, together with expected criteria.

c) Establish a deployed specific process for each project, tailored from the standard process in accordance with the needs of the project.

d) Establish and maintain information and data related to the use of the standard process for specific projects.

e) Understand the relative strengths and weaknesses of the organization's standard software processes.

f) Make changes to standard and defined processes in a controlled way.

g) Implement planned and monitored software process improvement activities in a coordinated manner across the organization.

Engineering Process Group Charter

The following is provided as an example of the type of information typically provided in an engineering process group (EPG) charter document. The information provided here is for illustrative purposes, the EPG charter of any organization should reflect the unique requirements of the developing organization.

Purpose. This section should describe the rationale and scope of the EPG charter. An example follows:

> This charter formally empowers the Engineering Process Group (EPG) to manage the released content of corporate software engineering processes at [Company Name].

Vision/Mission Statement. This section should describe the mission of the EPG to include major areas of responsibility. An example follows:

> The EPG provides management control of [Company Name] engineering processes. It
> a) Approves release into general use of engineering process assets
> b) Manages proposed improvements to the engineering process assets as necessary
> c) Provides tactical guidance on asset creation and maintenance priorities
> d) Adjudicates controversial or unprecedented process tailoring decisions as required

Authority and Responsibility. This section should provide a description of EPG authority and responsibility. An example is provided:

> The EPG has authority to define all systems and software engineering process assets in each released set comprising the [Company Name] system and software engineering process. The EPG is responsible for the management and implementation of that asset set.

Membership and Organization. A description of all voting and nonvoting members should be provided. The terms that each member is responsible for serving on the EPG should also be described. It is important to address what may be considered a quorum to convene meetings. An example of determining a quorum follows:

Quorum for the meeting shall consist of the Chair, SQA representative, and 60% of the voting membership.

Activities and Function. All activities should be described in this section of the charter document. These activities and functions may include a description of process development, procedures for fast-tracking processes, typical review and approval timelines, and supporting procedures.

Process Action Plan (PAP)

Process action plans usually result from appraisals and document how specific improvements targeting the weaknesses uncovered by an appraisal will be implemented. In cases in which it is determined that the improvement described in the process action plan should be tested on a small group before deploying it across the organization, a pilot plan is generated. Finally, when the improvement is to be deployed, a deployment plan is used. This plan describes when and how the improvement will be deployed across the organization.

The following sections should provide examples of the type of information typically provided in a process action plan (PAP) document. The information provided here is for illustrative purposes; any specific PAP of an organization should reflect the unique requirements of the developing organization.

Identification. This section should provide summary information that may be used when referring to the document. This should include the author, the date, and any unique tracking number.

Instructions. This section should provide directions for using the PAP, including submission procedures.

Issue. This section should provide a description of the process area being defined or improved. If this is an improvement suggestion, identify the process being used in the context of the proposal being submitted. If a specific tool, procedure, or other EPG asset is being addressed, describe it here as well.

Process Action Team. Identify the personnel to implement the actions.

Process Improvement. This section should provide the strategies, approaches, and actions to address the identified process improvements

Improvement. This section should describe the process or technological improvement being proposed to support resolution of the issue described above.

Tailoring Guidelines

Processes should be tailored for a specific project, as no two projects are the same. Tailoring is the adaptation of an organization's set of standard processes for use on a specific project by using tailoring guidelines defined by the organization. The purpose of tailoring the organizational process asset for a specific project is to ensure that the appropriate amount of effort is devoted to the appropriate activities to reduce project risk to an acceptable level while at the same time making most cost-effective use of engineering talent

IEEE Std 12207.0 Annex A, Tailoring Process, has four major steps:

1. Identify the project environment—strategy, activity, and requirements
2. Solicit inputs—from users, support team, and potential bidders
3. Select processes, activities, documentation, and responsibilities
4. Document tailoring decisions and rationale

Elaborations of these general steps in tailoring the organizational process asset for a project are:

1. Characterize the project environment:
   - Size of the development team
   - Strategy
   - Critical factors: project/customer requirements
   - Life cycle strategy: waterfall, evolutionary, spiral, and so on
   - Life cycle activity: prototyping, maintenance
   - Software characteristics: COTS, reuse, embedded firmware
   - Project policies, languages, hardware reserve, culture
   - Acquisition strategy: contract type, contractor involvement
2. Determine the cost targets and risk tolerance of the project.
3. Identify the organization process asset for which tailoring is being considered.
4. For each committed deliverable asset, use best practices and professional judgment to identify the *form* the asset should take, and *level of detail* necessary to achieve the purpose of each organization process asset.
5. Assess whether any organization process asset, or their forms or level of detail, are unaffordable given the project goals, cost targets, and level of tolerable risk.
6. Identify the level of detail needed for each process activity.
7. Document the tailoring planned to the organization process asset, and obtain approval. Typically it is unwise to tailor the asset intent or objectives. What normally is tailored are: number of phases/activities, roles, responsibilities, document formats, and formality/frequency of reports or reviews.
8. Document the planned processes, products, and reviews. Describe the completion criteria for each process.
9. After project completion, send example implementations, lessons learned, and improvements back to the organization to continually improve the organizational process assets.

TRAINING

The taining process provides and maintains trained personnel. As all software engineering processes are largely dependent upon knowledgeable and skilled personnel, it is imperative that training be planned and implemented early so that trained personnel are available as the software product is acquired, supplied, developed, operated, or maintained. The training process objectives are described in Table 5-8.

Table 5-8. Training process objectives

a) Identify the roles and skills required for the operations of the organization and the project.

b) Establish formal procedures by which talent is recruited, selected, and transitioned into assignments in the organization.

c) Design and conduct training to ensure that all individuals have the skills required to perform their assignments.

d) Identify and recruit or train, as appropriate, individuals with the required skills and competencies to perform the organizational and project roles.

e) Establish a work force with the skills to share information and coordinate their activities efficiently.

f) Define objective criteria against which unit and individual training performance can be measured, to provide performance feedback, and to enhance performance continuously.

Training Plan

This document template is based on IEEE Std 12207.0, Standard for Information Technology—Software Life Cycle Processes; IEEE Std 12207.1, Standard for Information Technology—Software Life Cycle Processes—Life Cycle Data; IEEE Std-12207.2, Standard for Information Technology Software Life Cycle Processes—Implementation Considerations [40]; and the U.S. DOE Training Plan Template* as primary reference material. Table 5-9 provides a suggested training plan document outline. Additional information is provided in Appendix A in the form of an example training request form and training log.

Table 5-9. Training plan document outline

Title Page
Revision Page
Table of Contents
1.0 Introduction
    1.1 Scope
    1.2 Objectives
    1.3 Background
    1.4 References
    1.5 Definitions and Acronyms
2.0 Training Requirements
    2.1 Roles and Responsibilities
    2.2 Training Evaluation
3.0 Training Strategy
    3.1 Training Sources
    3.2 Dependencies/Constraints/Limitations
4.0 Training Resources
    4.1 Vendor Selection
    4.2 Course Development
5.0 Training Environment
6.0 Training Materials

*Department of Energy, Training Plan Template, http://www.cio.energy.gov/documents/traintem.pdf.

The following provides section-by-section guidance in support of the creation of a training plan. Ensuring that all individuals are properly trained is integral to the success of any software development effort. Additional information is provided in the document template, *Training Plan.doc,* which is located on the companion CD-ROM. This training plan template is designed to facilitate the development of a planned approach to organizational or project-related training.

Introduction. This section should describe the purpose of the training plan and the organization of the document. As described by IEEE Std 12207.2,

> The Training Process is a process for providing and maintaining trained personnel. The acquisition, supply, development, operation, or maintenance of software products is largely dependent upon knowledgeable and skilled personnel. For example: developer personnel should have essential training in software management and software engineering. It is, therefore, imperative that personnel training be planned and implemented early so that trained personnel are available as the software product is acquired, supplied, developed, operated, or maintained [40].

Scope. This subsection should describe the scope and purpose of the training, such as initial training for system users, refresher training for the system maintenance staff, training for system administrators, and so on.

Objectives. This subsection should describe the objectives or anticipated benefits from the training. It is best to express all objectives as actions that the users will be expected to perform once they have been trained.

Background. This subsection should provide a general description of the project and an overview of the training requirements.

References. This subsection should identify sources of information used to develop this document to include reference to the existing organizational training policy. Reference to the organizational training plan, if the plan being developed is tactical, should be provided as well. Any existing engineering process group (EPG) charter should be referenced to provide insight into process oversight activities if needed.

Definitions and Acronyms. This subsection should provide a description of all definitions and acronyms used within this document that may not be commonly understood, or which are unique in their application within this document.

Training Requirements. This section describe the general work environment (including equipment), and the skills for which training is required (management, business, technology, etc). The training audience should also be identified (category of user: upper management, system administrator, administrative assistant, etc.). It may also identify individuals or positions needing specific training. Include the time frame in which training must be accomplished. Identify whether the training requirements are common, cross-project requirements or if the requirements are required in support of a unique project requirement.

Roles and Responsibilities. This subsection should identify the roles and responsibilities of the training staff and identify the individuals responsible for the management of the training development and implementation. It may also include the identification of other groups who may serve as consultants, such as members of the development team, experienced users, etc. The individuals responsible for keeping training records and for updating any associated organizational training repository should be identified. If this information is provided in the associated project plan, a reference to this plan may be provided here.

Training Evaluation. The effectiveness of training must continually be evaluated. This subsection should describe how training evaluation will be performed. Evaluation tools, surveys, forms, and so on, should be included. The revision process with regard to the modification of the course and course materials resulting from the evaluations should also be described.

Training Strategy. This section should describe the type of training (e.g., classroom, CBT, etc.) and the training schedule (duration, sites, and dates). Include adequacy of training facilities, accommodations, need to install system files, modem/communication issues, physical access to buildings, escorts needed within facilities, and so on. It is suggested that a training log be developed to document and track information associated with individuals receiving training.

Training Sources. This subsection should identify the source or provider for the training. Training may be internal (course developed in-house) or external (contracted to external training agencies).

Dependencies/Constraints/Limitations. This subsection should identify all known dependencies constraints, and or limitations that could potentially affect training on the project.

Training Resources. This section should include hardware/software, instructor availability, training time estimates, projected level of effort, system documentation, and other resources required to familiarize the trainer with the system, produce training materials, and provide the actual training. The identification and availability of other resource groups should also be included.

Vendor Selection. This subsection should describe the criteria for training vendor selection.

Course Development. This subsection should describe the requirements for internal course development. This should include all related life cycle activities associated with the development of the training.

Training Environment. Describe the equipment and conditions required for the training, including installations, facilities, and special databases (typically, there should be a separate, independent development/production environment). Also identify any actions required by other groups, such as trainees, to request training.

Training Materials. This section should describe the types of training materials required for the training. The training materials developed may include visuals for overhead projectors, handouts, workbooks, manuals, computerized displays, and demonstrations. The training materials and curriculum should accurately reflect the training objective.

Update/Revise Training Plan. Once the training plan is developed, it must be subjected to the same kind of configuration management process as the other system documentation. Training materials should remain current with system enhancements. Describe the change release process with regard to all training documentation.

6

LEAN SIX SIGMA FUNDAMENTALS

WHAT IS LEAN SIX SIGMA?

Almost all process improvement methods serve one of two purposes: (1) eliminating variation in quality or speed (which is the source of defects) or (2) improving process flow and speed [80]. Lean Six Sigma is no exception and directly supports both of these purposes. Lean Six Sigma is the combination of two techniques, Lean and Six Sigma. To understand Lean Six Sigma, you must first understand how these concepts are related, how they evolved, and how each of these concepts is defined.

LEAN PRINCIPLES

The core of Lean is based on the continuous pursuit of improving processes. Lean is a philosophy of eliminating all nonvalue-adding activities and of reducing waste within an organization [111]. Lean has been described as "the thinking way," or practical problem solving. This may sound simple in theory, but is much more complex in application. Lean should be considered as much more than a series of processes and techniques. Lean requires cultural change and includes the integration of vision, culture, and strategy to serve the customer (both internal and external) with high quality, low cost, and short delivery times. Lean must become a whole systems approach in order to create a new culture and operating philosophy for eliminating all nonvalue-adding activities from software production.

The primary source for Lean is Taiichi Ohno, an architect of Toyota's lean production system, commonly referred to as the "Toyota Production System" or "TPS." Toyota essentially refined the concepts of flow and pull. The principle of flow can be thought of as the flow of parts moving down a plant assembly line. Pull can best be described as the inventory restocking process. The foundations of lean processes combine flow and pull into the single unified concept of value flow, or the uninterrupted value flow at the pull of the customer. The goal of lean processes is to eliminate all activities that do not contribute to this customer-based value flow processes [61].

Value-adding activities are simply only those things that the customer is willing to pay for; everything else is classified as waste and should be eliminated or reduced. The Japanese word "Kaizen," which means "continuous improvement," is also closely associated with Lean. Kaizen focuses on continuously improving processes. Some of the main objectives of kaizen are to reduce waste, improve quality, reduce delivery time, assure a safer work area, and increase customer satisfaction.

Until the 1990s, automotive development, production, and supply-chain processes were sequential. In this environment, incorrect requirements and poor design decisions had long-lasting and far-reaching implications. The Lean revolution emphasized concurrent development, delayed decision making, and just-in-time (JIT) supply. When implemented correctly, JIT can lead to dramatic improvements in an organization's return on investment, quality, and efficiency.

Principles are those guiding ideas and insights that support a discipline, whereas practices assist in accomplishing the principle [72]. Principles tend to be universal and will transfer across domains, whereas practices will not. The challenge for those striving for software process improvement is how to translate Lean principles that have traditionally been applied in manufacturing environments to Lean practices in software engineering or development environments.

There are seven Lean principles that translate into practically applicable methods. These principles define the foundations of Lean thinking. In Table 6-1, these are described as they have been used traditionally in manufacturing and then again as they might be as applied to software engineering.

Lean principles place an emphasis on keeping the work in progress to a minimum. This helps to shorten the overall timeline and also helps to eliminate waste. The concept of delaying decisions is a hard one to grasp. Why delay decisions? This seems counterintuitive, especially since many of the traditional software life cycle models are driven by a firm definition of requirements, software design, and architecture early in the software development life cycle. When software development is truly being driven by business or customer requirements it is imperative to continually check against the true product development motivators.

The risk associated with the rigid and firm early definition of requirements, design, and system architecture is that the climates associated with customer and business requirements often change, and past decisions become incorrect. Hence, past development assumptions begin to generate production delays and waste. Previously defined requirements, architectural decisions, and each line of code that may have been written, based upon committed decisions, are now obsolete. This software development "waste" is analogous to the manufacturer's decaying "inventory" [74].

As previously mentioned, Lean focuses on the elimination of waste. The following have been identified as eight types of waste:*

---

*U.S. Army Business Transformation Center, http://www.army.mil/ArmyBTKC/focus/cpi/tools3.htm.

Table 6-1. Lean as translated from manufacturing to software engineering

Manufacturing	Software Engineering
1. Eliminate Waste	
Examine timeline	Examine timeline
Start from order receipt to payment collection	Start from order receipt (customer requirement) to software deployment
	Start and end of timeline may be modified
Focus on timeline reduction and removal of nonvalue-added wastes [102]	Focus on timeline reduction and removal of nonvalue-added wastes [105]
Forms of waste: inventory, number of actions to complete a step of work	Forms of waste: partially completed work, extra features ("gold plating"), complex code
2. Amplify Learning	
Detailed design during production	Detailed design during development
Customer feedback during production	Stakeholder feedback development
Modular design	Modular architecture
3. Delay Commitment	
Maintain a flexible plan	Maintain a flexible plan
Leave room for time to experiment with various design solutions	Leave room for time to experiment with various design solutions
4. Deliver Fast	
Must retain quality	Must retain quality
Develop standardized work products, workflow documentation, and specific job descriptions. Supported by training.	Develop standardized work products, workflow documentation, and specific job descriptions. Supported by training.
Goal: to achieve a standardized, repeatable process that can be continually improved	Goal: to achieve a standardized, repeatable process that can be continually improved
Supported by daily or weekly team meetings	Supported by daily or weekly team meetings
5. Empower the Team	
Every individual on the line has the right to halt production based upon issues with quality.	Each individual on the software development team should be equally empowered to improve the development process.
Teams are provided with targets and set their own production goals.	Teams are provided with general plans and goals and are trusted to self organize to meet these goals [105].
6. Build Integrity (Quality) in	
Inspection driven production	Test driven development
Inspect, stop the line, and fix the defect.	Unit test and acceptance testing, then code
Build in quality from the start, not inspect at the end.	Build in quality from the start, not test at the end.
7. See the Whole	
Focus on optimizing the entire production line. Cannot just focus on inventory reduction, or end of the line product inspection.	Focus on optimizing the entire development process. Cannot just focus on design or final product beta testing.

1. Talent—wasted human resources
2. Defects—items in the supply chain that need rework
3. Inventory—oversupply in the supply chain
4. Overproduction—oversupply at the end of the supply chain
5. Wait time—undersupply in the supply chain
6. Motion—excessive human movement
7. Transportation—unnecessary movement of items through the supply chain
8. Processing—anything accomplished that does not add production value

The essence of lean manufacturing is to compress the time from the receipt of a customer order all the way through to prompt receipt of satisfied customer payment. The results of this time compression are increased productivity, increased throughput, reduced costs, improved quality, and increased customer satisfaction.

So how do organizations implement Lean? There are a number of Lean techniques available. These include, but are not limited to, value stream mapping, visual workplace, setup reduction, cellular/flow manufacturing, pull systems, and total productive maintenance. It is imperative that implementing organizations view Lean from a total system perspective. If not, they risk incorrect implementation and/or noncontinuous process improvement. If analyzed and planned from the proper system viewpoint, the continuous implementation and improvement of the appropriate lean techniques can yield substantial gains.

SIX SIGMA PRINCIPLES

Six Sigma as a measurement standard in product variation can be traced back to the 1920s when Walter Shewhart showed that three sigma from the mean is the point where a process requires correction.* Six Sigma continued to evolve through a series of measurements standards, but in the late 1970s Dr. Mikel Harry, a senior staff engineer at Motorola's Government Electronics Group (GEG), began to develop what are now the foundations of Six Sigma through his experimentation with statistical analysis and its applications in problem solving. The term "Six Sigma" is credited to Bill Smith, who at the time was a Motorola engineer. He suggested that the company should require 50% design margins for all its key product performance specifications. This equates to a six sigma level of capability.[†]

Generally speaking, sigma represents the standard deviation of a data set. Standard deviation is a measure of the variability of the data. Being able to remove the variation (reduce the deviation) is essential to the reduction of cost and improvement in quality. This also directly impacts customer satisfaction. Six Sigma is the unrelenting pursuit of deviation reduction. Six Sigma is about producing better products and services at lower cost [61]. Table 6-2 provides a summary of the paradigm shift represented by the move to Six Sigma.

Motorola engineers, Dr. Harry and Richard Schroeder, are credited with creating what Six Sigma has evolved into today. They led the efforts to develop Six Sigma to its current form beginning in the mid 1980s after determining that they required a more robust way to measure defects. The engineers developed the Six Sigma standard and associated

*http://www.ssqi.com/six-sigma-library/origina.html.
[†]Six Sigma is a federally registered trademark of Motorola Corporation.

Table 6-2. Six Sigma, a paradigm shift [88]

	Old Models	Six Sigma
Example	Increasing yields, increasing % satisfied customers	Reducing defects, reducing complaints
Focus on	Ideas, the future, finding new solutions	Facts, data, past and present, finding root causes
Sources of data	Expert opinion	Current process and history
Use of metrics	Performance metrics; less detail: good transactions, performance summaries	Diagnostic metrics; more detail: defects within transactions, root cause analysis
Impact on reporting	Emphasis on good news	Focus on real issues and improvement

methodology and then implemented it within their organization. Following its implementation, they reported $16 billion in savings resulting from their Six Sigma improvement efforts. Today, Six Sigma is known as a business management discipline toolset that is focused on quality engineering as defined by customer needs. Six Sigma can be described in the following ways:

- A business philosophy and strategy
- An integrated fact-based management system
- A customer-focused quality strategy

Six Sigma offers a continuous improvement methodology that can be used to improve business processes and products. It forces organizations to define their vision of quality in numerical terms. Six Sigma provides us with data-driven problem solving.

It is important to distinguish between what has commonly become known as Six Sigma the statistical concept supporting process measures and Six Sigma the business philosophy with the goal of continual process improvement. As described above, these are merged into what is commonly referred to as Six Sigma, but within Six Sigma the concept and philosophy are distinctly separate, but complementary, concepts. Just to clarify, but hopefully not confuse, Six Sigma as a statistical concept measures a process in terms of defects, a defect being defined as any deviation from acceptable customer limits. Putting this in the perspective of a software engineering effort, not meeting schedule is a defect, not meeting a requirement is a defect, not meeting design is a defect, and obviously, not meeting a test specification is a defect.

Describing Six Sigma as a business philosophy may seem counterintuitive for those not familiar with the methodology. However, Six Sigma support continual process improvement and offers phases, tools, and techniques that help an organization improve their processes.

Lean Six Sigma, as applied to software engineering, does not employ traditional models of software development, but places importance on the basic underpinnings of the principles of software engineering. To be more specific, Lean Six Sigma emphasizes configuration management, software quality assurance, requirements management, software design, requirements definition and management, and so on [65]. *If an organization, and its employees, does not practice or understand the tenets of basic software engineering, trying to employ Lean Six Sigma methodologies will lead to disaster.*

Table 6-3. Comparisons between Lean and Six Sigma [70]

	Lean	Six Sigma
Goal	Create flow and eliminate waste	Improve process capability and eliminate variation
Application	Primarily manufacturing processes	All business processes
Approach	Teaching principles and "cookbook-style" implementation based on best practices	Teaching a generic problem-solving approach relying on statistics
Project Selection	Driven by value stream map	Driven by top-level business goals, tightly linked to lower-level objectives
Length of projects	1 week to 3 months	2 to 6 months
Infrastructure	Mostly ad hoc, no or little formal training	Dedicated resources, broad-based training
Training	Learning by doing	Courses and then application (learning by doing)

An organization must first have a functioning software engineering process with employees who understand and are trained in the supporting practices before any attempt can be made to increase software development speed. Quality improvement gains will be made simply by defining processes and practices in organizations in which there are none to be found.

As previously described, Lean and Six Sigma are complementary. Lean focuses on increasing the speed of a process, or in the elimination of any nonvalue-added steps or activities within a process, whereas Six Sigma focuses more on quality than speed. Lean defines and refines the customer value flow and Six Sigma helps ensure that the value continues to flow smoothly and to improve. Table 6-3 provides a comparison between lean processes and Six Sigma.

LEAN SIX SIGMA FUNDAMENTALS

Michael George, lead of the George Group, has summed up Lean Six Sigma in four key points; these have been reordered from his original presentation so that they might line up with typical software process improvement models* [80]:

Key # 1—Delight Your Customers. Customer satisfaction increases when software processes are improved (sped up) and process waste is eliminated. Customers are also delighted when they see an overall increase in product quality and increase in productivity.

Key # 2—Base Decisions on Data and Facts. Keep decisions data driven, while at the same time remembering to keep the data collection relative to the size and scope of the project. A common complaint is that too often the data collection and measure-

---

*In his original presentation they are ordered as follows: "Delight your customers with speed and quality; Improve your processes; Work together for maximum gain; Base decision on data and facts."

ment are either nonexistent and project decisions are subjective, or data collection and measurement are out of control and have become the main focus of the process, instead of the development effort.

Key # 3—Work Together for Maximum Gain. Ensure that efficient, and frequent, data-driven reviews are occurring. The feedback from these reviews should be used as input to the project schedule and deliverables as well as the direction of the software development process. All team members should have an equal voice and should be respected for their input in support of the software production and process efforts.

Key # 4—Improve Your Processes. Use feedback to make software process improvements where they will have the most immediate and dramatic impact. Make improvements gradually and iteratively. It is important that your organizational processes are documented and that team members are trained and kept abreast of any changes to policies and/or procedures.

Typical Lean Six Sigma Data

Many businesses are turning to Six Sigma and Lean in order to continually improve their processes, reduce costs, cut waste, and, ultimately, to remain competitive. Six Sigma is being employed to help businesses cut costs while improving quality, customer satisfaction, and cycle times. The Six Sigma methodology uses data and statistical analysis tools to identify, track, and reduce problem areas and defects in products and services. When paired with Lean, with its emphasis on waste reduction, the pair forms a strong and complimentary pair of process improvement principles that support four typical types of data, as shown in Table 6-4.

Many traditional improvement techniques focus only on the bottom data row, "Quality/defect," and do not employ the Lean Six Sigma (LSS) techniques with the other data rows. This traditional narrow focus can result in processes becoming slower and more expensive and result in unimproved customer satisfaction.

Table 6-4. Typical types of Lean Six Sigma Data [80]

Data type	Type of measure	Description
Customer satisfaction	Result	Customer value designed, implemented, and measured through surveys or interviews about the associated products or services.
Financial outcome	Result	Viewed in terms of the bottom line, dollars and cents, or overall financial improvement.
Speed/lead time	Result or process	The lead time is how long it takes for any individual work item to make it all the way from the beginning to the end of the production process. A measure taken at the end this would be a result measure. If a measure occurs during the production steps, it would be a process measure.
Quality/defect	Result or process	Provides a measure of the number of errors. If accomplished at the end of production, it is a results measure, with result or within process.

The Laws of Lean Six Sigma

The five laws of Lean Six Sigma have been developed in order to act as the guiding principles for organizations implementing the combined principles and practices of Lean Six Sigma.

Law # 0—The Law of the Market. The zeroth law is the Law of the Market.* Customer needs define quality and are the highest priority for improvement (there will be no sustained revenue growth without this) [80]. Aligning improvement efforts with customer requirements must take the highest priority. Care should be taken in determining all key measurable characteristics of the software development and software production processes, and emphasis should be placed on meeting customer requirements. Placing the customer needs first and having their requirements drive quality is also know as "Customer Critical to Quality (CTQ)" and should remain the highest priority for the improvement effort.

Without venturing into a mini economics lesson, there are two items that are associated with this law that are frequently referenced, but rarely explained. These are "Return on Invested Capital" (ROIC) and "Net Present Value" (NPV). Any company expending capital on an infrastructure investment should be concerned with their return on the investment in their business and the financial viability of the projects being targeted for process improvement. So the CTQ priority described above should be followed by an emphasis on ROIC and NPV.

This calculation is used to determine whether invested capital is being used effectively and is adding value. ROIC is usually defined as:

$$\frac{\text{Net Operating Profit} - \text{Adjusted Taxes}}{\text{Invested Capitol}}$$

The result is usually represented as a percentage. When the capital is greater than the cost of capital, it is providing value [179].

NPV is used to provide a financial snapshot of projects and has long been used in support of budget planning (capital budgeting) for new research and development efforts. It can also be used to provide a picture of the current financial status of a project. NPV provides a measure of the excess or shortfall of cash flow in present value terms, once all charges are met.

It is important to remember that NPV does not include opportunity cost, which can be described as the benefits that could be received from the opportunity should something else be accomplished instead. For example, after the decision to develop one software feature or component instead of another, what is gained by the decision to ramp up on one project versus another? This should be factored into NPV if possible so that, in the end, all projects with a positive NPV, and whose their NPV is greater that the opportunity cost, are profitable.

Law #1—The Law of Flexibility. This law supports the concept that the speed of any process is proportional to its ability to adapt to different circumstances, or its flexibility.

---

*These laws have been seen in the literature numbered as 1–5 or 0–4. The authors have chosen to list them here as 0–4 as the initial law is frequently referred to as the zeroth law.

As a general rule, the more the process is receptive to change, the better the progress of the project. To increase speed, look for ways to remove anything that causes losses in production or development waste. This can be directly related to how easily people switch between tasks.

Law #2—The Law of Focus. The Pareto principle is the primary motivator behind this second law, the Law of Focus. The Pareto principle states that 80% of the impact of a problem will show up in 20% of the causes. For example, if presented in a bar chart displaying the frequency of software defects, these defects are displayed in descending order and the Pareto principle will be evident. About 20% of the categories on the far left will cause about 80% of the problems.

To summarize the Law of Focus, 20% of activities within a process cause 80% of the problems and delay (time traps). It is much more effective to focus concentrated effort on the 20% that are the time traps.

Law #3—The Law of Velocity (Little's Law). As classically defined, the Law of Velocity states that "the velocity (speed) of any process is inversely proportional to the amount of work in progress (WIP)" [80]. WIP can be interpreted as unfinished tasks and, as such, impact the speed of a production cycle significantly. Therefore, as WIP increases, speed decreases, and as WIP decreases, speed increases.

Minimizing waste should also include an inventory analysis and reduction. Little's Law provides an equation for relating lead time, work in progress, and average completion rate (ACR) for any process. Named after the mathematician who proved the theory, Little's Law states:

$$\text{lead time} = \text{WIP (units)}/\text{ACR (units per time period)}$$

Knowing any two variables in the equation allows the calculation of the third. For example, reducing WIP while maintaining the same ACR reduces lead time, and improving the process to increase ACR while maintaining the same WIP also reduces lead time. If it is difficult to relate WIP to a given process, try using things in progress (TIP) instead [80]. For example, a software development group can complete five game levels per month (ACR), and there are 10 levels (TIP) in various stages in the development group. Applying Little's Law:

$$\text{lead time} = \text{TIP}/\text{ACR} = 10 \text{ levels}/5 \text{ levels/month} = 2 \text{ months}$$

Therefore, without changing the process, inventory or priorities, or accounting for variation, any new game level coming into the development group could reasonably be expected to be completed in 2 months.

Law #4—The Law of Complexity and Cost. The complexity of a product (or service) can result in more nonvalue-added costs* and WIP than either poor quality or slow speed processes. For example, trying to develop a software product that is heavily feature laden increases the complexity and at a certain point will add more burden to the development

---

*Value-added work is work that adds value in the eyes of the customer, whereas nonvalue-added work is work that produces delays, waste, rework, and constant checking.

process (nonvalue-added costs and WIP) than the poor quality or slower speed processes that might be associated with a similar but much less complicated software product. One early target when attempting waste reduction might be a reduction in the number, varieties, or features of a product under development. Inflexible production goes against the foundation of lean principles. Complex software development processes and the specifications that support them often ultimately render them ineffective.

Lean Six Sigma as Applied to Software

As previously described, Six Sigma aims to reduce defects. The key to defect reduction lies in understanding which part of the process should be improved. The two key components again are variation and waste—which two parts of the process are affected by variation and waste? Lean "thinkers" want to understand what is truly and absolutely necessary to create value. These are fairly easy concepts to grasp when discussed using classic production analogies.

Lean Six Sigma methodologies are applicable in many environments. It is extremely important that those implementing Lean Six Sigma methodologies in software development environments clearly understand the relationships between software and classic lean thinking. Table 6-5 shows how lean wastes relate to software development.

Table 6-5. Lean as related to software development [74]

Seven classic lean wastes	Seven wastes of software development
Overproduction	Extra features, functions not valued by customer
Inventory	Requirements not implemented
Extra processing steps	Extra steps, extra reviews
Motion	Finding information
Defects	Bugs not caught in testing
Waiting	Delays due to decisions
Transportation	Handoffs
Plus One	
Talent	Mismatch assignments to life cycle roles

Before venturing off into the land of Lean Six Sigma software process improvement, it is important that an organization understands the differences between both methodologies. It is also critical that the organization have a firm and established software development foundation and that its process champions understand the differences between Lean and Six Sigma, and how their organization intends to integrate them in support of their improvement strategy.

What Lean Six Sigma is Not

Lean Six Sigma does not

- Ensure that a product or service is without defects
- Provide specific instructions on how to build a quality system

- Provide for easy interpretation and, thus, implementation
- Provide a rulebook, prescribe formats, or prescribe explicit contents
- Prescribe media or specify technology
- Specify requirements for products or services
- Specify all of the product realization processes
- Offer details about its application to specific domains of expertise
- Prescribe measurement definitions
- Focus on associated financial issues
- Provide for conformity to or assessment of a standard

Lean Six Sigma does not provide processes or process descriptions. Actual processes are dependent upon:

- Application domain(s)
- Organization structure
- Organization size
- Organization culture
- Customer requirements or constraints

This text will provide the guidance on details about the application of Lean Six Sigma to the software engineering domains of expertise as a management tool to attest to the software engineering process and how it will be managed and reviewed.

SIX SIGMA LIFE CYCLE VARIANTS

Six Sigma has two major life cycle variants [100]: DMAIC (define, measure, analyze, improve, control) and DMADV (define, measure, analyze, design, verify). These are two data-driven quality strategies that software development teams might use in support of improving existing, or in defining new, products and processes.

DMAIC should be used by implementing teams to improve an existing product or process. Improvement is required when a product is not meeting customer expectations. DMADV should be used define a nonexistent product or process. DMADV is a data-driven quality strategy for designing new products and processes.

DMAIC is the most commonly use framework by organizations applying Six Sigma and Lean Six Sigma (LSS) principles, and is the framework that will be described here. Each step in the DMAIC process is required in order to ensure the best possible results.

7

LEAN SIX SIGMA TECHNIQUES AND DMAIC

DEFINE

This stage helps the team develop a shared understanding of the business priorities and confirm the opportunities with the customer. This phase sets the scope of the projects and defines how success will be measured.

During this stage, the team determines the project goals and deliverables and must agree on what the project is all about. They must:

- Determine sponsorship and determine project stakeholders
- Train participants
- Get customer data
- Review existing data
- Draft a map of process priorities
- Set up a plan and guidelines
- Baseline processes

Build Sponsorship and Determine Stakeholders

Consensus and agreement in support of improvement must be developed. This requires a gathering together of all stakeholders and the definition of their roles. Sometimes, this requires executive-level training in software engineering concepts and principles.

For a project to implement software process improvement and definition methodologies, the stakeholders need to be identified and committed to their roles. Several stakeholders have been identified as follows:

Practical Support for Lean Six Sigma Software Process Definition. By S. Land, D. Smith, and J. Walz

- **Senior Management** initiates the implementation/improvement project, commits to the investment, monitors the various stages toward completion, and guides the organization through its periodic review and planning cycles.
- **Customers and Investors,** or their representatives, have a variety interfacing roles, depending on their maturity, previous experiences, and type of contracting agreements. Several roles are discussed below.
- **Software Engineering Team Members** have technical and supporting roles as defined in their business improvement model and their software development life cycle model.

The definition of the software process improvement initiative must be accomplished. This should be described in terms of time frame, membership, organizational, structure, and project scope. The degree of formality, required infrastructure investments, and communication channels should also be defined.

Process improvement normally works best with two levels of committees or groups. The first is the Steering Committee, which includes the sponsors and meets infrequently through the year. The Steering Committee manages a portfolio of process improvement initiatives. The second is the Process Group, a process improvement team made up of people who are given responsibility and authority for improving a selected process in an organization; this team must have the backing of senior management represented by the Steering Committee.

Each process owner in the Process Group has responsibilities that cut across the organization, operate through influence rather than control, own the process design but not necessarily the resources needed to execute it, and must induce people of all levels to work together to achieve desired results. The process owner is further responsible for the process measurement and feedback systems, the process documentation, and the training of the process performers in the structure and conduct of the group. In essence, the process owner is the person ultimately responsible for improving a process. IEEE software engineering standards provide valuable support to the process team and each individual process owner. The standards can be used to help define and document the initial baseline of recommended processes and practices.

Set Priorities

Following Steering Committee review and approval of a process improvement initiative, the Process-Group-endorsed recommendations should be prioritized. The Process-Group-developed priorities of the recommended actions should be based on such things as:

- The availability of key resources
- Dependencies between the actions recommended
- Funding

Organizational realities influence priority setting. The Steering Committee has more control over the organizational realities than the Process Group, whose job is to bring these issues to the surface or issue caveats for Steering Committee support and decisions. Some common organizational realities are:

- Resources needed for change are limited
- Dependencies between recommended activities
- External factors may intervene
- The organization's strategic priorities must be honored

Once the Process Group has Steering Committee approval of the recommended actions, an approach should be developed that reflects both the established priorities and current realities. This approach should be a repeatable process and should consider the following:

- Strategy to achieve vision (set during the initiating phase)
- Specifics of installing the new technology
- New skills and knowledge required of the people who will be using the technology, and recommended training
- Organizational culture
- Potential pilot projects
- Sponsorship levels
- Market forces

Plan Actions

From the approach developed by the Process Group, they will next develop the action plans to implement the chosen approach. The action plans should describe the following:

- Deliverables, activities, and resources
- Decision points
- Milestones and schedule
- Risks and mitigation strategies
- Measures, tracking, and oversight

Many teams construct a project plan using programs such as Microsoft Project. This type of program may be suitable for the first three bulleted items above, but will not be sufficient for the last two.

Baseline Processes

The Process Group should define their process baseline, which is critical when implementing software processes that can be repeatable. Normally, baseline processes and descriptions can be managed in the Software Configuration Management (SCM) system as text documents.

Once a process baseline has been established, an action plan should be formulated. The action plan provides a map of the path forward—improvement of the baseline processes. Table 7-1 provides a sample from an action plan.

Take advantage of the information provided by the *IEEE Software Engineering Standards Collection* [42]. Many of these standards provide documentation templates and de-

Table 7-1. Example action plan

Process Area	Weakness or Area for Improvement	Short Description of How to Address	Project Point of Contact	Due or Resolution Date	Resources/ Support Required	Risks/ Contingency
Requirements Management	7.2.1.1 the establishment of a method for traceability of the requirements to the final product	Create requirements traceability matrix in SCM with links to and from Reviews	Jim Smith	4Q2008	SCM Req. doc. Req. Rev. Client Rev.	Review time and rework

scribe in detail what individual project support processes should contain. Think of the standards as an in-house software process consultant who has recommended, based upon years of experience, the proper methodologies and techniques to be used in support of software development.

Create a Solution

The solution arising from the approved action plan recommendations will result in defining and implementing new and/or revised procedures, templates, forms, and checklists, along with measurable objectives that support the organization's quality objectives.

The Process Group will bring together everything available to create a "best guess" solution specific to organizational needs, for example:

- Existing tools, processes, knowledge, and skills
- New knowledge and information
- Outside help

The solution should:

- Identify performance objectives
- Finalize the pilot/test group
- Construct the solution material and training with pilot/test group
- Develop draft plans for the pilot/test and implementation
- Technical working groups develop the solution with the pilot/test group

Software engineering processes create work products or artifacts that normally require team member and/or customer review and acceptance. Artifact templates, checklists, and record forms can facilitate these reviews and acceptance. In addition to policies, procedures, artifacts, and records are the process measures. Objective measures can cover time interval, investments, costs, output quantities, and rework indicators such as defect counts and feedback.

As previously described, the first phase is the define phase. It is during this phase that all stakeholders define the software engineering processes and what should be accomplished during the software process improvement initiative.

Tools commonly used during the define phase are:

- Affinity diagram
- Critical-to-quality (CTQ) tree
- Functional analysis diagram
- Kano Model
- Project charter
- Quality function deployment
- Stakeholder analysis
- Supplier, inputs, process, outputs, and customers (SIPOC) diagram
- Voice of the customer

Abstraction Tree (KJ or Affinity) Diagram

An affinity diagram (sometimes referred to as a "KJ," after the initials of the person who created this technique, Kawakita Jiro) is a special kind of brainstorming tool. This process has also been referred to as the creation of an abstraction tree. You use an affinity diagram to:

- Gather large numbers of ideas, opinions, or issues and group those items that are naturally related or organize large volumes of data into cohesive and logical groupings
- Identify, for each grouping, a single concept that ties the group together
- Encourage new patterns of thought when working with diverse groups of people or ideas

An affinity diagram is especially useful when:

- Chaos exists
- The team is drowning in a large volume of ideas
- Breakthrough thinking is required
- Broad issues or themes must be identified

Building an affinity diagram is a creative rather than a logical process that encourages participation because everyone's ideas find their way into the exercise.

Affinity diagramming should not be used if less than fifteen items of information have been identified. A decision process supporting priority ranking would be more appropriate for groups of items of this size.

Overviews of the five steps that support the process of affinity diagramming are provided in Figure 7-1. Detailed descriptions are provided below:

Step 1. Generate ideas. Members of the team use brainstorming to generate a list. The list items are documented. Traditionally, list items are written on Post-It™ notes

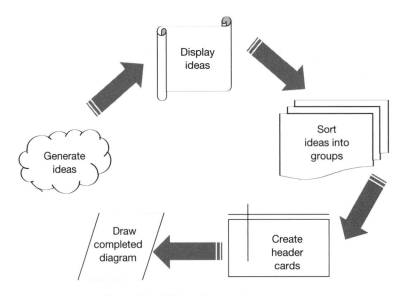

Figure 7-1. Affinity diagramming steps.

and these list items are then posted on walls or white boards for ease of later grouping and categorization.

Step 2. Display ideas. During this step, members of the group post the list items on a whiteboard, wall, or table in a *random* manner.

Step 3. Sort ideas. The list items are sorted into groupings. Typically, this occurs when members of the team begin by looking for related ideas and placing them in related columns or groups. Super- and subgroups may be created, depending upon how the list items best relate.

Step 4. Create idea headers. Based upon the grouping that have been created, the team member create headers that summarize the relationship between the logical groupings of the list items.

Step 5. Draw diagram. Team members will need to define an overall problem statement or a title for the diagram. They will need to ensure that all super- and subgroups for the appropriate list items have been captured and that all ideas have been properly reviewed and vetted. Figure 7-2 provides an example of what the structure of a completed affinity diagram might look like.

Functional Analysis Diagrams

Functional analysis diagrams may be used to classify and categorize the functionality associated with a product or process. This is a discovery technique and is used to uncover hidden requirements. It is important to remember when using functional analysis to remain "verb-dependent," that is, use verbs to precisely describe functions. Figure 7-3 provides an example of an initial functional analysis diagram. This is a very simplistic first pass of what might be found during the exploratory phase in support of the development of a new website.

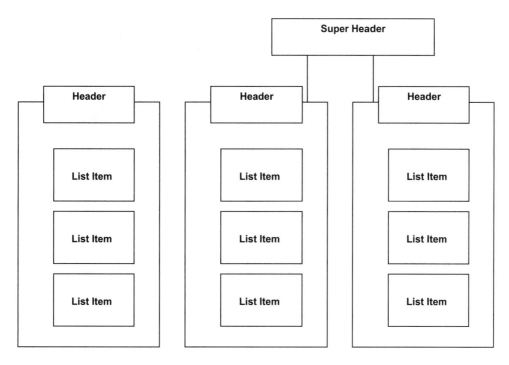

Figure 7-2. Example structure of a completed affinity diagram [129].

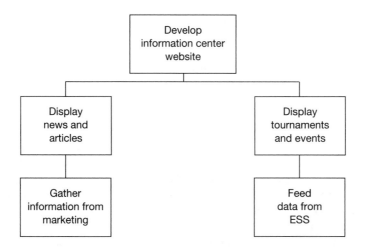

Figure 7-3. Example of an initial functional analysis diagram.

Of course, the requirements should be refined and another pass of the website functionality should be made. The development team should ask the following questions: "Does this communicate a comprehensive story? Do we have enough information to fully understand how to implement this product? Can we begin design?" If the answer to one or more of these questions is "No," then the branches of the diagram should be increased and additional research should be conducted into the product functionality and requirements. Figure 7-4 provides an example of the previously described functional analysis diagram with additional, further analysis of the functional analysis provided.

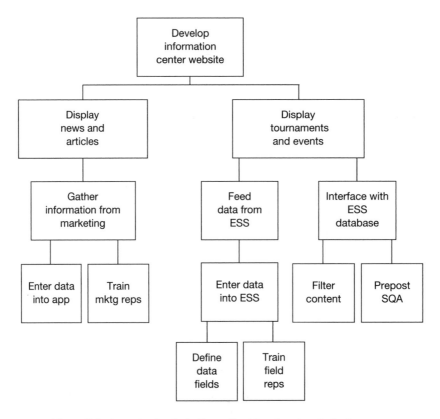

Figure 7-4. An example of a further-refined functional analysis diagram.

Critical-to-Quality (CTQ) Tree

The purpose of critical-to-quality (CTQ) trees is to convert customer requests into measurable requirements. The key to any successful software development effort is accurately identifying requirements. The concept behind the generation of a CTQ tree is that hard-to-define, general requirements are elicited from the target user community and these are then used to successfully capture the target market.

The first step in building a CTQ tree is to collect as much data about your target customers as possible. This information might be obtained for existing software products using help desk logs and user forums. When trying to define requirements for new products surveys, interviews and focus groups are just two of the methods that can be used to gather information to support the definition of new product requirements.

Once the data is gathered, it is examined. Positive and negative trends are then identified. These trends are utilized to create the CTQ tree to help convert the raw data into product requirements. Figure 7-5 provides an example of a CTQ tree describing how collected data has been converted to requirements in support of a second version of a software development effort.

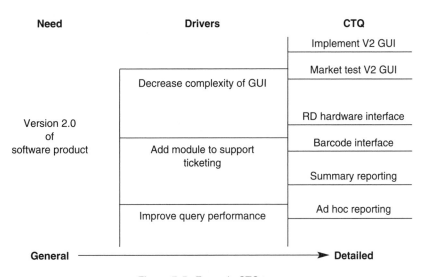

Figure 7-5. Example CTQ tree.

Kano Model

The Kano Model was developed in the 1970s by Professor Noriaki Kano. This model has also been called the "customer satisfaction model" as it applies quality concepts to customer preferences. The power in the application of this model is in how it can then be related to product development. A complete description of these quality types are provided in Table 7-2.

Table 7-2. Types of Kano quality [94]

Kano Original (English)	Customer Satisfaction	Description
Must-be (dissatisfier)	Basic factors	These are the minimum requirements. If these are fulfilled, they will not cause customer satisfaction because they are taken as minimal criteria.
One-Dimensional (critical)	Performance factors	These requirements are directly connected to the customer's business goals. Satisfaction is directly linked to the performance of these requirements.

(continued)

Table 7-2   *(continued)*

Kano Original (English)	Customer Satisfaction	Description
Attractive (satisfier)	Excitement factors	These are the requirements that the customers are looking forward to seeing in the product. When fulfilled, they will cause customer satisfaction.
Indifferent (neutral)	Indifferent attributes	These are attributes or requirements that, if fulfilled, are a low priority and to which the customer is ambivalent.
Reverse (expected)	Reverse attributes	These are "reverse" attributes and are the inverse of the customer's original requirement.
	Questionable attributes	Possible feature creep, unclear as to origin of requirement.

Prior to Dr. Kano's research, many of the previous definitions of quality were linear and one-dimensional in nature [122] (i.e., good or bad). He was able to demonstrate how quality could be integrated along two dimensions: (1) the degree to which a product or service performs, and (2) the degree to which the customer is satisfied. This integration is demonstrated in Figure 7-6. This Kano plot of quality along two axes led to initial unique definitions of quality: basic, performance, and excitement quality.

The Kano Model defines the categories of quality associated with customer requirements and emphasizes the importance of requirements definition. Additional information in support of the process of requirements elicitation and definition may be found in Chapter 3.

Project Charter

A charter is prepared for each Lean Six Sigma problem-solving team. It starts as the preliminary design for the team (as part of the sponsor or champion identifying and prioritizing the problem), and grows to become the contract between management and the team (at the conclusion of the define phase of the DMAIC process). The charter contains details about the project, and has two key subcomponents: the problem statement and the business case. Figure 7-7 provides an example project charter template.

The charter is used to keep the team focused on the sponsor's goal and aligned with organizational priorities. In organizations with a Six Sigma infrastructure, approval of the initial charter formally transfers the project from the champion to the project team. [102]

The charter is a dynamic document. Teams review it periodically and, as needed, update/revise it. All substantive changes should be approved by the champion/sponsor. A template in support of the project charter is provided on the companion CD-ROM with additional explanatory guidance. This template is entitled *LSS Project Charter.doc.*

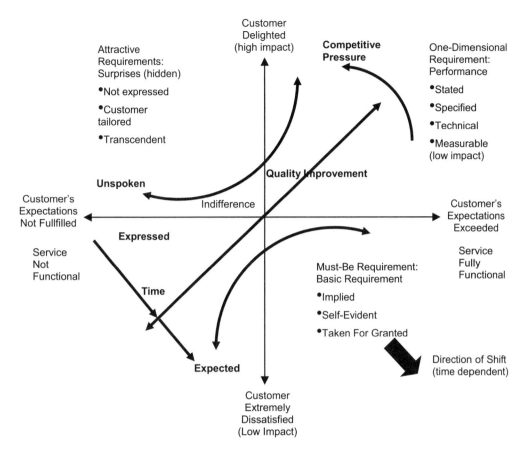

Figure 7-6. KANO quality plot.

Problem Statement. It is a Six Sigma maxim that a successful project flows from a well-written problem statement. Good problem statements concisely address as many of the following issues: *Who, What, When, Where, Effect, Gap, Measure,* and *Pain.* A good problem statement acts as springboard to the business case and makes the problem compelling to solve. It is *not* a solution disguised as a problem; it does not say what needs to be done, just states indisputable facts about the observed effects.

A sample problem statement is:

> In the first six months of 2007, an average of twenty-five working days was required to install and build baseline release 3.1 at customer locations. Our customer wants the system to be installed and operational within ten working days after the order is placed, and ten points of the contract Award Fee (worth $10,000/point) depend on meeting this threshold.

In many cases, the measure is not available initially—the extent of the problem is unknown. The team thus must plan to collect that data during the measure phase, and then update the problem statement based on the results.

Project details

Project
Description: _____
_____
_____

Business case: _____
_____

Problem/opportunity statement: _____
_____

Process start/stop events: _____
_____

Process boundaries: _____
In:
Out:

Project improvement criteria

Metric Type: Baseline   Current: Goal:

Metric Type: Baseline   Current: Goal:

Notes:
_____
_____
_____

Figure 7-7. Lean Six Sigma project charter.

Business Case. A Lean Six Sigma project must create benefits (direct/indirect, strategic, hard or soft savings*) now or in the future, or it is not worth doing. The business case quantifies these benefits, viewed from the customer's perspective. Business cases should reflect changes in metrics that are directly impacted by the scope of the Lean Six Sigma project.

When the charter is first drafted, the business case is often based on possible savings: "If we could reduce 'x' by this amount, then we would save 'y' dollars." By the end of the define and measure phases of the DMAIC, the team will have the data to update the business case to specify the range that "x" is expected to be reduced to and the actual savings that are forecasted.

Quality Function Deployment (House of Quality)

Quality function deployment (QFD) was developed by Yoji Akao and Shigeru Mizuno [69] and championed as a software engineering method by Richard Zultner [181] and Ralph Young [182]. QFD utilizes tools that result in graphs or matrixes that help organizations analyze large quantities of data. This data can be used to support group decision

---

*Hard savings are defined as identifiable changes in the cost of operations (which can include customer savings); for example, implementing the improvement reduces the number of computers leased for a project. Soft savings are defined as productivity increases or cost avoidance; for example, implementing the improvement will reduce by 10% the labor hours required to build software code.

making or to help improve processes or products. This data can then be reused for future product/service developments. One such tool is the House of Quality.

QFD had its origins in manufacturing, like many of the Lean Principles, and care should be taken when applying this technique to software development. Figure 7-8 pro-

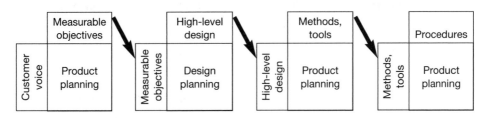

Figure 7-8. Alignment of QFD techniques to traditional software development [77].

vides an example of how QFD can be aligned with a traditional software development process. When applying any QFD technique, it is important to first align the requirements with a set of measurable business objectives. These measurable objectives should be turned into a high-level design. Then this high-level design should drive the selection of methods and tools for the projects and, subsequently, the procedures for the production of the software product [126].

Each QFD House of Quality has distinctive "rooms" and each room serves a specific purpose. An overview of a typical House of Quality QFD is shown in Figure 7-9. All requirements are listed (1) and are shown as related (5) to available development capabili-

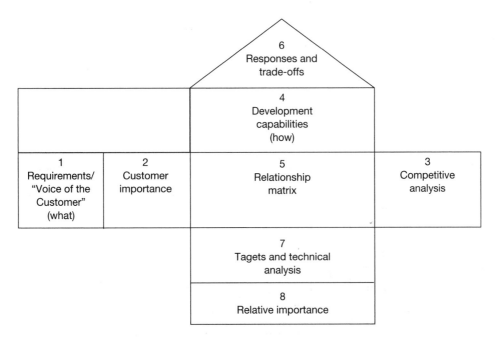

Figure 7-9. Typical areas for a House of Quality QFD [89].

ties (4). The customer's view of the requirements (2), relative importance (8), competitive analysis (3), targets (7) and trade-offs (6) are then factored into the House of Quality.

When filling out a House of Quality QFD, the first step is to identify the key customer requirements. One common technique is to use a Voice of the Customer analysis. Looking at Figure 7-10, which provides an example House of Quality QFD, these requirements should be listed in area 1. Each of the requirements should be analyzed. A method such as

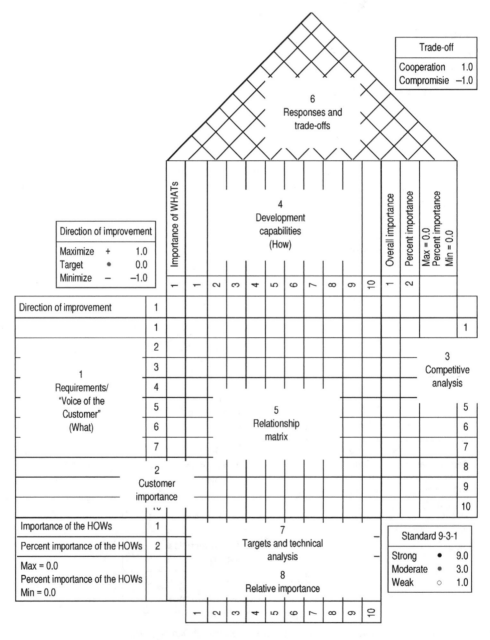

Figure 7-10. House of Quality diagram.

paired comparison analysis can be used to determine the weights associated with each requirement. This weight should be listed in area 2. This activity is followed by listing potential solutions for the requirements that have been previously identified. These will be listed in area 4; brainstorming techniques are usually used for this step in the production of the House of Quality dataset.

Building the data to support the relationship matrix, area 5, should be the next step. A relationship scale should be developed. This scale should provide a ranking from 0 to 9, where 0 indicates the weakest (or no) relationship and 9 indicating the strongest relationship between the requirement and potential solution.

The next calculation is to perform a technical analysis (area 7). This value is based upon the product of the values in customer importance (area 2) and the relationship matrix (area 5). In addition, positive and negative symbols are used to summarize areas of cooperation or compromise (area 6).

A QFD House of Quality template, entitled *QFD_HouseofQuality.xls,* is included on the companion CD-ROM.

Stakeholder Analysis

It is critical that the identification of all program stakeholders be of primary importance. When dealing with change management, communication is paramount, and in order to minimize the resistance to change it is essential that the stakeholders be identified and brought onto the program during the initial rollout phases. If the organization is a large one, a communication plan may be developed. Typical stakeholders include managers, people who work in the process group, customers, developers, and so on.

Supplier, Inputs, Process, Outputs, and Customers (SIPOC) Diagram

A SIPOC (supplier, inputs, process, outputs, and customers) diagram is a Six Sigma tool used by organizations primarily during two phases of DMAIC (define, measure, analyze, improve, and control). It is used primarily during the Define phase to identify all relevant process elements. It is also used during the measure phase to help determine how any given process is performing.

With the system or SIPOC model, organizations can create a picture of how work gets accomplished and then target improvement. When organizations can be described simply in terms of the suppliers, their inputs to the process, the process itself, the outputs the process creates, and the customers, you have perspective and boundaries for improvement.

For example, a programmer working to support the test and evaluation of a software development effort might describe his or her role on the project in the following manner: "I receive the Beta Build from the development team, a list of the software requirements, and the test cases. I begin to test, and when I find problems I enter the problem reports into the system online. These problem reports are then sent back to the development team and are also reported to the client in the monthly status report." This information could be described in the SIPOC model as follows:

S = Development staff

I = Beta Build, software requirements, test cases

P = Test process

O = Problem reports

C = Monthly status report

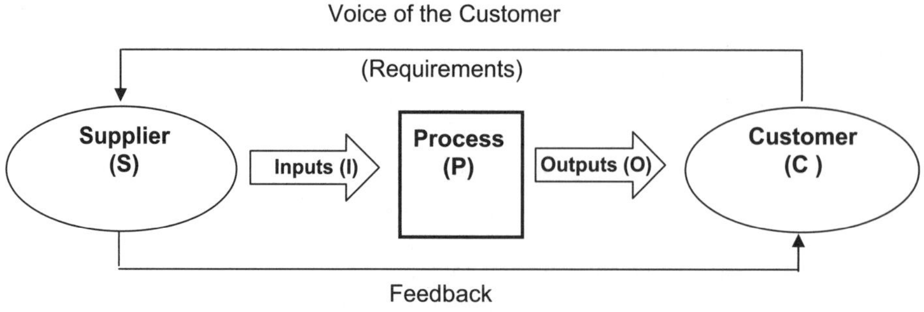

Voice of the Customer

(Requirements)

Feedback

Figure 7-11. SIPOC diagramming process.

SIPOC diagramming is very helpful when areas of the process require clarification; for example: identifying all suppliers and their relative importance, illumination of all process inputs and outputs, and the determination of all customer requirements.

Throughout the SIPOC process, suppliers (S) provide input (I) to the process. This is the process (P) that is being improved. This improvement results in outputs (O) that meet or exceed the customer (C) expectations (Figure 7-11).

To perform an SIPOC diagramming:

1. It is important that you begin with the targeted process. Map the process to four or five levels if possible.
2. Identify the process outputs.
3. Identify all customers receiving process outputs.
4. Identify process inputs.
5. Identify all process suppliers.
6. Optional. If used during the define phase of DMAIC, identify the high-level customer requirements, as these requirements will be verified during the DMAIC measure phase.
7. Verify the SIPOC diagram with all stakeholders.

A SPIOC template, entitled *SPIOC.xls,** is provided on the companion CD-ROM.

Voice of the Customer

The Voice of the Customer (VOC) is a LSS process in which requirements and ongoing feedback are solicited from the client so that they may receive the highest quality product and best service. With VOC, requirements are not frozen, but remain flexible and subject to change. This process supports constant innovation.

IEEE Standard 830, IEEE Recommended Practice for Software Requirements Specification [6], provides detailed guidance in support of the development of a software requirements management plan. This standard provides detail on what is required to effectively manage software requirements, how to elicit requirements, and how to document these requirements in a management plan. Additional guidance is also provided on the accompanying CDROM in the form of the template, *Software Requirements Management Plan.doc.*

*This template is from the makers of Breeze Tree Software, www.breezetree.com.

MEASURE

This stage helps the team develop an understanding of where they are and helps them develop an understanding of the entire development process. During this DMAIC phase, the team will:

- Gather data and combine with knowledge and experience for purposes of improvement
- Evaluate existing measurement system and improve if necessary
- Develop a measurement system if there isn't one
- Observe the process
- Gather data
- Map the process in more depth

The tools most commonly used tools in support of the measure phase are:

- Checksheet
- Control charts
- Failure modes and effect analysis (FMEA)
- Histograms
- Pareto charts
- Prioritization matrix
- Process cycle efficiency
- Run charts
- Time value analysis

Checksheet

The checksheet is one of the seven basic quality control tools and is used to make data collection easier and more accurate. It may range from a simple list to a complex form, with the consistent characteristic of having areas to make marks, or checks (as the name implies), to indicate completion of a step or record a quantity.

Checksheets take different forms depending on the data that needs to be collected. Common varieties include:

- Checklist. This is the simplest type of checksheet and contains a list of steps, actions, or conditions that are marked off as they are completed or verified. Checklists are very useful for defect prevention, and produce artifacts that can be useful proofs for quality assurance, CMMI, or ISO 9001 audits.
- Stratified Frequency Counts. These are used to collect the number of occurrences in different categories or measurement ranges. This form of checksheet is also referred to as a tally sheet.
- Data Sheet. This type of checksheet may be used to capture readings such as temperature, queue length, or utilization percentages, typically over specified time intervals.
- Data forms. When many different data elements need to be collected per instance, this type of checksheet is used. One sheet per occurrence is used, instead of the single summary sheet common in other varieties.

Guidelines for Checksheets. Einstein's famous guidance applies here: "Make everything as simple as possible, but not simpler." Complexity increases errors and can decrease completion rates, but the ability to stratify the collected data is often required. The guidelines are:

- Anticipate the key factors required for stratification, and collect the data according to those factors.
- Prefer single-page/single-screen checksheets.
- The larger the number or the more decentralized the collectors, the more labels, instructions, and definitions are required. These are useful when more than one person completes a checksheet, and increasingly essential as the number grows.
- Leave no question about the data. Include metadata such as date, time period, collector's name, and "place" (which may be location, organization, shift, team, process, machine, etc.) or "thing" under study. For paper checksheets, it is often useful to provide instructions and definitions on the back of the sheet.
- Organize the checksheet for ease of use in collection, and match the compilation method to the checksheet layout [102,103].

There are seven basic tools of quality control:

1. Cause-and-effect diagram
2. Checksheet
3. Flowchart
4. Histogram
5. Pareto chart
6. Scatter diagram
7. Quality chart

The checksheet is the second tool described in the basic toolset. Each of these tools is being described separately. It is important for the reader to understand their efficacy when used together and value when applied separately.

Control Charts

The control chart was invented by Walter A. Shewhart while working for Bell Labs in the 1920s. His experiments showed that data from physical processes was not as tidy as first thought. This data did not produce a normal distribution or "bell" curve (Gaussian distribution). He concluded that although every process displays some type of variation, some processes display controlled variation natural to the process, whereas others display uncontrolled variation, which is to say it cannot be assigned directly to definable causes as related to the process [99].

Control charts have three basic components (Figure 7-12):

1. A centerline comprised of the average of all samples
2. Upper and lower statistical control limits
3. Performance data plotted over time

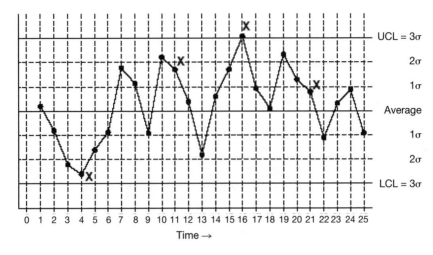

Figure 7-12. Example control chart [120].

There are two popular variations, the common-cause, also referred to as "attribute data," and the special-cause, also commonly referred to as "variable data." Table 7-3 compares the two and provides easily understood examples.

Types of special-cause/Variable control charts [120] are:

- X and R chart (also called averages and range chart)
- X and S chart
- Chart of individuals (also called X chart, X–R chart, IX or MR chart, Xm R chart, moving range chart)
- Moving average or moving range chart (also called MA or MR chart)
- Target chart (also called difference chart, deviation chart, and nominal chart)
- Cumulative sum chart (CUSUM)

Table 7-3. Comparison of the two popular variations of control charts

Descriptor	Common-cause (Attribute data)	Special-cause (Variable data)
Occurrence	Consistent	Unanticipated
Variance	Predictable	Unpredicted
Experience base	Historical basis with irregular variation	No historical basis
Significance	Lack of significance to individual highs and lows	Evidence of inherent change in the system
Examples	Inadequately trained employee; poor program design	Software or hardware failure; abnormal web hits (site crash); e-mail virus
Measured	Continuous scale; fractional	Counts; not fractional
Sample data	Time, weight, distance, and/or temperature	Numbers of errors

- Exponentially weighted moving average chart (EWMA)
- Multivariate chart (also called Hotelling T2)

Types of common-cause/attribute control charts are:

- Proportion chart (p chart)
- np chart
- Count chart (c chart)
- u chart

Charts that support both common- and special-cause data are:

- Short-run charts (also called stabilized charts or Z charts)
- Group charts (also called multiple-characteristic charts)

FMEA (Failure Modes and Effects Analysis)/EMEA (Error Modes and Effects Analysis)

Failure modes and effects analysis (FMEA), also called error modes and effects analysis (EMEA), is a technique that is used to identify and categorize the failures associated with the production of software. Following the identification of software failures, these failures are then ranked and prioritized. The possible causes of the failures are identified and a prevention plan should be developed. Appendix A provides examples of IEEE artifacts that may be used in support the FMEA process. These include the forms and reports used in support of SQA inspections and walk-throughs that are listed in support of validation.

In summary, FMEA is intended to document a failure, its mode, and its effect, by analysis.

Histograms

Histograms, also called frequency plots or frequency distributions, are vertical bar chart depictions of a set of data. They show the shape of the data's distribution, and are visual displays of its central tendency and dispersion. The central tendency is measured numerically by mean, mode, and median, whereas dispersion is shown by range and standard deviation (Figure 7-13).

One of the seven basic quality control tools, histograms can be very useful for visualizing a set of data. These are often used in the measure and analysis phases of DMAIC. They allow visual comparisons between sets of data, and comparisons of data to the standard distributions (e.g., normal, Poisson, etc.) that have to be verified before using numerical tests such as ANOVA [80,102].

There are seven basic tools for quality control:

1. Cause-and-effect diagram
2. Check sheet
3. Flowchart
4. Histogram
5. Pareto chart

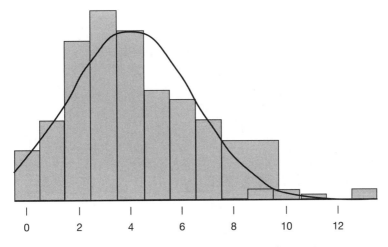

Figure 7-13. Histogram example [68].

6. Scatter diagram
7. Quality chart

The histogram is the fourth tool described in the basic toolset. Each of these tools is being described separately. It is important for the reader to understand their efficacy when used together and value when applied separately.

Pareto Charts

In the early 1900s, an Italian economist, Vilfredo Pareto, determined that the distribution of wealth in different countries was allocated among a fairly consistent minority. During this period of time, about 20% of the people, a fairly small proportion, controlled the large majority, about 80%, of any given society's wealth. This distribution has been termed the "Pareto effect". Hence, charts used to graphically summarize and to display graphical distributions are called Pareto charts (Table 7-4 and Figure 7-14).

Table 7-4. Data supporting Pareto chart for reported software errors

Priority	Category	Frequency	Module
1	1	22	12
1	2	15	13
2	2	26	12
2	2	7	15
3	3	2	12
3	3	18	15
4	2	9	12
2	2	2	13
4	1	6	13
2	1	15	15
1	3	5	15

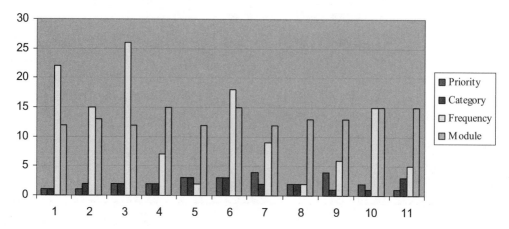

Figure 7-14. Pareto chart showing an example of reported software errors by priority, category, frequency, and software module.

As applied to process and product quality, the Pareto effect speaks directly to the dynamic that 80% of the problems identified can often be traced to 20% (or less) of causal issues. Pareto charts are useful tools that can be used to support problem solving and issue resolution.

There are seven basic tools of quality control:

1. Cause-and-effect diagram
2. Checksheet
3. Flowchart
4. Histogram
5. Pareto chart
6. Scatter diagram
7. Quality chart

The Pareto chart is the fifth tool described in the basic toolset. Each of these tools is being described separately. It is important for the reader to understand their efficacy when used together and value when applied separately.

Prioritization Matrix

A prioritization matrix is a tool that can be used to help prioritize problems for resolution. This is a useful approach when dealing with a large list of items. It is a useful approach as it removes ambiguity and associates weighted scores with each issue. Prioritization matrixes are very helpful in gaining group consensus and setting actionable priorities.

The prioritization matrix is said to:

- Quickly surface basic disagreements so that they may be resolved up front.
- Force a team to focus on the best things to do, not everything they could do, dramatically increasing the chances for implementation success.

- Limit hidden agendas by surfacing the criteria as a necessary part of the process.
- Increase the chance of follow-through because consensus is sought at each step in the process (from criteria to conclusions) [71].

An example of a prioritization matrix is provided in Table 7-5. Software features for a proposed web upgrade are listed down the left side of the table and the ranking criteria are listed across the top of the table. Members of the team would then fill in the matrix. This could be done individually and the numbers could be averaged, or the matrix could be filled in during a group discussion. In the matrix in Table 7-5, the rank values are numbers 1 to 10, low to high, respectively.

Table 7-5. Prioritization matrix showing proposed features for a web upgrade

	Criteria			
Features	Frequency	Importance	Feasibility	Priority
Use stylesheets for layout	10	5	8	23
Provide searchable content	10	9	7	26
Provide currency converter	4	6	6	16
Blog capability	8	4	7	19

A template prioritization matrix, entitled *PrioritizationMatrix.xls,* is included on the companion CD-ROM [73].

Process Cycle Efficiency (PCE)

Process cycle efficiency (PCE) is a calculation that relates the amount of value-added time to the total cycle time within any given process. In order to perform this calculation, a common understanding of the following steps must be shared:

1. Value-added. This step in the process adds form and function to the end product and, therefore, adds value to the customer.
2. Nonvalue-added. This step does not add form or function to the manufacturing of the product and, therefore, does not add value to the customer.
3. Nonvalue-added but necessary. This step does not add value but is a necessary step in the final value-added product as perceived by the customer.

In order to calculate PCE, the entire process must first be mapped and each of the value-added, nonvalue-added, and nonvalued-added but necessary steps should be identified. A time dimension should be associated with each of the process steps. Following the mapping, the amount of total time is calculated as a percentage. Divide the value-added time by the cycle time for the process to determine the PCE. Figure 7-15 provides an example of the PCE of a software-problem-reporting process. This is a simplified overview, of course, but it shows the value-added and nonvalue-added steps broken down, and the average amount of time spent on each of those process steps overall.

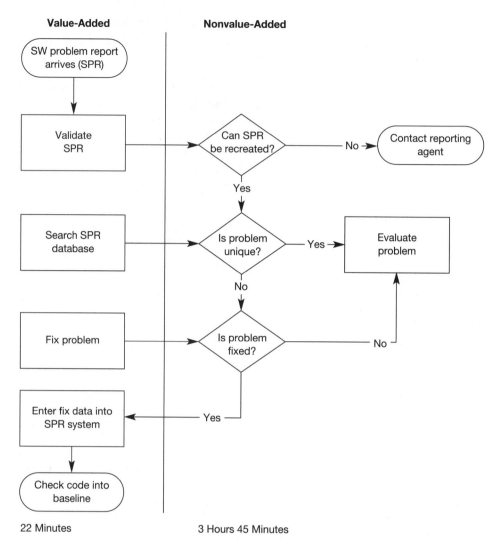

Figure 7-15. Example PCE for a software-problem-reporting process.

Run Charts

Run charts were initially designed by Walter Shewhart, a statistician at Bell Telephone Laboratories in New York. Shewhart was the originator of the control chart. Using control charts, he established the foundations for bringing processes under statistical control. Run charts evolved from the development of these control charts. Shewhart's theories are the basis of what as known as statistical quality control (SQC).

Run charts display process performance over time. Upward and downward trends, cycles, and large aberrations in either the product or process being examined may be immediately identified. In a run chart, events, shown on the y axis, are graphed against a time period on the x axis. In Figure 7-16, a run chart in support of the code review process of a

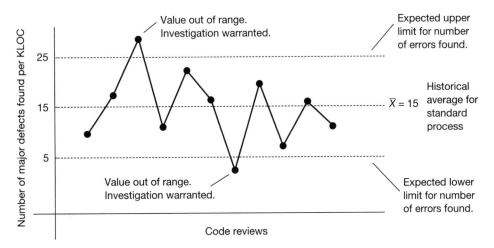

Figure 7-16. Example run chart for software code reviews (http://programminglarge.com/software_quality_management/code-review-cc.gif).

software development project has been plotted. This shows at a quick glance that there may be problem with either the product or the process.

Time Value Analysis

Placing the value-added and nonvalue-added data in a chart so that it might be readily analyzed, so that the value-added and nonvalue-added amounts might be easily compared, is a time value analysis. This type of chart visually separates the value-added from nonvalue-added time in a process (Figure 7-17).

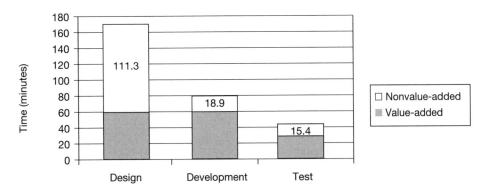

Figure 7-17. Time value analysis of three phases of a software development effort.

ANALYZE

At this point, the team will begin to make sense of all the information collected in the measure phase and they will begin to use that data to confirm the source of delays, waste, and poor quality. This is done in order to determine and eliminate problems at their source. During this phase the team will:

- Evaluate the data
- Verify the problem statement
- Confirm root causes
- Target lost and wasted time
- Propose process improvement opportunities

The Process Group should audit the organization according to applicable desire state requirements. Next, the Process Group determines what was actually accomplished by the improvement effort:

- In what ways did it or did it not accomplish its intended purpose?
- What worked well?
- What could be done more effectively or efficiently?

The team will begin to make incremental changes in the process that will eliminate process defects, waste, cost, and so on. These changes should always be linked to customer need as identified during the define phase.

The tools most commonly used in support of the analyze phase are:

- 5 Whys Analysis
- Affinity diagrams (discussed in the define phase)
- Analysis of variance (ANOVA)
- Brainstorming
- Cause-and-effect diagram
- Control charts (discussed in the measure phase)
- Flow diagram
- Multivoting
- Pareto charts (discussed in the measure phase)
- Pugh Concept Selection Matrix
- Regression analysis
- Root cause analysis
- Scatter plots
- Stratification
- Value stream mapping
- Waste identification and elimination

5 Whys Analysis

5 Whys Analysis is a problem solving technique that facilitates root cause analysis. It was developed by Toyota in the 1970s. Application of the strategy involves taking any problem and repeatedly asking the question "Why" five times. Using this method of repeated questioning, the symptoms associated with a problem are questioned until the root cause of the problem is ultimately identified. Although this technique is called "5 Whys," it is important to not constrain the questions to five, but rather to question until the root cause of the problem is identified. This technique is described in Table 7-6.

Table 7-6. The Five Whys. Repeating "Why" Five Times [61]

1.	Why did the release candidate not get built on time?	The developers could not create the build.
2.	Why couldn't the developers create the build?	There was an issue with the source code.
3.	Why was there an issue with the source code?	The files were all checked out and locked in the source control system.
4.	Why were the files all checked out and locked in the source control system?	A developer on another project had the files checked out, because the permissions in the source control system were not set up correctly.
5.	Why were the permissions not set up correctly in the source control system?	There is no process for associating developer permissions and the creation of new projects in the source control system.

Analysis of Variance (ANOVA)

Analysis of variance (ANOVA) is typically used in the analyze phase of DMAIC to compare multiple sets of data simultaneously. It is used to test hypotheses about the mean (average) of a dependent variable across datasets. Whereas the t-test is used to compare the means between two groups, ANOVA is used to compare means between three or more groups. ANOVA is a parametric test that requires sets of continuous data (time, weight, distances, etc.) from populations that are normally distributed.

ANOVA provides for the comparison of averages between datasets of trial data. ANOVA helps to determine if one or more of the different trials is better or worse by a statistically significant amount. ANOVA can also used to compare the variances of the datasets, to check the assumptions that the variances are the same when comparing averages.

There are a variety of ways in which ANOVA may be applied: one-factor (or one-way) ANOVA, two-factor (or two-way) ANOVA, and repeated-measures ANOVA. The factors are the independent variables, each of which must be measured on a categorical scale, that is, levels of the independent variable must define separate groups.

ANOVA is most easily done with a statistical analysis tool like Minitab®,* which produces the output shown in Figure 7-18.

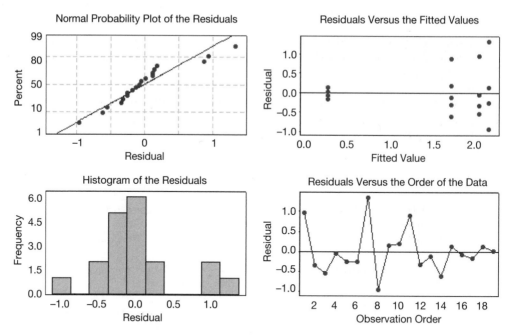

Figure 7-18. Example ANOVA datasets.

Brainstorming

This is simply listing all ideas put forth by a group in response to a given problem or question. In 1939, a team led by advertising executive Alex Osborn coined the term "brainstorm." According to Osborn, "Brainstorm means using the brain to storm a creative problem and to do so "in commando fashion, each 'stormer' audaciously attacking the same objective" [183]. Creativity is encouraged by not allowing ideas to be evaluated or discussed until everyone has run dry. Any and all ideas are considered legitimate, and often the most far-fetched are the most fertile. Structured brainstorming produces numerous creative ideas about any given "central question." Done right, it taps the human brain's capacity for lateral thinking and free association. Brainstorms help answer specific questions such as:

- What opportunities face us this year?
- What factors are constraining performance in Department X?
- What could be causing problem Y?
- What can we do to solve problem Z?

However, a brainstorm cannot help you positively identify causes of problems, rank ideas in a meaningful order, select important ideas, or check solutions.

*For information on Minitab®, visit www.minitab.com.

Cause-and-Effect (C&E) Diagrams

Kaoru Ishikawa first established the use of the cause-and-effect (C&E) diagram while pioneering the quality management processes in the Kawasaki shipyards. The C&E diagram is used to explore all the potential or real causes (or inputs) that result in a single effect (or output). Causes should be arranged according to their level of importance, resulting in a depiction of relationships or hierarchy of events. The goal is to help to determine root causes, identify problem areas, compare the relative importance of different causes, and then develop a solution.

The approach is simple. Beginning with an observable and verifiable effect, the question is asked as to why the situation exists. The goal is to craft a factual answer that is within the project's scope—to direct the "why" down toward the root cause. The cause-and-effect diagram is also commonly referred to as the "fishbone" diagram because of its resemblance to a fish skeleton once fully assembled, as shown in Figure 7-19.

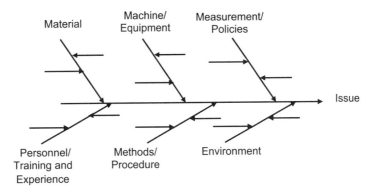

Figure 7-19. Typical categories associated with cause-and-effect diagrams.

When creating a C&E diagram, begin by clearly stating the issue and by writing it down on the right side of the diagram. If the issue is not clearly stated and understood, the problem solving session will not resolve the problem. The main line through the middle of the drawing is the "Effect" arrow. The point of the session is to ask questions that help identify causes as related to each effect category labeled in Figure 7-19.

The problem solving session should begin by discussing one of the main effect categories. The group should focus on one category, asking "why" questions that relate to the effect category as probable causes are identified and documented. These probable causes should be recorded on the spines of each associated effect category. This process should be repeated until all effect categories are exhausted.

The group should then review all probable causes and prioritize them. The last thing to do would be to discuss probable courses of action for the top-priority items and assign team action items. Figure 7-20 provides an example of what a completed C&E diagram might look like.

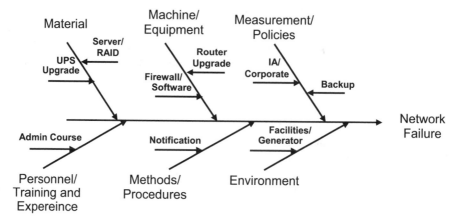

Figure 7-20. Example of a completed cause-and-effect diagram.

There are seven basic tools of quality control:

1. Cause-and-effect diagram
2. Checksheet
3. Flowchart
4. Histogram
5. Pareto chart
6. Scatter diagram
7. Quality chart

The C&E diagram is the first from the basic toolset. Each of these tools will be covered separately, but it is important for the reader to understand their efficacy when used together and value when applied separately.

Flowcharts—y-to-x Flowdown Diagrams

The use of diagrams, or flowcharts, can be very helpful when trying to demonstrate process functionality. Diagrams can help uncover problems with logic (e.g., dead ends, indirect paths), miscommunication, redundancy, and process boundaries. Diagrams can also be used to help define a common process baseline, or a starting point for process improvement or process improvement discussion.

The y-to-x diagram is initiated with a results measure (the y). It begins with the statement of the desired business outcome that is defined in a quantitative way, and then asks a question, "What are the drivers, or causes, of y"? Each node should describe a cause, and these should be described in measurable terms. These measures can be thought of in terms of constants or as something that defines a scope or category. The numbers of failures found during a test cycle is an example of one type of constant. An example of a category might be the categories used to define software failures (e.g., minor to severe) during the same test cycle.

Figure 7-21 provides an example of a y-to-x flowdown diagram that might be used to initiate a discussion between a client and organization supporting the test and evaluation of their product. This discussion would include a review of all associated data, how the measures align with the customer's business goals, and the support of required test and evaluation.

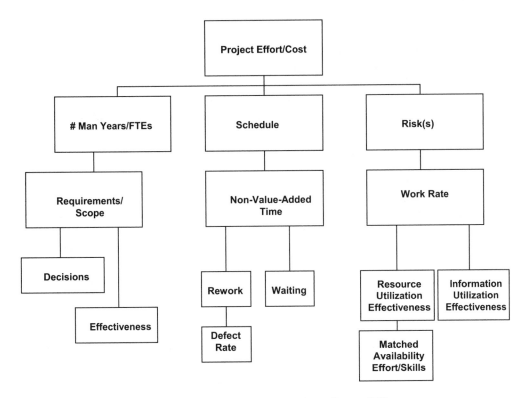

Figure 7-21. Example y-to-x flowdown diagram [88].

There are seven basic tools of quality control:

1. Cause-and-effect diagram
2. Checksheet
3. Flowchart
4. Histogram
5. Pareto chart
6. Scatter diagram
7. Quality chart

The flowchart is the third tool described in the basic toolset. Each of these tools is being described separately. It is important for the reader to understand their efficacy when used together and value when applied separately.

Multivoting

Multivoting is an effective technique for a group to reduce a large number of items to a manageable few. It also a way to group items by rough priority. It is particularly effective at taking the output from a brainstorming session and identifying the "key few" among the many ideas generated, while minimizing "win–lose" confrontations and maximizing group consensus.

The multivoting process consists of the following steps:

1. Number all of the items, so they can be referred to by number during discussions.
2. Walk through the list from beginning to end, checking for duplicates and common understanding of each item. Add context to items as necessary, to ensure that everyone understands what is included in each item and what is not. If an item is an exact duplicate of a previous one, delete it, but do not renumber items after it. If it is a restatement of a previous item, add it (with its original number) as part of the context of the first occurrence, but do not do this just to group similar items. Depending on the number of people, the type of items, and the number of items, this can take a couple of minutes or a couple of hours.
3. For the first round of voting, everyone gets an unlimited number of votes, but each person can only cast one vote per item (i.e., vote "yes" or "no"). They must clarify what they are voting for, typically, whether or not the item is valid. (Brainstorming sessions often encourage the generation of offbeat, silly, or outlandish items, to help the team think "out of the box" and have some fun.) This step eliminates items no one supports. If every item in the list is likely to get at least one vote, skip this round of voting and go directly to the next round (step 5).
4. Collect and compile the votes. If the list is softcopy (for an example, see the template entitled *multivote.xls,* provided on the companion CD-ROM), sort the items by number of votes received, in descending order. Drop the items that received the fewest votes. (Note: always drop items that received zero votes. If there is a team consensus, you can draw the line higher, especially if there is a gap between items with many votes and items with few votes. However, be careful to avoid the case in which a single team member has insight into a particular area, and may be the only one voting for a significant issue that other team members just are not aware of.)
5. Count the number of items that are still in the list ("above the line"). Give each participant approximately half of that number of votes. Reiterate what they are voting for (greatest impact, most troublesome, most important, etc.). Participants do not have to use all of their votes, but unused votes do not carry forward into the next round.
6. Repeat steps 4 and 5 until either:
   a. There are few items left in the list. "Few" is typically three to seven, depending on your needs. Never multivote down to a single item.
   b. Enough items receive unanimous votes (where "enough" is the same number as "few" above).

Variations:

- If necessary to decrease the number of rounds of multivoting in step 5, reduce the number of votes from half of the count to one-third of the count.

- Allow participants to cast multiple votes for a given item in step 4. But they still must stay within the given number of votes for the round. So in a round with 10 votes, each of the following would be allowed:

    Cast all 10 votes for a single item

    Cast 5 votes for one item, 3 for another, and 1 each for two other items

    Cast 2 votes each for five different items

    Cast 1 vote each for ten different items

    This variation helps avoid the case described in step 4 [80, 68, 102].

Pugh Concept Selection Matrix

The Pugh Concept Selection Matrix, more commonly referred to as the Pugh Matrix, is a systematic approach for selecting among alternatives. This technique was developed by design professor Stuart Pugh at the University of Strathclyde, Glasgow, Scotland in the 1980s to select the best from among several design concepts. This technique is an extension of traditional trade study/decision analysis grids and is useful whenever it is necessary to evaluate several alternatives against multiple criteria. The Pugh Matrix not only identifies the best defined alternatives but this technique also helps synthesize the best elements of the lesser alternatives into the top alternative to produce a new, optimal alternative.

Figure 7-22 shows a Pugh Matrix in which five alternatives are being evaluated against six criteria. The steps to construct and evaluate a Pugh Matrix are:

1. List the criteria that the alternatives will be evaluated against (cost to implement, etc.) in the numbered rows. The criteria can come from goals, requirements, specifications, "voice of the customer" data (the customer's wants and needs), or other items brainstormed by a team.
2. Determine weighting. The weights are the numeric values that are added or subtracted for alternatives that are better or worse than the baseline alternative. The default is all weights equal to 1, which gives each criterion even weighting. A weight of 2 gives a criterion twice as much weight. Fractional weights are allowed.
3. List the name of the baseline alternative at the top of the BL column. The baseline may be the current situation, the preferred alternative, or the highest scoring alternative from a previous run through the Pugh Matrix. When there is no obvious choice for the baseline, any alternative will do.
4. List other alternatives in columns to the right of the baseline.
5. Compare each alternative against the baseline, and for each performance criterion:
    a. Enter a "+" if the alternative is better than the baseline for the given criterion.
    b. Enter a "−" if the alternative is worse than the baseline for the given criterion.
    c. Enter an "s" if the alternative is the same as the baseline for the given criterion.
6. Analyze the sums at the bottom to identify the best solution(s). Higher positive-weighted net scores are better.
7. Work to improve those alternatives that scored best by incorporating strong ideas from other alternatives.
8. Iterate steps 3 through 7 until a clearly superior alternative emerges [80].

The Pugh Matrix is often used by Lean Six Sigma teams in the improve phase when

Alternatives

Criteria	Weight	BL	Alt 1	Alt 2	Alt 3	Alt 4
1 Cost to implement	1	S	+	S	–	+
2 Staffing required	1	S	S	S	S	+
3 Effectiveness	2	S	S	+	–	S
4	1	S	+	S	+	–
5	2	S	+	+	–	S
6	1	S	S	+	+	+
Sum of positives		0	2	2	1	3
Sum of negative		0	1	1	2	1
Sum of sames		6	3	3	1	2
Weighted sum of positives		0	2	4	1	3
Weighted sum of negatives		0	2	1	3	1
Weighted net		0	0	3	–5	2

Legend:

+	Better than baseline
–	Worse than baseline
S	Same as baseline

Figure 7-22. Example Pugh Matrix for selecting problem solutions.

selecting among alternative solutions for root causes. It is useful on software projects when designing products, services, or processes. In either case, include customers and other stakeholders when feasible.

An electronic version of the Pugh Matrix template as described, entitled *Pugh_Matrix.xls,* is provided on the companion CD-ROM.

Regression Analysis

Regression analysis is a statistical forecasting model that is used to predict and model fundamental relationships, and test scientific hypotheses. Regression analysis is used to describe and evaluate the relationship between any given variable, usually called the dependent variable, and one or more other variables, which are referred to as independent variables. When working with functions, the dependent variable is equated to the function output and independent variables to the function inputs. Regression analysis is used to help predict the value of one unknown variable when the information associated with one or more other variables is readily available.

The goal of regression analysis is to determine the values of parameters for a function that cause the function to best fit a set of data observations that you provide. Simply put, it is a useful tool for investigating the relationships between variables; the causal effect of one variable upon another is examined. To employ the regression analysis method, organizations assemble the data and employ regression to estimate the effect the independent variables have upon the dependent variable that they influence [172].

The first step is the identification of the dependent variable, or function output. Next, multiple regression analysis will be carried out. The primary focus will be placed on the independent variables, or function inputs. The model of these regression analyses will then be reviewed and the data will be used to identify any relationships between the variables. The following is an example of a simple linear regression.

Description:
Predict the amount of time required for a test cycle, using the amount of time spent per development cycle, the number of requirements, the number of developers, and their relative skill levels.

Using the description above:

Dependent variable (output) = test cycle time

Independent variables (inputs) = development cycle time
number of requirements
number of developers
relative developer skill level

Root Cause Analysis

Root cause analysis is the heart of the analyze phase of the DMAIC process. The purpose of the analyze phase is to identify potential root causes, and then quantify and verify the vital few that have the most impact on the problem so that effective countermeasures may be identified. The tools most commonly used for root cause analysis are 5-Whys, cause-and-effect diagrams, and failure modes and effects analysis (FMEA); the tool of choice depends on the type of problem and the availability of the related data.

In Toyota's Lean approach, 5-Whys is the immediate response to any problem. Taiichi Ohno noted that problems usually have a direct cause that is the source of the problem, but the root cause typically lies hidden beyond the source, upstream of the place where the problem occurs. Fixing the direct cause solves the problem for the moment, but does not prevent it from reoccurring, which requires finding and fixing the root cause. Both individuals and teams can use 5-Whys to find the root cause whenever there is single, primary, path of cause and effect [102].

Cause-and-effect diagrams, also know as Ishikawa diagrams or fishbone diagrams, are used when data* about causes is available, and multiple root causes may be present. When the problem is potential product or process defects, FMEA is used to find the "vital few" out of the many possible causes of defects. FMEA begins with generation of the range of possible failure modes, which are each rated (typically 1 to 5, or 1 to 10) according to their severity, probability of occurrence, and likelihood of escaping detection and

---

*Preferably quantitative; anecdotal data is less definitive but can be effective.

correction. The ratings are multiplied to produce a risk number, which is used to prioritize the most severe causes [62].

Once the tools have identified the potential root causes, these are then quantifed and verified with data. The cause-and-effect relationship should be confirmed (using, histograms, dotplots, boxplots, time series plots, control charts, scatter plots, or Pareto diagrams). Where possible, justify and determine the strength of confirmation statistically using hypothesis testing [80, 96].

Scatter Plots (Diagrams)

Scatter plots (also called scatter diagrams) are used to investigate the possible relationship between two variables that relate to the same event. Figure 7-23 provides an example of a scatter plot of temperatures over a period of years, from 1900 to 1920. A straight line of best fit, using the least squares method, is also often included.

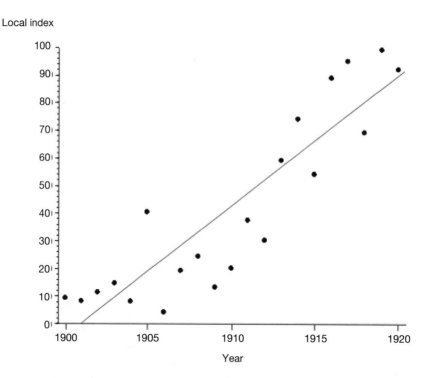

Figure 7-23. Scatter plot of temperature ranges.

When evaluating scatter plot data, there are several things to look for. Positive relationships appear if the points cluster in a band running from lower left to upper right (as x increases, y increases). Negative relationships appear if the points cluster in a band running from upper left to lower right (if x increases, y decreases). The more points that cluster around the imaginary line of "best fit," the stronger the relationship that will exist between the two variables, and if it is hard to see where the line would be drawn there is probably no significant correlation between the variables.

There are seven basic tools of quality control are:

1. Cause-and-effect diagram
2. Checksheet
3. Flowchart
4. Histogram
5. Pareto chart
6. Scatter diagram
7. Quality chart

The scatter diagram is the sixth tool described in the basic toolset. Each of these tools is being described separately. It is important for the reader to understand their efficacy when used together and value when applied separately.

Stratification

Stratification is the separation of a dataset into subgroups, looking for meaningful patterns. Stratification is not a specific tool; it is a technique that is a fundamental part of data analysis, so significant that one author even lists it as one of the seven basic quality tools [120]. Stratification is performed by using tools (dot plots, scatter diagrams, Pareto charts, histograms, etc.) to "slice and dice" the data to find related subgroups that tell you something significant about the data. In the analysis phase, this often consists of plotting the data against various attributes (such as time, location, size, etc.), looking for meaningful patterns.

When performing root cause analysis, use stratification to narrow the search for the root cause. Brainstorm characteristics or factors that influence or are related to the issue under study, and either manipulate available data or collect the data necessary to confirm or refute the influence of the factors. When collecting data, such as with a checksheet, try to anticipate the characteristics that might be useful for stratification, and ensure that they are recorded along with the data. It is often easy to identify most of the factors that will affect the results, but rarely possible to be sure that you have identified *all* significant factors in advance. Pilot the data collection and analysis before conducting large or expensive data collections, and use the pilot data to look for lurking variables that should be explicitly captured in the full data collection [80].

In statistical process control, the need for stratification is indicated by "hugging"— when a control chart shows most of the points near the centerline (within ± 1.5 sigma). The problem here is not that the points are too near the centerline but that the control limits are driven too wide by the range of variance between observations from different subgroups (or completely different populations). If you can find the right variable(s), you can stratify the observations into separate groups, each more predictable and with a control chart that shows greatly reduced variation.

A common use of stratification for projects using the CMMI® involves the analysis of defect data from peer reviews. Whether you have a control chart with hugging, a causal analysis and resolution cycle on the peer review data, or an attempt to identify optimal values for number of reviewers, reviewer preparation time, or size of product to be reviewed at a time, stratification is usually necessary. Even with data from similar processes, the stratification variable may differ [63].

One team found that they could separate effective reviews from ineffective ones by collecting the time spent by reviewers before the meeting. Despite a single published process, there were actually two different types of reviews taking place on this project. In one, most defect finding took place before the meeting; in the other, defects were found by walking through the product during the meeting. Once the defects-found-per-review-hour data was stratified by review time, it was clear that the "review in the meeting" variant was responsible for the ineffective results. Another team found that code reviews needed to be stratified by module size in order to get repeatable results; small modules could register high defect densities from a single defect, since the denominator was so small. Mid-size modules consistently had a low defect density, and then the density rose with the complexity of vary large modules [102].

Value-Stream Mapping

Once candidate projects have been identified and team members have been properly prepared, the continuous improvement process can begin. This typically starts with a top-level analysis of all the candidate projects, called a "value stream analysis." This is done to in order to develop a clear understanding of the current state of all development processes and to determine where to most effectively apply Lean Six Sigma techniques. Figure 7-24 provides an example process diagram.

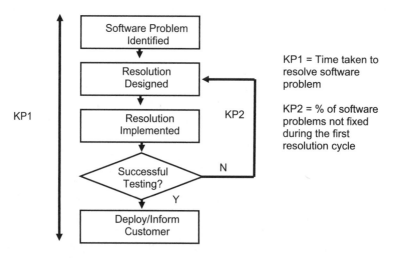

Figure 7-24. Example process diagram (see http://www.transacthr.com).

Value stream mapping attempts to document all of the process elements associated with the software development projects. The ultimate goal for the organization should be to map each process so that it can be analyzed for waste. It is important to remember that all activities add cost, whereas only a few add value. There are three simple types of questions that can be asked while working through this phase to help with the analysis:

1. Would the customer care about this activity? Are they ultimately willing to pay for it?

2. Is the activity currently being carried out correctly?

3. Does the activity currently have any impact on the development activity? Is it a required part of the development process? What happens if it disappears?

The next step should be to develop a roadmap or plan. This roadmap should describe what the organization views as achievable goals. Finally, an action plan should be developed identifying all of the activities that must take place to achieve the goals. It is important that the plan include the goals and all supporting milestones, actors, and selected projects. Figure 7-25 provides a process mapping example.

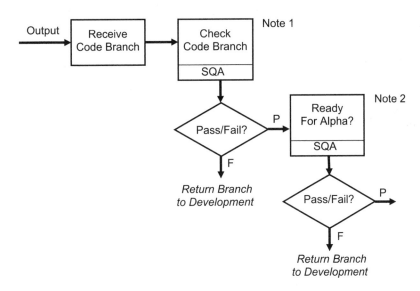

Figure 7-25. Process mapping example. *Note 1:* Checking Code Branch. Feature does not violate established CM or coding standards. Package and Asset naming conventions are adhered to, as per approved naming convention documentation. Asset packages do not contain grossly redundant assets. Map-exclusive assets are stored within the map package. *Note 2:* Branch is ready for alpha or beta testing and has reached a major milestone in development. Features do not obscure or remove functionality, except by design.

Waste Identification and Elimination (Seven Elements of Waste)

A major focus of Lean is learning to see waste (Japanese: muda) in our processes, and then eliminating the root causes of that waste. Taiichi Ohno at Toyota developed a taxonomy of types of waste, to help in identifying waste in process execution. This waste classification has proved useful in a broad spectrum of environments, from manufacturing to health care, from software development to delivery of services like help desks or network management, with just one modification: the addition of an eighth element, which was implicit in the Toyota culture but needs to be explicitly considered elsewhere. The wastes can be organized to fit the mnemonic DOWNTIME:

1. Defects (correction)
2. Over production

3. Waiting
4. Not utilizing people's talent
5. Transportation (conveyance of material)
6. Inventory
7. Movement (motion of people)
8. Excess processing (processing)

Defects (Correction). Defects require rework of the product or service to fulfill customer requirements. Bugs; problem reports; conversion errors; incomplete, ambiguous, or inaccurate specifications; and missing tests or verifications are all examples of defects.

Overproduction. Overproduction is producing more, sooner, or faster than is needed by the customer. It can be considered the worst waste since it contributes to all the others. Examples include too much detail in documents, unnecessary information, redundant development, overdissemination, extra features, and pushing data (by plan or control) rather than pulling it on request. In the office, overproduction includes printing documents before or in greater quantities than needed.

Waiting. Waiting is idle time in which no value-added activities take place—wait time for people or products. This includes the time things spend in a queue or inbox waiting to be processed, or schedule slippage due to upstream tasks.

Not Utilizing People's Talent. Not utilizing the talent of staff in the right jobs or place within the organization. This occurs when performers do not own their process, or have rigid roles with little or no cross-training. Command and control management command, limited employee authority, and inadequate business tools are also symptoms of this waste.

Transportation (Conveyance of Material). Transportation waste includes any unnecessary material movement. No matter how short the distances involved and no matter what methods are used, acts of transport have inherently no value in terms of production. Applied to information technology, this includes the flow of electronic information and problems with incompatibility, communication failure, multiple sources, and security issues. It applies to e-mail attachments and movement of paperwork, especially hand-offs and approvals.

Inventory. Inventory wastes are supplies or materials (beyond the minimum necessary buffer or safety stock) that are not delivered just in time. Any form of batch processing represents inventory waste, as do filled inboxes or multiple copies of files that are not configuration managed. Unfinished code is a similar manifestation of software inventory.

Movement (Motion of People). Movement waste covers any movement of people which does not add value to the product, such as searching, walking, moving, reaching, carrying, bending, adjusting, storing, and retrieving. This includes looking for documents and data, switching between computer keyboards/monitors, handling things more than once, travel to other offices or conference rooms, or poor facility layout.

Excess Processing (Processing). Excess processing is any effort that adds no value to a product or service, such as work that can be combined with other processes, enhancements that are transparent to the customer, reentering data, making extra copies, or any unnecessary reports, transactions, accounting, or budgeting. Unneeded processes running on a computer also constitute processing waste.

IMPROVE

During the DMAIC improve phase, the Process Group will revisit and modify the solution to incorporate new knowledge and understanding. Iterations of pilot/test and refinement may be necessary before arriving at a solution that is deemed satisfactory to meet the expected process measurement goals. The appropriateness of training the process participants should be examined.

The tools most commonly used in the improve phase are:

- 5S's Method
- Balanced work flow
- Brainstorming (discussed in the analyze phase)
- Cellular/flow manufacturing
- Flow charting (discussed in the analyze phase)
- FMEA (discussed in the measure phase)
- Jidoka (autonomation)
- Kaizen events
- Little's Law
- Process stability
- Production smoothing
- Setup reduction
- Stakeholder analysis (discussed in the define phase)
- Takt time
- Total productive maintenance
- Visual workplace

5S's Method

There is a method, called 5S, that supports the organization and standardization of workplace procedures. The goal of this in support of the Lean methodology is to utilize the elements of the 5S philosophy to help eliminate waste and nonvalue-added activity while improving quality. The idea is that following the application of the 5S steps, the targeted process will be streamlined and more efficient.

The 5S notation has evolved from the traditional Japanese notion of housekeeping. A good way to think of the application of the 5S philosophy is as a stepped-based process. The original translations and an overview of the process are provided in Table 7-7.

Table 7-7. The 5S philosophy [117]

Step	Japanese	English	Definition
1	Seri	Sort	Put things in order; remove what is not needed and keep what is needed.
2	Seiton	Straighten	Proper arrangement; place things in such a way that they can be easily reached when needed.
3	Seiso	Shine	Clean; keep things clean and polished; no clutter.
4	Seiketsu	Standardize	Purity; maintain cleanliness after cleaning; continue standards.
5	Shitsuke	Sustain	Commitment; this is not a part of the original "4S," but is a typical teaching; adherence to program and established standards.

Balanced Work Flow

Balanced work flow has it origins in lean manufacturing techniques and principles. Balanced work flow is integral to standardizing work and the associated processes and just-in-time production. Standardizing work processes includes designing the cycle time to be equal to takt time.

An example problem in support of work balancing is presented here in its simplest terms. In the typical home washer and dryer, the washer usually has a cycle time that is much faster than the dryer. If you continue to load and unload the washer at the end of its cycle, inventory of wet clothes piles up ahead of the dryer.

Typical problems with software engineering life cycles are:

- The requirement engineers "wait" for the customer
- The designers "wait" for the requirement engineers
- The testers and implementers "wait" for the designers
- The customer "waits" for the testers and implementers

These "wait" areas are examples of unbalanced work flow within the life cycle that stress the engineering and management processes with peak loads that show up as congestion and delays. It would be ideal if there were a more consistent, or balanced work flow, but cyclic feast and famine are the rule in many areas of engineering rather than the exception.

The software engineering life cycle can be designed for balanced work flow to smooth work and reduce inventory and bottlenecks. Work flow design to be balanced can include a pull system, possibly using a "kanban" pull system.

There are several approaches:

- Pooling staff. Cross-training staff to step in during times of peak loads. Experienced engineers can move from one role to another role in different processes in a well-defined and -managed life cycle. The role changes have other positive effects on the engineering team as dependencies are better understood.
- Triaging. Sorting work requests, enhancement, and features into categories that reflect different levels of effort required. Once triaging categories have been identified, then different strategies can be developed to deal with each category

- Systematic determination and documentation of processes and their work products or artifacts can allow better estimates of effort and better allocation of experienced staff.
- Software engineering tools can automate or support engineering work flows by minimizing
  Searching for tools and work products or artifacts
  Excess communications or movements
  Hand-backs or double handling
  Record keeping or awkward ergonomics
- Working constantly with the customer's advocate to develop small but important features on a short cycle.

Cellular/Flow Manufacturing

An efficient flow-through manufacturing (development) system will eliminate waste, minimize work in process, optimize floor space, reduce lead time, and improve customer response time, leading to reduced costs and greater production capacity.

Cellular/flow manufacturing is the linking of manual and machine operations, while leveraging the concept of process balancing, to produce the most efficient combination of resources that maximize value-added content while minimizing waste. When processes are balanced, product development flows continuously, extra (parts) movement is minimized, wait time between operations is reduced, inventory is reduced, and productivity increases. The bottom line is that when all production processes are balanced, production flows continuously and customer demands are easily met.

Jidoka (Autonomation)

Jikoda, also called autonomation, is the quality control process that relies on work stoppage when quality defects are detected. When this concept was introduced, the idea was revolutionary. Most production error detection was done through human observation. Today, the concept of routinely detecting production errors without the assistance of automation would be seen as production waste.

Jikoda can be called upon to shut down a process when:

- The required number of items has been produced
- A defective item has been detected
- A type of physical blockage occurs and causes a stop in the production process

The Jidoka steps are detection, a stop in production, an immediate evaluation and correction of the problem, and subsequent improvement of the process.

One common use of Jikoda as applied to software production would be the idea of continuous integration and the automation of the software build and unit test processes. In many software development environments, software teams continuously integrate their individual pieces of code. As each developer works separately, changing their individual pieces of code, they unit test and then check their source code into a staging area. It is then that they perform a full build to ensure that no incompatibilities exist and that no other code breaks because of their changes. If problems are identified, these are fixed prior to code being migrated into the central code baseline.

Kaizen Events

Kaizen events, also referred to as a Kaizen blitz, are used to quickly map and then subsequently improve an existing process. These events are typically applied to smaller-scale, localized, efforts. The idea is that whereas large-scale organizational improvement may be attractive, it is slow and involves associated risks. For example, what if the organizational change being invoked involves many people and large sums of capitol but is not successful? Kaizen events allow for the deployment of incremental improvement on a smaller scale. The risk can be minimized for an organization, while the overall benefit may be accumulated over time. The risk is that since the change is not organizational, it may not be permanent or become a part of the corporate culture. This risk, of course, can be minimized if the Kaizen events are regular occurrences intended for the purposes of continuous improvement (Figure 7-26).

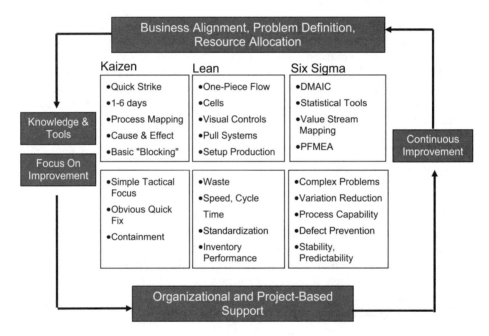

Figure 7-26. Overview of the applications of Kaizen, Lean, and Six Sigma [72].

Little's Law

Little's Law, also commonly referred to as inventory reduction, relates three critical performance measures that are associated with any type of production. Little's Law provides an equation for relating lead time, work in process (WIP), and average completion rate (ACR) for any process. Little's Law states:

$$\text{Lead time} = \text{WIP (units)}/\text{ACR (units per time period)}$$

As you can see, an improvement in the process used to support an increase in the ACR while maintaining the same WIP, or a reduction in WIP while maintaining the same ACR, will reduce the required lead time.

Little's Law is very useful when applied to software development processes. For example, this law can be used to validate software testing and the testing tool environment [82]. This formula can be applied to help identify and subsequently eliminate stress-testing bottlenecks.

Process Stability

Lean Six Sigma requires a robust management and process measurement system foundation. The management system is composed of interrelated processes. The process is stable if the same results can be produced in a predictable manner. This stable performance occurs in comparable instances. Thus, process stability is the ability of the process to perform in a predictable manner over time. Process stability is also known as the reduction of variation. These processes are said to be stable when process measures have both constant means and constant variances over time, and also have a constant distribution [62].

The software engineering process measures include:

- Life cycle data collected during life cycle phases
- Defects discovered during these life cycle phases
- Project size and number of source statements
- Effort expended and number of people employed during these life cycle phases
- Elapsed time
- Computer resources utilization

Process stability can be achieved through a quantitatively managed and defined process that is controlled using statistical and other quantitative techniques. The product quality, service quality, and process-performance attributes are measurable and controlled throughout the project [168]. For organizations using CMMI-DEV, process stability is part of organizational process performance process area, the purpose of which is to establish and maintain a quantitative understanding of the performance of the organization's set of standard processes in support of quality and process-performance objectives, and to provide the process-performance data, baselines, and models to quantitatively manage the organization's projects.

The most common focus of process stability is improving product quality by analyzing the relationship between product quality and process stability. Once process stability is achieved, then process capability can be improved. Process capability is delivering products that meet requirements from processes whose performance meets the business needs of the organization. Both software engineering process stability and capability can be measured by statistical process control (SPC) techniques. These SPC techniques include scatter plot or control charts and run charts with a sufficient number of independent samples to provide an analysis of process stability.

Another focus for process stability is the organization's processes change management and adoption by the staff. Many organizations make the revised process elements available via the Web and expect that the projects will simply start using them, a "Field of Dreams" approach. To change the staff's behavior, consider an approach similar to deploying software packages to clients:

- Periodic "process releases"
- Release notes
- Testimonial data from the pilot projects

- Process training
- Internal consultation
- Publicity

Production Smoothing (Heijunka)

Production smoothing, also referred to by the Japanese term heijunka, refers to the lean tool in which the total volume of items used in support of production are kept as constant as possible throughout the manufacturing life cycle. Yasuhiro Monden, a renowned expert on the Toyota production system,* stated:

> The smoothing of production is the most important condition for production by Kanban and for minimizing idle time in regard to manpower, equipment, and work-in-process. Production smoothing is the cornerstone of the Toyota Production System. [184]

The goal is to maintain as close to the ideal inventory quantities of required stock so that actual production quantities are met (Figure 7-27.).

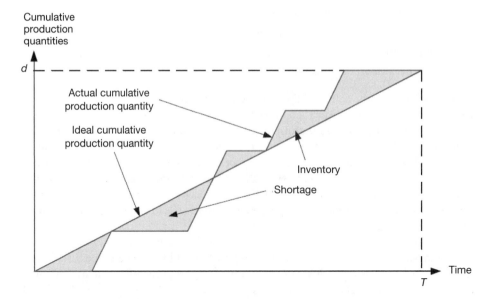

Figure 7-27. Ideal and actual production quantities [125].

When the term production leveling is used, it can be used to describe one of two types of leveling. One is the leveling of production by volume. The other is the leveling production by product type or mix.

Production Leveling by Volume. This technique is appropriate when you have a production process that produces a single product. The product remains the same, but the demand for that product varies. For any given week, the number of units may vary. For the purposes of illustration, we will say that we are producing and shipping diskettes. De-

---

*His book, *The Toyota Production System,* was awarded the 1984 Nikkei Prize by the *Nikkei Economic Journal.*

mands range from 100 to 400 diskettes, but the average is 300 per week. The notion of production leveling is simple. You level the production to meet this average, supporting long-term demand, rather than catering to the whims of the highly variable daily market.

This example has been highly simplified and, as stated previously, production leveling is critical to overall lean production. Additional factors should be considered, such as:

- How much inventory should be carried and is this inventory proportional to the variability of the demand?
- How stable is the production process and how does this stability affect the numbers of units produced?
- How do the frequencies of shipments affect production demands?

These are just a few of many and they tend to be unique to each production environment. The key is to try to factor in all the variables that will affect long-term demand for the product in question.

Production Leveling by Product Type. This technique is appropriate when you have a production process that produces a variety of the same base product. However, there are also the added consumer demands to consider. A company that produces laptops offers Models A, B, C, and D, and weekly demand averages three of Model A, two of Model B, two of Model C, and five of Model D. A mass producer, seeking economies of scale and wishing to minimize changeovers between products, would probably build these products in the following weekly sequence:

AAABBCCDDDDD

A lean producer, mindful of the benefits of smoothing output by mix as well as volume, would strive to build in the repeating sequence:

ABCDDABCDDAD

The two approaches are contrasted in Figure 7-28 [93].

The Heijunka Box. The heijunka box is a simple tool that can be used to help support production leveling by volume or product type. A typical heijunka box has horizontal rows for each member of a product family, and vertical columns for identical time intervals of production. Production control kanban are placed in the slots created, in proportion to the number of items to be built [185].

For lean process managers who accept the notion that leveling by volume and mix produces benefits throughout the value stream, the problem remains how to control production so that true heijunka is consistently achieved. Toyota came up with a simple answer many years ago in the form of the heijunka box (or leveling box.)

A typical heijunka box has horizontal rows for each member of a product family, in this case five (Figure 7-29). It has vertical columns for identical time intervals of production, in this case twenty minutes. Heijunka refers to the specific concept of achieving production smoothing; the heijunka box is the name of a specific tool used in achieving the aims of production smoothing.

Historically, the heijunka box was a box with production tasks inserted (Figure 7-29). Colored cards representing individual jobs, which can vary in duration and complexity,

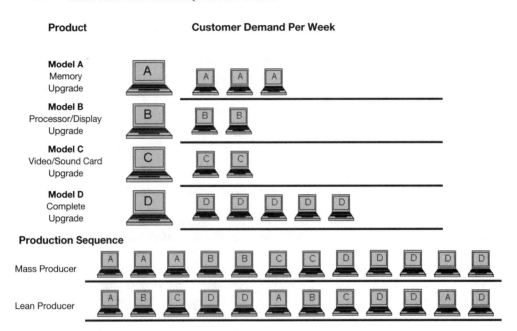

Figure 7-28. Economies of lean production sequencing.

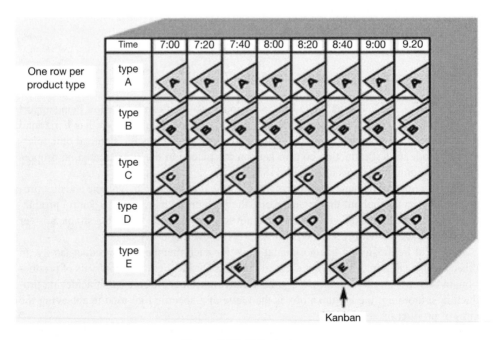

Figure 7-29. Heijunka box.

are placed on the heijunka box to provide a visual representation of the upcoming production runs. However, as applied to software and systems engineering, a heijunka box is more generally represented as a production schedule. This schedule makes it easy to see what types of jobs are queued for production. For additional information in support of software project planning, refer to Chapter 3.

Setup Time Reduction

Setup time reduction is the process of reducing changeover time and is driven by the need to change over a given production process to produce a different product in the most efficient manner. The goal is to eliminate downtime due to these production setups and changeovers. Reducing setup time is the lean manufacturing technique allowing the mixing of production/products without slowing output or creating nonvalue-adding activity. Quick changeover will increase productivity, reduce lead time, lower total costs, and increase flexibility to adapt to a changing market and/or product mix.

The Lean tool for tackling setup time is the four-step rapid setup method. The principle of this method is to eliminate anything that interrupts or hinders productivity. The following steps provide a high-level description of the four-step rapid setup method:

Step # 1. Identify and chart any process-related activity that can be described as one or more of the following activities:
- Delays the start of value-added work
- Causes interruptions to value-added work
- Is similar to an existing process task

Step #2. Offload any of the interruptive/delaying tasks.

Step # 3. Streamline or automate any interruptive/delaying tasks that cannot be offloaded.

Step # 4. Bring the process under statistical control, meaning the amount of variation in lead time is within predictable limits of $\pm$ 3 sigma.

Takt Time

Takt time can most simply be described as the maximum amount of time allowed to produce a product in order to meet demand.* Takt is derived from the German word taktzeit, which means "clock cycle." Takt is also the term used to describe the pace the conductor sets when directing an orchestra. German engineers were supporting the rebuilding of Japanese industry in the 1930s and used the analogies of cycle time and work pace, or takt; the term and concept stuck. Every Lean organization knows its takt time.

As previously stated, takt time is the maximum amount of time allowed to produce a product in order to meet demand. It naturally follows that the pace of production flow would then be set based on this takt time. Production flow is expected to fall within a pace that is less than or equal to the takt time. The key is to ensure that the total of each step during the production process is less than or equal to the cycle time [97].

A common misconception is that takt time can be measured, it cannot. It is also often confused with cycle time, which as described above is the time it takes to complete one task. Takt time must be calculated:

---

*See http://www.isixsigma.com/dictionary/Takt_Time-455.htm.

$$T = \frac{T_a}{T_d}$$

where t = takt time, T_a = net available time to work, and T_d = total customer demand. T_a = the amount of time available for work to be done, excluding any expected downtime.

For example, there are a total of 9 hours in a software quality inspector's day (gross time), less 1 hour for lunch, 30 minutes for breaks (2 × 15 mins), 10 minutes for a production meeting, and 10 minutes for miscellaneous:

Net available time to work = (9 hours × 60 minutes) – 30 – 30 – 10 – 10 = 460 minutes

If customer demand is 460 software units a day and you were running one shift, then your line would need to make one per minute to be able to keep up with customer demand. However, these are extremely close margins and the risk for failure in this scenario is high. This will work only as long as 100% efficiency is maintained, and there may be work stoppages for many reasons. Typically, production would be run at a proportionally faster rate to allow for contingencies.

Total Productive Maintenance (TPM)

Total productive maintenance (TPM) is a company-wide equipment maintenance program concept primarily focused on minimizing production downtime due to required unscheduled maintenance activities. TPM covers the entire equipment life cycle and requires participation by every employee. One of the key elements of TPM is the independence of maintenance. Each individual is responsible for the maintenance of his or her own equipment. Production and maintenance staff work together in support of continuous production support and improvement.

TPM was introduced to achieve the following:

- Avoid waste
- Keep quality high
- Reduce cost
- Keep production time low
- Eliminate defects

TPM cannot be discussed without mentioning total quality management (TQM) as TPM evolved from TQM. Dr. W. Edwards Deming, building on the earlier statistical work of Walter Shewhart, supported the military reconstruction efforts in Japan following World War II. Deming used these statistical techniques and methods and applied them to improving the concepts of manufacturing and production quality control. These manufacturing concepts became known as TQM.

Preventive maintenance was practiced as part of any TQM program. However, this practice required the rigid application of maintenance schedules, scheduled downtime, and repair by external maintenance personnel with little or no involvement by the machine operator. TPM began as a concept in the 1960s and by the late 1980s was generally widely accepted. Today, few organizations adhere to the original TQM concepts. These concepts have been evolved to include the notion of integrated systems maintenance [112].

TPM has been compared to TQM. The similarities are striking as each requires management commitment, employees are empowered to initiate corrective action, and long-range planning must occur as each may take a year or more to implement and are ongoing processes [113]. The differences between TQM and TPM are summarized in Table 7-8.

Table 7-8. Differences between TQM and TPM

Category	TQM	TPM
Object	Quality (output and effects)	Equipment (input and cause)
Means of attaining goal	Organizational process improvement. Management-centric.	Maintenance process improvement. Employee-centric.
Target	Customer satisfaction	Minimizing production downtime

TPM has five goals. These goals are shown in Figure 7-30 and are described as follows:

1. The identification and subsequent elimination of key problems. All losses associated with an organization's productions processes are examined.
2. Achieving independent maintenance by allowing people who operate the equipment to take responsibility for some of the equipment maintenance.
3. Having a systematic and planned approach to maintenance activities.
4. Achieve maintenance prevention through early equipment management. The goal is to track maintenance problems back to their cause so that they can be eliminated completely.
5. Ensure that staff are continuously and appropriately trained. Ensure that staff also have defined maintenance responsibilities.

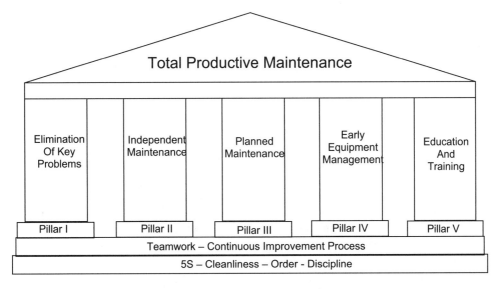

Figure 7-30. TPM based upon the pillars of the 5S.

TPM can be easily adapted to a software engineering environment, moving from an environment of classic manufacturing, where the same part is produced and machines are used, to one where unique software products are developed and digital computers are used to produce the "soft" product. The key is engaging the machine operator in the maintenance process. This is happening in many organizations. Gone are the days where the systems administrator walked around with the most recent Windows® upgrade. Today, end users are expected to download and install systems updates and to keep their systems updated with the latest and greatest updates. The ratio for systems support for corporations used to be 1:10; this is no longer the case, partly due to TPM.

TPM as applied to software and systems engineering extends to software configuration management and the build tree—the responsibility for and ownership of each programmer and member of the team to ensure that code meets standards and is checked in appropriately.

The TPM model can also be extrapolated to the code base, to eliminate "maintenance" (or catastrophe and downtime) before it is required. TPM can be applied to the refactoring of baselined code. Code can be examined for improvements; these improvements can be made before they are required. The idea is that all participants take ownership of the process and participate in the prevention of waste, or downtime, due to negligence and oversight.

The following software and systems engineering standards may be used in support of TPM implementation efforts: IEEE Std.1219, IEEE Standard for Software Maintenance [22]; IEEE Std 12207.0, Standard for Information Technology—Life Cycle Processes [40]; and U.S. DOD Data Item Description DI-IPSC-81429, Software Transition Plan [139]. IEEE Std. 1219 describes a process model for software maintenance, suggested supporting measures and metrics, maintenance guidelines and suggestions in support of planning, and definitions for the various types of maintenance.

Visual Workplace

The visual workplace is a lean concept that typically takes the pillars of the 5S and displays these throughout the work environment. 5S stands for sort, set in order, shine, standardize, and sustain. A visual workplace is a work environment that is based on visual clues. In the classic lean production environment, the visual workplace has been used to transform factory floors into places where messages in support of individual productivity, product quality, and safety are delivered.

These visual messages help turn the production lines into places of training, self-regulation, and improvement. This happens 24/7, all because of the ever-present visual information. The visual workplace should try to supply answers to the following six core questions: where? what? when? who? how many? and how?

It is important to think of the differences between the classic manufacturing environment and software development when applying the principles of the visual workplace. Software engineers are focused on the inner space of their digital environment and it makes sense to pay attention to this visual workplace when targeting the placing of messaging. Reminders regarding proper coding conventions that are delivered via e-mail, automatic validation of code with noncompliance notification, automatic meeting reminders, and schedules available from a centralized server are all examples of the application of the visual workplace to this digital environment.

CONTROL

Teams should ensure that any gains will last and that enacted process improvements will provide added benefit. During the control phase, the Process Group should review, catalog, and save reusable intellectual assets in the SCM for future organizational process assets. The Process Group should develop recommendations concerning management of future change efforts:

- Improving an organization's ability to perform
- Addressing a different aspect of the organization's business

The requirements of any software process improvement methodology are broad. No single IEEE standard can be used in isolation to support process improvement requirements or the implementation of basic software engineering processes and practices. Rather, a subset of the available IEEE software engineering standards should be employed in combination to provide effective support for these activities. IEEE software engineering standards can be used to provide detailed, prescriptive support for software engineering process definition and improvement activities.

Tools most commonly used in the control phase are:

- Control charts (discussed in the measure phase)
- Evolutionary operation (EVOP)
- Flow diagrams (discussed in the analyze phase)
- Measurement systems and their assessment
- One-piece flow
- Pareto charts (discussed in the measure phase)
- Plan–Do–Check–Act (PDCA)
- Poka-yoke (error proofing)
- Quality control process charts
- Standardized work processes
- Statistical process control

Evolutionary Operation (EVOP)

Anyone working in software development should be interested in supporting the definition and improvement of the processes that support the development of their products. Evolutionary operation (EVOP) is a methodology that supports continuous and incremental process improvement during normal production or software development operations.

The National Institute of Standards and Technology (NIST) defines EVOP in this way:

An EVOP is a special type of on-line experiment with several distinguishing features:

1. The experimental material is production material intended to be delivered to customers.
2. In each experimental cycle, the standard production recipe is changed.
3. The experimental factor levels are less extreme than in conventional off-line experiments.
4. The experiment is run over a longer term with more material than in conventional off-line experiments. [186]

The main thrust of EVOP is controlled change and the avoidance of unwanted and harmful change. EVOP may be best utilized on a pilot or existing production process but should be deployed by well-trained personnel through a series of small, routinely applied changes. This controlled approach to process change management is an effective way to manage software process improvement activities and observe proposed changes to software development activities.

One-Piece Flow

One-piece flow, also known as continuous flow, is the idealized goal of Lean and is achieved only when all waste is removed from the value stream and only the value-added work remains. One-piece flow is an ideal that should be strived for, but this ideal begins to break down as the transfer time begins to approach the work time.

One-piece flow is not appropriate for all types of production. It best supports production environments in which development occurs in a logical sequence, for uniform products, and the steps are interdependent. As applied in a software production environment, this technique could be best applied to the areas of verification and validation. Automated test procedures could bring about significant increases in the value stream. For additional information regarding test planning and test plan development, refer to Chapter 4.

Measurement Systems and their Assessment

The purpose of measurement, or the collection of program metrics, is to establish, document, and maintain consistent methods for measuring the quality of all software products. Any metrics and measurement plan must define the minimum standardized set for information gathering over the software life cycle, as these metrics will serve as software and system measures and indicators that critical technical characteristics and operational issues have been achieved.

An example of metrics as defined and employed on a specific software project is shown in Table 7-9. These metrics fall into three general categories. Management metrics

Table 7-9. Example measurements from a Software project

Metric	Producer	Objective	Measurement
		Management Category	
Manpower	Program Manager	Track actual and estimated man-hours usage	Hours used vs. estimated
Development progress	Project Manager	Track planned vs. coded	Number of coded Units
Cost	NA	Track software expenditures	NA
Schedule	Project Manager	Track schedule adherence	Milestone/event slippage
Computer resource utilization	NA	Track planned and actual resource use	NA
Software engineering	Delta EPG	Quantify developer software engineering environment	Computed # of KPAs

(continued)

Table 7-9. *(continued)*

environment		maturity	
Requirements Category			
Requirements traceability	Project Manager	Track requirements to code	% of requirements traced throughout program life cycle
Requirements stability	Project Manager	Track changes to requirements	# and % of requirements changed/added per project

deal with contracting, programmatic, and overall management issues. Requirements measures pertain to the specification, translation, and volatility of requirements. Quality measures deal with testing and other software technical characteristics Only examples of management requirements measures are provided in Table 7-9.

For additional information on the creation of a software measurement and measures plan, refer to Chapter 4.

PDCA

PDCA is an acronym for the Plan—Do—Check—Act cycle of continuous improvement, shown in Figure 7-31. It is commonly referred to as the "Deming Wheel," since it was popularized by W. Edwards Deming.* In his later teachings, Deming replaced "Check" with "Study" to reinforce the importance of verifying results with data.

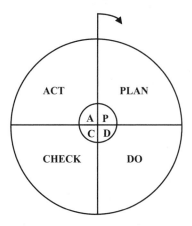

Figure 7-31. Plan—Do—Check—Act continuous improvement cycle.

PDCA can be viewed as both a philosophy of work and as a specific tool used in support of the improve and control phases of DMAIC. In support of the common management guidance, "plan the work and work the plan," PDCA adds the additional steps of checking results of the work against goals and then acting based upon the outcome to prevent errors, replanning if the goals are not met, and continuous improvement if they are. It is important to note that PDCA applies equally to efforts whether they be the conduct of a year-long Lean Six Sigma project or the conduct of an individual meeting.

*Deming himself attributed it to Walter Shewhart and called it the Shewhart Cycle.

Poka-yoke (Error Proofing)

The term poka-yoke is derived from two Japanese words, avoiding (yokeru) and inadvertent errors (poka) and means to error or mistake proof. This concept was originally called bake-yoke, which literally means fool (or idiot) proofing. The idea is place constraints on an operation so that it can only be successfully completed, or that the chances of failure are significantly minimized.

As applied to the process of production, poka-yokes are any mechanism that can be applied to help ensure that each process step is as error free as possible, thereby helping to mistake-proof the entire production process. Poka-yokes are mechanisms that prevent the incorrect execution of a process step. The basic principles of Poka-yoke advocate developing tools, techniques, and processes so that it is impossible, or very difficult, for people to make mistakes.

An example of this in the software industry is the insertion of a CD-ROM in a reader. Improper insertion will not cause any injury to the media or the reader. Another example is autorun install programs that automatically launch when installing a new software product. It is also common practice to use checklists in support of software quality inspections and walk-throughs to help standardize and eliminate errors. Software programs that utilize poka-yoke techniques to eliminate users errors are applications like FrontPage®, which mitigates or highlight errors as programmers code HTML, and Microsoft Word®, which autocorrects spelling errors as authors create documents.

There are two categories of poka-yokes; those used to help prevent defects from occurring and those to help detect defects. Table 7-10 describes these two types.

Table 7-10. Prevention- and detection-based poke-yokes

Prevention-based methods	
Control	Senses an impending problem, stops production or process so that corrective action can be taken.
Warning	A warning of an impending problem occurs, but production is not halted.
Detection-based methods	
Contact	The removal of defects through direct contact or observation based upon conformance specifications.
Fixed value	Defects are detected through the measurement of the numbers of parts or the detection of critical conditions through monitoring devices. A missing part or pressure out of range would signal a production problem.
Motion step	Helps to ensure that all process steps are performed and in the proper order.

Quality Control Process Chart

A quality control (QC) process chart, more commonly called a control chart, is a tool helps determine whether a process is under statistical control. The patterns revealed in the control chart can help determine the source(s) of any process variation and help to bring the errant process back under control.

Control charts are also referred to as "Shewhart charts" or "process-behavior charts," and are specific types of run charts. Walter A. Shewhart was an engineer at Western Electric and Bell Telephone Laboratories in the 1920s and is known as the father of modern quality control. He relentlessly pursued ways in which statistical theory might support the needs of industry and is most widely known for the control chart. This simple but highly effective tool represented what Shewhart felt was an initial step toward "the formulation of a scientific basis for securing economic control."*

Control charts plot a quality characteristic and typically contain a center line representing the average value of the quality characteristic, the upper control limit (UCL), and the lower control limit (LCL). If the points associated with the quality characteristic plot within these control limits, the process is assumed to be in control (Figure 7-32).

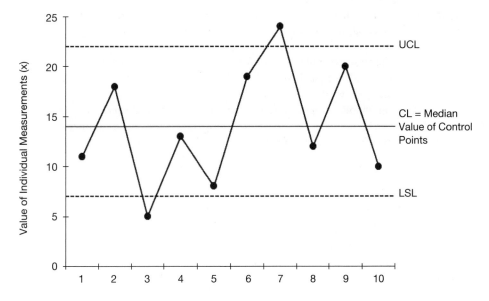

Figure 7-32. Control chart example with UCL and LCL exceeded.

Before constructing a control chart, the target process must first be identified, the sampling plan must be agreed upon, the data collected, and all calculations used to populate the control chart and control limits must be calculated [117].

There are seven basic tools of quality control:

8. Cause-and-Effect Diagram
9. Check Sheet
10. Flowchart
11. Histogram
12. Pareto Chart
13. Scatter Diagram
14. Quality Chart

The quality chart is the seventh, and final, tool described from the basic toolset. Each of these tools is being described separately. It is important for the reader to understand their efficacy when used together and value when applied separately.

*See http://www.asq.org/about-asq/who-we-are/bio_shewhart.html.

Standardized Work/Processes

In Chapter 2, we discuss the benefits of software engineering standards. Standardized work reaps those same benefits, not just for the products we deliver, but also for the processes we use to produce them. Process standardization is fundamental to every continuous improvement approach, because there is no hope of institutionalizing an improvement unless there is a context, a framework to hold the change. Standardized work provides this framework.

But standard work does not mean rigid and unchanging. Liker [96] examines the technical structure (low or high bureaucracy) and social structure (coercive vs. enabling) of organizations and notes that although Toyota took the notion of standard work from Ford Motor, the implementations today are very different. Ford's "coercive bureaucracy" uses standards to control people, catch them breaking the rules, and punish them to get them back in line. By contrast, enabling systems are simply the best practice methods, designed and improved upon with the participation of the work force. The standards actually help people control their own work.

Standardization enables high-quality production of goods and services on a reliable, predictable, and sustainable basis, ensuring that important elements of a process are performed consistently in the most effective manner. Changes are made only when data shows that a new alternative is better. Use of standardized practices will:

- Reduce variation
- Make the output of the production process more predictable
- Provide "know-why" for operators and managers now on the job
- Provide a basis for training new people
- Provide a trail for tracing problems
- Provide a means to capture and retain knowledge
- Give direction in the case of unusual conditions

In lean usage, standard work means a process description (sequence of task steps), definition of the takt time (expected cycle time for the process), and standardized materials for the task. The process detail in the description balances the need to be specific enough to be useful, yet flexible enough to allow improvement. And most importantly, the person doing the work is the one who improves the process.

Statistical Process Control

Statistical Process Control (SPC) is used to monitor the consistency of processes used to manufacture a product as designed. SPC is different from more traditional inspection-based quality approaches, as this method for process monitoring applies resources to defect prevention rather than defect detection.

Statistical Process Control was engineered by Walter A. Shewhart in the 1920s. Working at Bell Labs, he created the basis for the control chart, which is the most common technique for determining whether a software process is under statistical control. Shewhart's research was later applied by W. Edwards Deming. Deming introduced these SPC methods to the Japanese industry following World War II in support of the reconstruction. SPC and Deming's methods laid the groundwork for what was later to become know as total quality management (TQM).

The following SPC techniques are used commonly in addition to control charts [108]:

- Histograms
- Pareto analysis
- Radar charts
- Regression analysis
- Run charts
- Scatter diagrams
- Pie charts

SPC will not improve a poorly designed product's reliability, but can be used to maintain the consistency of how the product is made and, therefore, of the manufactured product itself and its as-designed reliability. It is easy to see how SPC fits naturally into the repeatable classic manufacturing environment, in which the same type of component is produced in an assembly line fashion. The important question is how would SPC apply to software and systems engineering?

SPC can be applied to all aspects of the software development life cycle. However, in practice, it makes sense to target the most "reusable" portions as candidates for SPC methods. Specifically, SPC can be used to monitor and control the requirements management process, design, inspections, and testing (Table 7-11).

Table 7-11. Example statistical applications in software engineering [75]

Phase	Use of Statistics
Requirements	Specify the number of changes to requirements throughout the software life cycle.
Design	Use Pareto analysis to identify fault-prone module(s).
Coding	Statistical control charts applied to inspections.
Testing	Coverage metrics provide attributes. Testing based on specified operational profile.

When looking for information in support of software and systems reliability, it is important to remember the following IEEE standards that provide invaluable information:

- IEEE Std. 982.1, IEEE Standard Dictionary of Measures to Produce Reliable Software
- IEEE 1061, IEEE Standard for a Software Quality Metrics Methodology
- IEEE Std. 14143.1, Implementation Note for IEEE Adoption of ISO/IEC 14143-1:1998 Information Technology—Software Measurement—Functional Size Measurement—Part 1: Definition of Concepts
- ISO/IEC 15939, Information Technology—Systems and Software Engineering—Measurement Process

8

APPLYING LEAN SIX SIGMA

WHY SHOULD MY ORGANIZATION IMPLEMENT LEAN SIX SIGMA?

As previously described, defects are principal to the Six Sigma model. Defects are isolated and eliminated, and thereby lower the overall costs of rework during production and postproduction maintenance. Six Sigma places primary importance on customers, business results and statistical analysis. Lean principles aim to uncover and reduce waste. Software defects are only one of the targets of Lean waste reduction. Lean also aims to make discoveries and targeted improvements in support of eliminating wasted time and effort. The implementation of both Six Sigma and Lean techniques will address and change similar management and technical staff behaviors. The organizational change management approach used in both techniques will be quicker and less costly than a sequential approach and provides more benefits than an "either/or" approach. Figure 8-1 provides an overview of the strong partnership provided when these two process improvement methodologies are combined.

Keys to Success

When companies choose Lean Six Sigma (LSS) implementation for improvement, there are many benefits. A few of these are:

- Customers are confident about the consistent performance of the software development organization.
- Transparency of operations within the organization.
- Lowering of costs and shortening of cycle times, through the effective use of resources.
- Improved, consistent, and predictable results, resulting in an overall increase in production and growth.

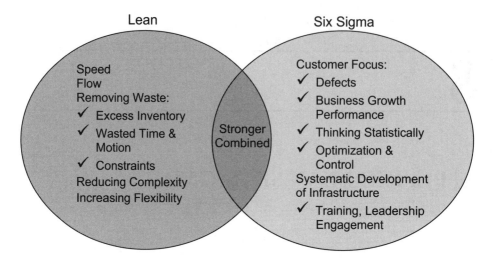

Figure 8-1. Lean and Six Sigma—strong partners [88].

- Provision of opportunities for focused and prioritized improvement initiatives.
- Integration and alignment of processes that enable the achievement of planned results, resulting in an increase in organizational confidence.
- Ability to focus effort on process effectiveness and efficiency.

It is not enough to simply focus on the benefits or results from the implementation of a software process improvement initiative. When considering traveling the road to software process improvement, you must also consider all the items that will affect success or failure. A few of these critical success factors are:

- Senior leadership support. Is there adequate and appropriate support from the senior leadership within the organization? Senior leadership can learn from this initiative to support future efforts.
- Process improvement infrastructure. Does your organization have the necessary, preexisting corporate policies, procedures, methods, tools, measurements, and training required to support your process improvement effort? Infrastructure improvements will benefit future projects.
- Integration with preexisting initiatives or goals. You must ensure that you resolve any conflicts between your improvement effort and any existing processes or practices. Coordination, alignment, communications, and training will maintain staff momentum.
- Adequate and appropriate staffing. Are you going to require full-time dedicated staff for this effort? Is funding and training for these individuals available? Process skills can complement development skills.
- Adequate time and support for training. Individuals must have a firm grasp of the principles or the time to prepare. One root cause of rework is the lack of training.
- Program accountability and participant recognition. Program participants must be recognized for their participation. Improvement initiatives need the same structure

as managed projects: deliverable schedules, measurements, and recognition and rewards.

- Knowledge transfer. How effectively is knowledge management occurring within your organization? Project knowledge and best practices are organizational assets that need to be shared.

- Customer participation. It is critical that your customer remain engaged throughout the production process. Knowledgeable and engaged customers are a key investment for success.

- Process change management. How is your organization going to support the software process improvement process and the process of change management? Change management is critical. Senior leadership understands the planning, investment, and benefit of change management, so get their involvement.

Figure 8-2 provides an overview of some of the perceived benefits to the software development organization. These are dependent upon several critical "keys to success."

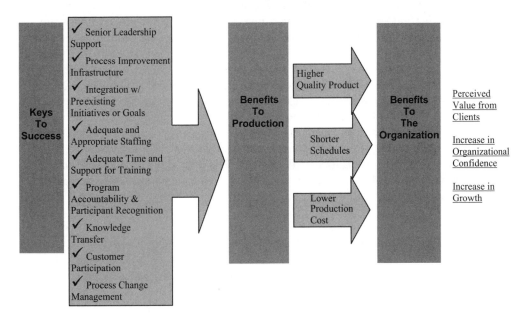

Figure 8-2. Key to Lean Six Sigma implementation success [66].

The Change Context

It is important to understand the context driving change. Often, the dominant forces for Lean Six Sigma implementation are external/customer satisfaction and/or internal business improvement; these require focusing on different issues. The latter requires more change activities, with a focus on return on investment (ROI). The former requires a focus on minor process changes for alignment, consolidation, training, and documentation. If management has determined that there is a lack of customer confidence, then this is often a primary motivator for change. However, these two concepts are closely mingled and often when organizations aim for improvements in customer satisfaction they find that business improvements are required in order to reach this goal.

This book assumes that your software engineering organization is either (1) developing products that internally support customers within your company or (2) developing products and/or providing services to external customers. In the first case, your organization may be a small part of the company, representing a relatively small percentage of annual revenue; an example might be an information technology (IT) group supporting a company's IT infrastructure. In the second case, the developing organization may be a large contributor to a company's annual business bottom line. However, in both scenarios each organization is important to the corporate mission. Each organization should have defined processes and practices and should target and eliminate waste using Lean Six Sigma techniques.

In either case, the company's strategic plan or annual business plan should identify directions, objectives, and risks for the software engineering organization. Many companies use strategic planning for the start of their risk planning and management. The human resources (HR) department has several responsibilities impacting the organization, including job descriptions, staffing goals, training, and recognition plans. After plans and resources, the portfolio management of the software development projects allows smooth startup, transition, and ending of projects. The focus should be on the project success criteria.

Project Planning starts with Software Development Life Cycle (SDLC) selection and the tailoring of existing software engineering processes. This is followed by the development of project schedules and the subsequent management of the SDLC. The project members work their roles within the processes and create, review, revise, and transmit project artifacts to other project members or stakeholders. Project artifacts are stored and managed as part of the project configuration management system. The final artifacts are customer deliverables, installations, and/or training. It is critical that the processes that are fundamental to basic software engineering exist prior to the initiation of any development effort or Lean Six Sigma activity.

WHAT IF MY ORGANIZATION IMPLEMENTS LEAN SIX SIGMA?

Implementing Lean Six Sigma principles in many cases requires significant organizational cultural change. It requires the movement from a static, reactionary, development environment to an integrated, adaptable Lean Six Sigma environment supporting significantly improved software development techniques. Table 8-1 presents the key themes of Lean Six Sigma again for review.

Table 8-1. Key themes of Lean Six Sigma

✓ Customers are important.
✓ Speed, quality, and low cost are linked.
✓ You need to eliminate variation and defects and focus on process
   flow if you want to deliver quality, speed, and low cost.
✓ Data is critical to making sound business decisions.
✓ People have to work together to make the kinds of improvements that
   customers will notice.

Lean and Six Sigma initiatives produce a synergy that works to eliminate waste while ensuring overall process quality.

Identify Your Target(s)

Where to begin? As with any software process improvement effort, the first challenge is determining an appropriate target area for your improvement event. The selection of candidate projects is critical and the following combination of lean principles and Six Sigma concepts should all be considered when introducing Lean Six Sigma organizational process improvement:

- Quality/variability
- Customer focus
- Speed
- Cost reduction

It is also important to remember that any project selected should be well scoped, have internal management support, and be linked to business strategic objectives (e.g., the business case for these projects and the benefits from any process gains must be easy to describe and immediately obvious). Projects fall into one of three categories:

1. Just Do Its. Simple projects with single ownership, usually completed within a month.
2. RIEs—Rapid Improvement Events. Planning events supported by several cross-functional teams. They occur in support of structured seven-week improvement cycles (3 weeks for preparation, 1 week for execution, and 3 weeks for follow-up).
3. Long-Range Projects. Projects supported by Six Sigma experts (Black Belts); ownership may cross departments. Projects are typically longer than 3 months.

Staff Responsibilities and Training

Training your process champions may be the single most important part of any process improvement initiative. However, training is only effective if it is applied. It is important that training be part of a planned process improvement initiative based upon tangible results. The principles must be applied to real projects and real results must be expected.

Training for Lean. Classically, lean principles were taught by a mentor on the job or factory floor. There are courses available that now teach these principles. The principles used in support of the lean methodology are typically taught as separate workshops. Each workshop usually combines a short training session with a lean principle and the direct application of that principle. Students are then advised to take what they have learned during the workshop and apply their knowledge.

Six Sigma Training. Six Sigma training is broken down into phases that support the DMAIC (define–measure–analyze–improve–control) process. Individuals attending training are provided with time in between each training session to apply the tools that they have learned during each training session. Ideally, trainees then go back to their projects and apply their newly gained knowledge. During the next phase of training, these same students would return with a renewed respect and ready for the next Six Sigma training phase. Table 8-2 describes the individuals within a Six Sigma organization, their roles, and the typical training that they might receive.

In the true spirit of Lean Six Sigma it is important to have individuals who are trained experts but it is critical to remember that successful lean process improvement efforts are team driven. Lean methods emphasize process improvement through the application of the simplest methods, not the most complex statistical formulae. Identify the greatest process challenges that will bring the most production gain and work toward these incrementally. Six Sigma techniques should be team driven through the application of common logic. Teaching or applying Six Sigma as gospel minimizes the possible process benefits and is a mistake made by many implementing organizations. It is important to go through the discipline but only through practical application; remember to ask what is right for your organization or project.

Table 8-2. Six Sigma: Special staffing and typical training programs

Role	Description	Training	Program Participation
Champion	Executive management	1–2 hour overview briefing	Responsibility for managing and guiding Six Sigma efforts and assuring that those efforts support and drive corporate priorities.
Supporters	Company employee	2–4 hour awareness briefing	Goal is familiarization; project participation is not a requirement. These individuals are not candidates for future participation but should understand the program and its priorities.
White Belt	Company employee	1–2 days awareness course	Goal is familiarization; project participation is not a requirement. These individuals are candidates for future participation.
Yellow/Green Belt	Yellow—project participant Green—project lead	1–2 weeks introductory methods/tools course	Individuals get to practice using implementation methods and techniques.
Black Belt	Lead projects and serve as coaches and resources to other projects	4–5 weeks leadership and problem solving	Skill building tools/methods course.
Master Black Belt or Enrichment Course	Have led a number of project teams and have a proven track record of delivering results. Train and coach Black Belts and monitor team progress, also aid teams as needed.	Varies	Advanced training in one or more specialties; sophisticated problem solving.

Follow-up

How does an organization know whether process improvement has been successful? Whether it has been sustained or is continually improving? Keeping track of key improvement measures in the form of performance metrics is a critical part of program maintenance. Follow-up events should be discussed and planned as a part of the program implementation and not as an afterthought, and the results should be shared with all employees.

GOAL-DRIVEN IMPLEMENTATION

Any approach requires a strategy to be used by the organization to facilitate process institutionalization. This strategy should include targeted projects, key personnel, training, and, most importantly, schedules reflecting specific software process improvement targets. Goal-driven process improvement is the most effective. Identify both short- and long-term goals stating concise objectives and time periods and associate these goals as schedule milestones. Table 8-3 provides a sample timeline for goal-driven process improvement.

Table 8-3. Example goal-driven implementation time line

0–3 months
- Identify individuals responsible for software process improvement.
- Identify project managers who will be participating.
- Identify list of candidate projects.
- Solidify backing of senior management.
- Look at existing processes and make sure they are appropriate and reflect current business needs (small vs. large projects) using identified goals and IEEE software engineering standards.
- Define the formats for your process plans (software configuration management plan, software requirements management plan, software quality assurance plan) using IEEE software engineering standards and measure them against the project requirements.
- Get project members to provide feedback on process plans; review and incorporate feedback.
- Conduct a gap analysis against the existing requirements.

3–6 months
- Create process document templates for project documentation based upon defined processes; projects will use these to develop their own plans (e.g., software development plan, software requirements specification).
- Conduct weekly/monthly status reports/reviews to gauge and report progress and provide areas for improvement.

6–9 months
- Conduct Lean-Six-Sigma-based reviews of the projects. It would be ideal to also include members from unselected projects to participate in these reviews, with reporting senior management.
- Provide feedback regarding project review; include requirements for improvement.

9–12 months
- Conduct internal appraisals or assessments, with reporting to senior management.
- Provide feedback regarding project review, include requirements for improvement.

LEAN SIX SIGMA AND CONFORMANCE

It is important to understand the context driving change. Often, the dominant forces for change are business development or customer satisfaction. This chapter will assume business development is the main driver, or force, for change.

The crisis in software has been well documented. In order to compete successfully, software professionals have found it necessary to improve both the quality of their software products and to also improve their ability to work within time and budget constraints.

These improvements depend just as strongly on process as well as technology. Over the years, many have focused on this problem, as illustrated by the number of standards and process methodologies that have evolved (Figure 8-3) that directly support software process improvement or assessment.

From a general problem-solving point of view, project management may be described simply as identifying what is to be done, deciding how to do it, monitoring what is being done, and then evaluating the outcome. From a process improvement point of view, and stating things extremely simply, apply the lessons that you learn and make improvements during the next project cycle.

There are currently no formal, declared, conformance measures that support LSS implementation. However, some organizations employing LSS determine their overall process improvement capability with one or more of the following:

The Lean Aerospace Initiative's Lean Enterprise Self-Assessment Tool (LESAT)

TickIT Guide for Software (ISO 9001)

ISO/IEC 90003 Guidelines for the Application of ISO 9001:2000

SCAMPI (A, B, C)/CMMI

The Balanced Scorecard methodology [187] is the recommended best practice for LSS initiatives. This methodology is widely recognized as an effective performance management tool for tracking progress against stated goals and objectives in a "balanced" manner.

The Balanced Scorecard approach establishes performance measurement criteria for business improvement initiatives, examining both direct and indirect benefits and performance measures balanced across four key categories, linked by vision and strategy (Figure 8-4). The four key categories are:

1. Financial
2. Internal business process
3. Customers
4. Employee learning and growth

There are several program minimums that are essential in supporting the role of any software process improvement initiative. These have been mentioned previously, but will be restated here in summary:

- There must be strong executive leadership and support
- The effort must be associated with clear and measurable organizational goals:

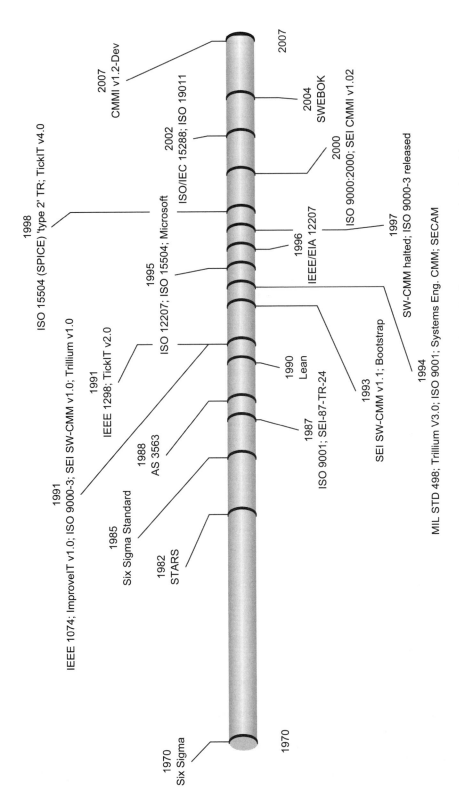

Figure 8-3. Time line of major standards supporting process methodologies and software-process-related models.

241

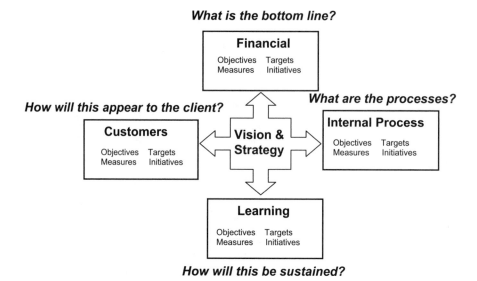

Figure 8-4. The Balanced Scorecard.

To include actionable business case and performance measures

Program benefits and returns on investment must be clearly defined and approved by all key stakeholders

- A well-designed execution and continuous improvement plan must exist and should be readily available to all program participants:

    Transformation (change) management must be an integral part of the program

    The execution strategy and implementation plan must be well defined and clear to all stakeholders

    Program milestones must be designed to deliver program goals and objectives

- A training program must exist for all program participants:

    An integrated training environment should exist, including integrated record tracking

- Clearly defined roles and responsibilities for all program participants:

    Key stakeholders, internal and external, must be identified early in the program

IMPLEMENTATION PITFALLS

What is the overall goal of the software process improvement effort? Who is the customer? What needs to be collected to support the organization? In addition, when looking at the organization, what needs to be aggregated across projects in order to provide meaningful data to the organization? Are all stakeholders committed? These are just a few of the questions that every organization should ask prior to undertaking any process improvement initiative.

Organizations should look for ways to operate leanly and efficiently. Are the selected sets of activities appropriate? Data collection simply for the sake of data is at best inefficient; this data is often ignored, wasting valuable time and money. When data is misused, it can have devastating effects. When applied correctly, LSS can help the project or program manager do a better job at implementing more realistic software plans, more precisely track and report progress against those plans, and take corrective action when required.

Being Overly Prescriptive

Implementing software engineering processes and practices in support of specific process improvement requirements verbatim can be costly and may not reflect the specific process or business requirements of an organization. This approach can result in an overly prescriptive process that will increase cost and slow product cycles. This can rapidly destroy the credibility of the process improvement implementation with management and the software development staff.

The key to implementing effective measurement programs is that the members of the implementing organization must endorse the program as implemented. The program must be perceived as adding value, the measures must be viewed by all stakeholders as being realistic, easily implemented, and readily understood by all participants. Carefully analyze what areas of process improvement will have the most positive impact on existing programs; implement these first and management will see the tangible results and continue their support. Small projects may require less formality in planning than large projects, but all components of the standard should be addressed by every software project. Components may be included in the project-level documentation, or they may be merged into a system-level or business-level plan, depending upon the complexity of the project.

Documentation, Documentation

Be careful not to generate policies and procedures to simply satisfy the implementation requirements of your process improvement methodology. The goal should be improved performance, not process simply for the sake of process. Generating documentation for the sake of documentation is a waste of time and resources.

Policies and procedures should be developed at the organizational level and used by all projects, only requiring documentation where there is deviation from the standard. When developing and implementing software metrics programs, just like any other program or initiative, the implanting organization should focus on specific goals and determine the most appropriate metric and its proper application. These metrics should then be systematically collected and analyzed, and their use should support both project and organizational decision making.

The pairing of IEEE standards, which are prescriptive and describe software engineering minimums, with these models can reduce this risk significantly.

Lack of Incentives

Many times the success of the process improvement initiatives is the responsibility of the process improvement team, program manager, or even members of the development team. Members of the process improvement team should act as mentors, guiding the software projects through continual process improvement. Each selected project should be required to follow the recommendations of these groups. Process improvement is the most effective and the results more permanent when the projects are stakeholders in the software improvement process. Appropriate incentives should be determined for program participants and these should be clearly communicated throughout the effort.

Overemphasis on Certification and Not on the Lean

As organizations look to the combination of Six Sigma and Lean as the source for their software process improvement principles, the authors would like to emphasize the importance of remaining results focused. As with any process improvement initiative, there is the temptation to look for the "quick fix" and the danger with Six Sigma is that this is often linked with the numbers and types of Six Sigma "belts" that can be quickly hired or trained. Adequate and appropriate training is an integral part of any successful software process improvement initiative; the keywords here are "adequate" and "appropriate." Organizations should be aiming for cultural change. This can only be accomplished by selecting the correct projects and by supporting the change from within those projects.

CONCLUSION

Process improvement can be intimidating. Many times, the task of process improvement comes in the form of a directive from senior management, or as a customer requirement, leaving those assigned with a feeling of helplessness. However, all those practicing as software engineers should desire to leave behind the chaotic activities associated with uncontrolled software processes and the associated required heroic efforts. When software engineering processes are under basic management control and there is an established management discipline, then this provides benefits to all involved. When used in conjunction with IEEE software engineering standards, the customer may be assured of a lower risk of failure, the organization is provided with accurate insight into the effort, management can more effectively identify and elevate development issues, and team members can work to efficiently managed baselines.

12207 SOFTWARE PROCESS WORK PRODUCTS

ACQUISITION

Make/Buy Decision Matrix

This matrix may be used to support the analysis of commercial off-the-shelf acquisition versus in-house software development solutions. This matrix should include a listing of all project requirements and their estimated cost. It should also include a list of all candidate applications that come closest to meeting the specifications as well as a list of reusable components. This matrix should be used to compare and analyze key functions and cost, addressing the qualitative as well as quantitative issues surrounding the selection process. An example decision matrix is provided on the companion CD-ROM as and is entitled *Decision Matrix.xls*.

Table A-1 provides some sample questions to support the analysis of each of the make/buy/mine/commission options.

Table A-1. Questions in support of make/buy/mine/commission [167]

Sample questions for the "make" option:
- Are developers with appropriate expertise available?
- How would the developers be utilized if not on this project?
- If additional personnel need to be hired, will they be available within the needed time frame?
- How successful has the organization been in developing similar assets?
- What specific flexibilities are gained by developing products in-house as opposed to purchasing them?
- What development tools and environments are available? Are they suitable? How skilled is the targeted workforce in their use?
- What are the costs of development tools and training, if needed?

Table A-1. *Continued*

Sample questions for a "commission" option (*cont.*):
- What are the other specific costs of developing the asset in-house?
- What are the other specific benefits of developing the asset in-house?

Sample questions for the "buy" option:
- What assets are commercially available?
- How well does the COTS software conform to the product-line architecture?
- How closely does the available COTS software satisfy the product-line requirements?
- Are small changes in COTS software a viable option?
- Is source code available with the software? What documentation comes with the software?
- What are the integration challenges?
- What rights to redistribute are purchased with the COTS software?
- What are the other specific costs associated with purchasing the software?
- What are the other specific benefits associated with purchasing the software?
- How stable is the vendor?
- How often are upgrades produced? How relevant are the upgrades to the product line?
- How responsive is the vendor to user requests for improvements?
- How strong is the vendor support?

Sample questions for a "data mining" option:
- What legacy systems are available from which to mine assets?
- How close is the functionality of the legacy software to the functionality that is required?
- What is the defect track record for the software and nonsoftware assets?
- How well is the legacy system documented?
- What mining strategies are appropriate?
- How expensive are those strategies?
- What experience does the organization have in mining assets?
- What mining tools are available? Are they appropriate? How skilled is the workforce in their use?
- What are the costs of mining tools and training, if needed?
- What are the other specific costs associated with mining the asset?
- What are the other specific benefits associated with mining the asset?
- What changes need to be made to the legacy asset to perform the mining?
- What are the costs and risks of those changes?
- What types of noncode assets are available?
- What modifications or additions need to be made to them?

Sample questions for a "commission" option:
- What contractors are available to develop the asset?
- What is the track record of the contractor in terms of schedule and budget?
- Is the acquiring organization skilled in supervising contracted work?
- Are the requirements defined to the extent that the asset can be subcontracted?
- Are interface specifications well defined and stable?
- What experience does the contractor have with the principles of product-line development?
- Who needs to own the asset? Who maintains it?
- What are the other specific costs associated with commissioning the asset?
- What are the other specific benefits associated with commissioning the asset?
- What are the costs of maintaining the commissioned asset?
- Does commissioning the asset involve divulging to the contractor any technology or information that it is in the acquiring organization's interest to keep in-house?

Alternative Solution Screening Criteria Matrix

This aternative solution screening criteria matrix may be used to support the analysis of alternative solutions. This matrix should include a listing of all project requirements and their estimated cost. It should also include a list of all candidate applications that come closest to meeting the specifications, as well as a list of reusable components. This matrix should be used to compare and analyze key functions and cost, addressing the qualitative as well as quantitative issues surrounding the selection process. An example decision matrix is provided on the companion CD-ROM as and is entitled *Decision Matrix.xls*. Suggestions in support of effective screening practices are provided in Table A-2.

Table A-2. Tips for effective solution screening [41]

Strike a reasoned balance between efficiency and completeness	Efficient screening often requires employing high-level criteria that can quickly exclude many COTS components. For example, a common and often appropriate screening strategy suggests that examining only the top few (e.g., 3–5) competitive components in terms of sales or customer-installed base. However, there are often reasons to additionally consider smaller, niche players that provide a unique or tailored capability more in line with system expectations. This suggests that there may be multiple pathways for inclusion of a component.
Avoid premature focus on detailed architecture and design	Although high-level architecture is an appropriate screening criterion, there is a tendency to prematurely focus on detailed architecture and design as screening criteria (for example, specifying the architectural characteristics unique to one component). Where architectural or design decisions have already been made that must be reflected by the chosen component, then it is important that criteria reflect these decisions. However, to fight against the tendency to think within the box, consider the risks and potential work-arounds if a highly similar capability is delivered in a different manner.
Optimize the order in which you apply screening criteria	Criteria will vary along dimensions of ability to discriminate and effort required to obtain data. By their nature, some criteria will be highly discriminatory in determining which components are appropriate. Other criteria, while important, will eliminate fewer components from consideration. Still other criteria will require more work than the norm to accumulate data. Consider both how discriminating a criterion is and the effort to evaluate components against the criterion when determining an order for applying screening criteria.
Market share/growth are good screens for long-lived systems	A defining characteristic of many systems is expected lifespan. Often, systems are used for over 20 years. COTS components in these systems should come from financially sound companies with good prospects to remain in business.

Cost–Benefit Ratio

A cost–benefit ratio is a calculation that depicts the total financial return for each dollar invested. Many projects have ready access to staff "hours" rather than accounting figures, which is good enough for decision analysis and resolution. The simplified example that follows is based on a fictitious case of developing a Web-based user support program that is estimated to cost 1,000 hours to develop and deliver, and which should save $6,400 staff hours in the first year due to a reduction in field service calls. Table A-3 describes the calculation.

Table A-3. Cost–benefit ratio calculation

	Benefit (first yr hours)	Total investment (hr)	Cost–benefit ratio	Return after 1 year each dollar invested returned over:
Planned	6,400	1,000	6.4	$6
Actual	6,000	800	7.5	$7

SUPPLY

Recommendations for Software Acquisition

IEEE Std 1062-1998, IEEE Recommended Practice for Software Acquisition, describes the software acquisition process. IEEE Std 12207.0 also recommends a set of objectives in support of the acquisition process. Table A-4 describes these combined recommendations and should be used to support the complete life cycle of the acquisition process.

Table A-5 lists the acquisition checklists provided in IEEE Std 1062. Items listed also support conformance with IEEE 12207.0. Items in boldface are included in this appendix.

Table A-4. Recommendations for software acquisition process

Rec. 1	Planning organizational strategy. Review acquirer's objectives and develop a strategy for acquiring software [4].
Rec. 2	Implementing an organization's process. Establish a software acquisition process that fits the organization's needs for obtaining a quality software product. Include appropriate contracting practices [4].
Rec. 3	Determining the software requirements. Define the software being acquired and prepare quality and maintenance plans for accepting software supplied by the supplier [4].
Rec. 4	Identifying potential suppliers. Select potential candidates who will provide documentation for their software, demonstrate their software, and provide formal proposals. Failure to perform any of these actions is basis to reject a potential supplier. Review supplier performance data from previous contracts [4].
Rec. 5	Preparing contract requirements. Describe the quality of the work to be done in terms of acceptable performance and acceptance criteria, and prepare contract provisions that tie payments to deliverables [4]. Develop a contract that clearly expresses the expectation, responsibilities, and liabilities of both the acquirer and the supplier. Review contract with legal counsel [40].
Rec. 6	Evaluating proposals and selecting the supplier. Evaluate supplier proposals, select a qualified supplier, and negotiate the contract. Negotiate with an alternate supplier, if necessary [4]. Obtain products and/or services that satisfy the customer need. Qualify potential suppliers through an assessment of their capability to perform the required software [40].
Rec. 7	Managing supplier performance. Monitor supplier's progress to ensure all milestones are

(continued)

Table A-4 *(continued)*

met and to approve work segments. Provide all acquirer deliverables to the supplier when required [4]. Manage the acquisition so that specified constraints and goals are met. Establish a statement of work to be performed under contract. Regularly exchange progress information with the supplier [40].

Rec. 8   Accepting the software. Perform adequate testing and establish a process for certifying that all discrepancies have been corrected and that all acceptance criteria have been satisfied [4, 40].

Rec. 9   Using the software. Conduct a follow-up analysis of the software acquisition contract to evaluate contracting practices, record lessons learned, and evaluate user satisfaction with the product. Retain supplier performance data [4]. Establish a means by which the acquirer will assume responsibility for the acquired software product or service [40].

Table A-5. IEEE Std 1062 acquisition checklist summary

Checklist 1:	**Organizational strategy**
Checklist 2:	Define the software
Checklist 3:	**Supplier evaluation**
Checklist 4:	Supplier and acquirer obligations
Checklist 5:	Quality and maintenance plans
Checklist 6:	User survey
Checklist 7:	**Supplier performance standards**
Checklist 8:	Contract payments
Checklist 9:	Monitor supplier progress
Checklist 10:	Software evaluation
Checklist 11:	Software test
Checklist 12:	Software acceptance

Organizational Acquisition Strategy Checklist

IEEE Std 1062, IEEE Recommended Practice for Software Acquisition, provides a example checklist in support of the definition of an organizational acquisition strategy as described in Table A-6. Any checklist used by an organization to determine acquisition strategy should reflect the requirements of the acquiring organization. An electronic version of this document, entitled *Acquisition Strategy Checklist.doc,* is provided on the companion CD-ROM.

Table A-6. Acquisition strategy checklist [4]

1. Who will provide software support?	Supplier ☐	Acquirer ☐
2. Is maintenance documentation necessary?	Yes ☐	No ☐
3. Will user training be provided by the supplier?	Yes ☐	No ☐
4. Will acquirer's personnel need training?	Yes ☐	No ☐
5. When software conversion or modification is planned:		
a. Will supplier manuals sufficiently describe the supplier's software?	Yes ☐	No ☐

(continued)

Table A-6. *(continued)*

b. Will specification be necessary to describe the conversion or modification requirements and the implementation details of the conversion or modification?	Yes ☐	No ☐
c. Who will provide these specifications? Supplier ? Acquirer ?		
d. Who should approve these specifications? _____		
6. Will source code be provided by the supplier so that modifications can be made?	Yes ☐	No ☐
7. Are supplier publications suitable for end users?	Yes ☐	No ☐
a. Will unique publications be necessary?	Yes ☐	No ☐
b. Will unique publications require formal acceptance?	Yes ☐	No ☐
c. Are there copyright or royalty issues?	Yes ☐	No ☐
8. Will the software be evaluated and certified?	Yes ☐	No ☐
a. Is a survey of the supplier's existing customers sufficient?	Yes ☐	No ☐
b. Are reviews and audits desirable?	Yes ☐	No ☐
c. Is a testing period preferable to demonstrate that the software and its associated documentation are usable in their intended environment?	Yes ☐	No ☐
d. Where will the testing be performed? _____		
e. Who will perform the testing? _____		
f. When will the software be ready for acceptance? _____		
9. Will supplier support be necessary during initial installations of the software by the end users?	Yes ☐	No ☐
10. Will subsequent releases of the software be made?	Yes ☐	No ☐
a. If so, how many? _____ Compatible upgrades?	Yes ☐	No ☐
11. Will the acquired software require rework whenever operating system changes occur?	Yes ☐	No ☐
a. If so, how will the rework be accomplished? _____		
12. Will the acquired software commit the acquiring organization to a software product that could possibly be discontinued?	Yes ☐	No ☐
13. What are the risks/options if software is not required? _____		

Supplier Evaluation Criteria

IEEE Std 1062, IEEE Recommended Practice for Software Acquisition, provides a checklist in support supplier evaluation. This evaluation criterion is provided in Table A-7. The information provided here provides an illustrative example and should be tailored to reflect organizational process needs. An electronic version of this checklist, entitled *Supplier Checklist.doc,* is provided on the companion CD-ROM.

Table A-7. Supplier evaluation checklist [4]

Financial soundness
1. Can a current financial statement be obtained for examination?
2. Is an independent financial rating available?
3. Has the company or any of its principals ever been involved in bankruptcy or litigation?
4. How long has the company been in business?
5. What is the company's history?

(continued)

Table A-7. *(continued)*

Experience and capabilities
1. On a separate page, list by job function the number of people in the company.
2. On a separate page, list the names of sales and technical representatives and support person-nel. Can they be interviewed?
3. List the supplier's software products that are sold and the number of installations of each.
4. Is a list of users available?

Development and control processes
1. Are software development practices and standards used?
2. Are software development practices and standards adequate?
3. Are the currently used practices written down?
4. Are documentation guidelines available?
5. How is testing accomplished?

Technical assistance
1. What assistance is provided at the time of installation?
2. Can staff training be conducted on-site?
3. To what extent can the software and documentation be modified to meet user requirements?
4. Who will make changes to the software and documentation?
5. Will modification invalidate the warranty?
6. Are any enhancements planned or in process?
7. Will future enhancement be made available?

Quality practices
1. Are the development and control processes followed?
2. Are requirements, design, and code reviews used?
3. If requirements, design, and code reviews are used, are they effective?
4. Is a total quality program in place?
5. If a total quality program is in place is it documented?
6. Does the quality program assure that the product meets specifications?
7. Is a corrective action process established to handle error corrections and technical questions?
8. Is a configuration management process established?

Maintenance service
1. Is there a guarantee in writing about the level and quality of maintenance services provided?
2. Will ongoing updates and error conditions with appropriate documentation be supplied?
3. Who will implement the updates and error corrections?
4. How and where will the updates be implemented?
5. What turnaround time can be expected for corrections?

Product usage
1. Can a demonstration of the software be made at the user site?
2. Are there restrictions on the purposes for which the product may be used?
3. What is the delay between order placement and product delivery?
4. Can documentation be obtained for examination?
5. How many versions of the software are there?
6. Are error corrections and enhancements release-dependent?

Product warranty
1. Is there an unconditional warranty period?
2. If not, is there a warranty?
3. Does successful execution of an agreed-upon acceptance test initiate the warranty period?
4. Does the warranty period provide for a specified level of software product performance for a given period at the premises where it is installed?
5. How long is the warranty period?   *(continued)*

Table A-7. *(continued)*

Costs

1. What pricing arrangements are available?
2. What are the license terms and renewal provisions?
3. What is included in the acquisition price or license fee?
4. What costs, if any, are associated with an unconditional warranty period?
5. What is the cost of maintenance after the warranty period?
6. What are the costs of modifications?
7. What is the cost of enhancement?
8. Are updates and error corrections provided at no cost?

Contracts

1. Is a standard contract used?
2. Can a contract be obtained for examination?
3. Are contract terms negotiable?
4. Are there royalty issues?
5. What objections, if any, are there to attaching a copy of these questions with responses to a contract?

Supplier Performance Standards

IEEE Std 1062, IEEE Recommended Practice for Software Acquisition, provides a checklist in support of determining satisfactory supplier performance. This evaluation, provided in Table A-8, is provided as an example and should be based upon all known requirements and constraints unique to the development effort. An electronic version of this checklist, called *Supplier Performance Standards.doc.,* is provided on the companion CD-ROM.

Table A-8. Supplier performance standards [4]

Performance criteria

1. Approach to meet software's functional requirements is defined
2. Growth potential or expansion requirement of the system is defined.
3. Supplier meets time constraints for deliverables.
4. Test and acceptance criteria that are to be met are defined.
5. Programming language standards and practices to be followed are defined.
6. Documentation standards to be followed are defined.
7. Ease of modification is addressed.
8. Maximum computer resources allowed, such as memory size and number of terminals, are defined.
9. Throughput requirements are defined.

Evaluation and test

1. Software possesses all the functional capabilities required.
2. Software performs each functional capability as verified by the following method(s): Documentation evaluation, demonstration, user survey, test.
3. Software errors revealed are documented.
4. Software performs all system-level capabilities as verified by a system test.

Correction of discrepancies

1. Supplier documents all identified discrepancies.

(continued)

Table A-8. *(continued)*

2. Supplier establishes discrepancy correction and reporting.
3. Supplier indicates warranty provisions for providing prompt and appropriate corrections.

Acceptance criteria
1. All discrepancies are corrected.
2. Prompt and appropriate corrections are provided.
3. Satisfactory compliance to contract specifications is demonstrated by evaluations and test.
4. Satisfactory compliance to contract specifications is demonstrated by field tests.
5. All deliverable items are provided.
6. Corrective procedures are established for correction of errors found after delivery.
7. Satisfactory training is provided.
8. Satisfactory assistance during initial installation(s) is provided.

DEVELOPMENT

Requirements Traceability

Requirements traceability throughout the software development life cycle is not directly addressed by the IEEE software and systems engineering standards set. This traceability can be accomplished by the addition of a traceability matrix. The following example (Table A-9) supports backward and forward traceability for validation testing. Additional columns should be added to support traceability to software, or system, design and development. The conversion of this type of matrix to a database tracking system is a common practice for developing organizations. There are many commercially available tools that support requirements tracking.

Table A-9. Requirements traceability matrix example

CER #	Requirement Name	Priority	Risk	Requirements Document Paragraph	Validation Method(s)	Formal Test Paragraph	Status
1	Assignments and Terminations Module (A&T) Performance	2	M	3.1.1.1.1.1	Test Inspection Demonstration	4.1.2.3 4.1.2.6 4.1.2.7 4.1.2.9	Open

Software Development Standards Description

The definition of standards information item content associated with the development of software is integral to the success of any software development effort. The following outline is provided by section 6.17 of IEEE Std 12207.1, Software Life Cycle Processes Life Cycle Data [40].

a) Generic description information
     Date of issue and status
     Scope
     Issuing organization
     References

Context
Notation for description
Body
Summary
Glossary
Change history

b) Description of methods used to allocate system requirements, to develop the software requirements, architecture, and design, and to implement the source code and executable object code;

c) Notations used to describe the system requirements and architecture;

d) Notations used to describe the software requirements, architecture, design modules, design limitations, and code, including identification of the programming language(s) or subset used and reference to the definition of the language syntax, control and data behavior, and side effects;

e) Naming conventions for requirements, design, and source code, including source code, executable object code files, and data;

f) Methods of design and coding and constraints on design and code constructs and expressions, including design and code complexity restrictions and quality criteria for assessing requirements and design data and code;

g) Presentation conventions and content standards for requirements data, design data, source code, and test data

h) Description of methods and tools used to develop safety monitoring software (if applicable);

i) Description of the methods and tools used to define traceability between system requirements, system architecture, software requirements, software architecture, design, code, and test elements;

j) Description of methods, tools, and standards for testing.

System Architectural Design Description

When a software product is part of a system, then the software life cycle should specify the system architectural design activity. The resulting system architectural description should identify items of hardware, software, and manual operations, and all system requirements allocated to the items. The outline in Table A-10 is provided in section 6.25 of IEEE Std 12207.1, Software Life Cycle Processes Life Cycle Data [40].

Table A-10. System architectural design document outline

a) Description
b) System overview and identification
c) Hardware item identification
d) Software item identification
e) Manual operations identification
f) Concept of execution
g) Rationale for allocation of hardware items, software items, and manual operations

Software Architectural Design Description

The resulting software architectural description describes the top-level structure and identifies the software components, which are allocated to all the requirements and further refined to facilitate detailed design.

The information provided here in support of design and development is designed to facilitate the definition of a software architectural description. This information was developed using IEEE Std 1471 and has been adapted to support Lean Six Sigma requirements. The modification of the recommended software architectural description table of contents to support the goals of Lean Six Sigma more directly is shown in Table A-11; the outline is provided in section 6.12 of IEEE Std 12207.1, Software Life Cycle Processes Life Cycle Data [40].

Table A-11. Software architectural design document outline

a) Generic description information
    Date of issue and status
    Scope
    Issuing organization
    References
    Context
    Notation for description
    Body
    Summary
    Glossary
    Change history
b) System overview and identification
c) Software item architectural design, including
    1) Software architecture general description
    2) Software component definition
    3) Identification of software requirements allocated to each software component
    4) Software component concept of execution
    5) Resource limitations and the strategy for managing each resource and its limitation
d) Rationale for software architecture and component definition decisions, including database and user interface design

Database Design Description

Many software efforts require the description of the database components. The outline in Table A-12 is provided in section 6.4 of IEEE Std 12207.1, Software Life Cycle Processes Life Cycle Data [40].

Table A-12. Database design description document outline

a) Generic description information
    Date of issue and status
    Scope
    Issuing organization
    References
    Context

(continued)

Table A-12. *(continued)*

Notation for description
Body
Summary
Glossary
Change history
b) Database overview and identification
c) Design of the database, including descriptions of applicable design levels (e.g., conceptual, internal, logical, physical)
d) Reference to design description of software used for database access or manipulation
e) Rationale for database design

Software Architecture Design Success Factors and Pitfalls [132]

Through a series of software architecture workshops, Bredemeyer Consulting* has identified what they call the top critical success factors for an architecting effort. These are identified in Table A-13. Conversely, the top pitfalls associated with an architecting effort have also been identified. These are shown in Table A-14.

Table A-13. Software architecting success factors

1. The strategic business objective(s) of key sponsors must be addressed. The lead architect and team must have a sound understanding of functional requirements. There must be a good match between the technology and the business strategy.
2. The project must have a good lead architect with a well-defined role who:
    Exhibits good domain knowledge
    Is a good communicator/listener
    Is competent in project management
    Has a clear and compelling vision
    Is able to champion the cause
3. The architecting effort must contribute immediate value to the developers. The architecture must be based upon clear specifications.
4. There must be architectural advocates at all levels of the organization, ensuring that there is strong management sponsorship and that required talent and resources are available.
5. Architecture must be woven into the culture to the point that the validation of requirements occurs during each step of the process.
6. The customer must be involved and expectations are clearly defined.
7. Strong interpersonal and team communication and ownership must be present.

Table A-14. Software architecting pitfalls

1. The lead architect, or supporting senior management, does not effectively lead the effort. Roles and responsibilities are inadequately defined.
2. Thinking is at too low a level. The "big picture" or integration requirements are not considered.
3. There is poor communication inside/outside the architecture team.
4. There is inadequate support to include inadequate resources and/or talent.
5. There is a lack of customer focus. The architecture team does not understand or respond to functional requirements.
6. Requirements are unclear, not well defined, not signed off on, or changing.

*www.bredemeyer.com.

UML Modeling

Figure A-1 uses the UML notation to illustrate the inheritance, aggregation, and reference relationships between work product sets, their work products, their versions, the languages they are documented in, and the producers that produce them.

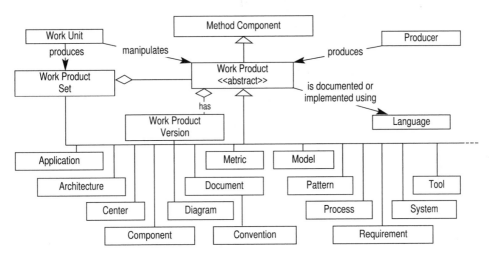

Figure A-1. UML notation.

Unit Test Report

This unit test report template is designed to facilitate the definition of processes and procedures relating to unit test activities. This template was developed using IEEE Std 829-1998, IEEE Standard for Software Test Documentation, and IEEE Std 1008-1987, Software Unit Testing, which has been adapted to support Lean Six Sigma requirements.

The items listed in the subsections under each section in the sample outline in Table A-15 are suggested for inclusion in the unit test report. It is important to note that significant additional information is available from IEEE Std 829 and IEEE Std 1008. The companion CD-ROM also provides an electronic template, *Unit Test Report.doc,* in support of this document.

Table A-15. Unit test report document outline

1. Introduction
2. Identification and Purpose
3. Approach/Strategy
4. Overview
5. Test Configuration/Environmental
   5.1   Software Environment
   5.2   Test Suite
   5.3   Test Stub
   5.4   Test Driver
   5.5   Test Harness

(continued)

Table A-15 *(continued)*

5.6 Preservation of Testing Products
6. Test Cases/Items
   6.1 Test Case Identifier
   6.2 Test Sequence for Interdependencies
   6.3 Items Tested, Indicating their Version/Revision Level
   6.4 Specific Test Data
   6.5 Test Conditions Met
   6.6 Test Procedure/Script/Task Identifier User
   6.7 Test Tool Used
   6.8 Results/Outputs/Summary of Faults
7. Evaluation
   7.1 Termination Requirements Met
   7.2 Test Effectiveness
   7.3 Variances from Requirement # and/or Design #
   7.4 Risks and Contingencies
8. Summary of Activities
9. Notes
10. Approvals

Unit Test Report Document Guidance

The following provides section-by-section guidance in support of the creation of a unit test report. This guidance should be used to help define a unit test process and should reflect the actual processes and procedures of the implementing organization. Additional information is provided in the document template, *Unit Test Report.doc,* that is located on the companion CD-ROM.

Introduction. This section should provide a brief introductory overview of the project and related testing activities described in this document.

Identification and Purpose. This section should provide information that uniquely identifies all test cases, items, and component descriptions involved in the testing effort. This can also be provided in the form of a unique project plan unit test identifier. The following text provides an example:

> The responsible tester was [Tester Name], [responsible organization], Role [coder, designer, test team member]. The unit test was completed on [Date/time of test conclusion].
> This Unit Test Report details the unit testing conducted for the test items/component description in [Project Name] ([Project Abbreviation]) Version [xx], Statement of Work, [date], Task Order [number], Contract Number [number], and Amendments.
> The goal of [project acronym] development is to [goal]. This development will allow [purpose].

Approach/Strategy. This unit test is the first or bottom level of all tests planned according to the development life cycle. The unit test consists of functional testing, which is the testing of the application component(s) against a subset of the operational requirements. The unit test is classified as white-box testing, whose goal is to ensure that code is constructed properly and does not contain any hidden weaknesses.

This section should describe the unit test approach. Identify the risk areas tested. Specify constraints on characteristic determination, test design, or test implementation. Identify existing sources of input, output, and state data. Identify general techniques for data validation. Identify general techniques to be used for output recording, collection, reduction, and validation. Describe provisions for application software that directly interfaces with the units to be tested.

Overview. This section should provide a summary of all test items/components tested. The need for each item and its history may be addressed here as well. References to associated project documents should be cited here. The following provides an example:

The list of individual program units (subprogram, object, package, or module) identifiers and their version/revision levels and configuration management Item(s) are:

Subprogram079 v15   Config12345
Subprogram081 v10   Config12346]

Specifics regarding the testing of these modules are identified in the [xxxx]
Any specific exclusion should be discussed.
This document describes the Unit Test Report (UTR) for the [Project Abbreviation] System. [Project Abbreviation] documentation related to this UTR are:

- Project Plan Milestone/Task #
- Software Requirements Traceability Matrix (RTM) line item descriptions:
   Software Requirements Specification (SRS)
   Software Development Plan (SDP)
   Software Design Document (SDD)
- User's Manual/System Administrator's Manual
- Data Dictionary
- Test Problem Report (TPR) #
- Change Enhancement Request (CER) #

This UTR describes the deliverables to be generated in the testing of the [Project Abbreviation]. This report complies with the procedures defined in the [Project Abbreviation] Software Configuration Management (SCM) Plan and the [Project Abbreviation] Software Quality Assurance (SQA) Plan. Any nonconformance to these plans will be documented as such in this UTR.

This document is based on the [Project Abbreviation] Unit Test Report template with tailoring appropriate to the processes associated with the creation of the [Project Abbreviation]. The information contained in this UTR has been created for [Customer Name] and is to be considered to be "For Official Use Only."

Test Configuration/Environmental. This section should provide a high-level summary of test configuration/environmental items. Depending on system complexity, this could include:

- Software environment
- Test suite
- Test stub
- Test driver
- Test harness
- Preservation of testing products

Software Environment. This subsection should provide a complete list of all communications and system software, operating system version, physical characteristics of the facilities including the hardware, and database version. System and software items used in support of testing should include versions. If the System is kept in an online repository reference, the storage location may be cited instead of listing all items here. The following are examples:

> System and software used in the testing of [Project Abbreviation]: [System and Software List]
>
> For details of client and server system specifications see the [Project Abbreviation] System Requirements Specifications (SysRS).
>
> Unit tests were to be performed at [company name], [site location] in the test system [test system number], using OS X, 10.3.2.

Test Suite. This subsection should describe the set of test cases that serve a particular testing goal.

Test Stub. This subsection should describe the temporary, minimal implementation of a component to increase controllability and observability in testing.

Test Driver. This subsection should describe the class or utility program that applies test cases to the component.

Test Harness. This subsection should describe the system of test drivers and other tools to support test execution.

Preservation of Testing Products. This subsection should describe the location of resulting test data, any revised test specifications, and any revised test data.

Test Cases/Items. This section should describe each of the test cases/items. Depending on system complexity, this could include

- Test case identifier
- Test sequence for interdependencies
- Items tested, indicating their version/revision level
- Specific test data
- Test conditions met
- Test procedure/script/task identifier used
- Test tool used
- Results/outputs summary of faults

Test Case Identifier. This subsection should describe the unique identifier of the test case, allowing others to find and repeat the same test case. For example, Test Case 051.

Test Sequence for Interdependencies. This subsection should describe where the sequence of this test case fits in the hiearcrchy or architecture of other test cases identifiers. The following is provided as an example:

Test Case 052 was the second of eight test cases following Test Cases 04x and preceeding Test Cases 06x.

Items Tested, Indicating their Version/Revision Level. This subsection should describe the test items/component tested by this test case. The following is provided as an example:

Subprogram079 v15   Config12345
Subprogram081 v10   Config12346

Specific Test Data. This subsection should describe the input data and states used by this test case (e.g., queries, transactions, batch files, references, software data structures, or invalid inputs). For example, Subprogram081, testfile123456.

Test Conditions Met. This subsection should describe the resulting test condition for the test case, where the test condition compares the values of two formulae. If the comparison proves true, then the condition is true; otherwise, it is false. Some typical conditions may be:

- Check for correct handling of erroneous inputs
- Check for maximum capacity
- User interaction behavior consistency
- Retrieving data
- Saving data
- Display screen and printing format consistency
- Measure time of reaction to user input

The following is provided as an example:

Subprogram081 True, created 256 byte testfile123457

Test Procedure/Script/Task identifier Used. This subsection should describe the tasks conducted for this test case. Also note if there are any constraints on the test procedures, or special procedural requirements, or unusual resources used. The following is provided as an example:

Task062   initiate Subprogram079, followed by Subprogram081

Test Tool Used. This subsection should describe the test case tool use from the test environment. The following is provided as an example:

Test Suite Alpha, Test Driver A1 of Test Harness A

Results/Outputs/Summary of Faults. This subsection should describe the results of using this test case. For each failure, have the failure analyzed and record the fault information:

- Case 1: Test Specification or Test Data
- Case 2: Test Procedure Execution
- Case 3: Test Environment

- Case 4: Unit Implementation
- Case 5: Unit Design

The following is provided as an example:

Task062   revised to v8 (Case 2)
Testfile123456   revised to v3 (Case 1)
Subprogram081   revised to v8, v9, & v10 (Case 4)]

Evaluation. This section contains the following subsections.

Termination Requirements Met. This subsection should provide a list of the termination requirements met by the unit testing, starting with a summary evaluation of the test cases/items.

Test Effectiveness. This subsection should identify the areas (for example, features, procedures, states, functions, data characteristics, and instructions) covered by the unit test set and the degree of coverage required for each area. Coverage can be stated as a percentage (1 to 100%) for the major objective units:
- Code
- Instructions
- Branches
- Path
- Predicate
- Boundary value
- Features
- Procedures
- States
- Functions

Variances from Requirement # and/or Design #. This subsection should report any variances of the test items from their design and/or requirements specifications. If this occurred, specify the:

- Reason for each variance
- Identification of defects that are not efficiently identified during previous peer reviews
- Resulting test problem report (TPR) #s

Risks and Contingencies. This subsection should identify any unresolved incidents and summarize the test item limitations.

Summary of Activities. This section should provide actual measurements for calculation of both test efficiency and test productivity ratios:

- Hours to complete testing, not counting test suspension
- Hours for setup

- Hours of test environment use
- Size of testing scope

Notes. This section should provide any additional information not previously covered in the unit test report.

Approvals. This section should include the approvals specified in the Project Plan.

System Integration Test Report

This System Integration Test Report template is designed to facilitate the definition of processes and procedures relating to system integration test activities. This template was developed using IEEE Std 829, IEEE Standard for Software Test Documentation, and IEEE Std 1008, Software Unit Testing, which has been adapted to support Lean Six Sigma requirements.

The subsections under each section in Table A-16 are suggested items for inclusion in the system integration test report. It is important to note that significant additional information is available from IEEE Std 829-1998 and IEEE Std 1008, as the standards provide additional tutorial information. Table A-16 provides a sample document outline. An electronic version of this document, *System Integration Test Report.doc,* is provided on the companion CD-ROM.

Table A-16. System integration test report document outline

1. Introduction
2. Identification and Purpose
3. Approach/Strategy
4. Overview
5. Test Configuration
    5.1 Test Environment
    5.2 Test Identification
    5.3 Results/Outputs/Summary of Faults
6. Test Schedules
7. Risk Management/Requirements Traceability
8. Notes
9. Approvals

System Integration Test Report Document Guidance

Introduction. This section should provide a brief introductory overview of the project and related testing activities.

Identification and Purpose. This section should provide information that uniquely identifies the system and the associated testing effort. This can also be provided in the form of a unique project plan milestone identifier. The following text provides an example:

The responsible lead tester was [Tester Name], [responsible organization], Role [lead test team member], and system integration test was completed on [date/time of system integration test conclusion], as planned in the System Integration Test Plan.

This System Integration Test Report is to detail the system integration testing conducted for the [Project Name] ([Project Abbreviation]) Version [xx], Statement of Work, [date], Task Order [to number], Contract No. [Contract #], and amendments.

The goal of [project acronym] development is to [goal]. This system will allow [purpose].

Specifics regarding the implementation of these system modules are identified in the [Project Abbreviation] System Requirements Specification (SysRS) with line item descriptions in the accompanying Requirements Traceability Matrix (RTM) and references to the Interface Control Documentation (ICD) and the System Integration Test Plan (SysITP).

Approach/Strategy. This system integration test follows unit testing of all modules/components scheduled for integration, as planned according to the system integration test plan. The result was the creation of either a partial or complete system. These tests resulted in discovery of problems that arose from modules/components interactions and these problems were localized to specific components for correction and retesting. The system integration test included white-box testing, whose goal is to ensure that modules/components were constructed properly and do not contain any hidden weaknesses.

This section should summarize the deviations from the system integration test plan objectives, kinds (software, databases, hardware, COTS, and Prototype Usability), and approach (e.g., top-down, bottom-up, object-oriented, and/or interface testing).

Overview. This section should provide a summary of all system modules/components and system features tested. Any specific exclusion should be discussed. The following provides an example:

The list of individual program units (subprogram, object, package, or module) identifiers and their version/revision levels and configuration management item(s) # are:

Subprogram078 v19	Config12343
Subprogram079 v15	Config12345
Subprogram080 v13	Config12344
Subprogram081 v10	Config12346

This document describes the System Integration Test Report (SysITR) for the [Project Abbreviation] System. [Project Abbreviation] documentation related to this SysITR is:

Project Plan Milestone/Task #
SystemRequirements Specification (SRS),
System Requirements Traceability Matrix (RTM)
Test Problem Report (TPR) #
Change Enhancement Request (CER) #

This SysITR describes the deliverables to be generated in the system integration testing of the [Project Abbreviation]. This report complies with the procedures defined in the [Project Abbreviation] Software Configuration Management (SCM) Plan and the [Project Abbreviation] Software Quality Assurance (SQA) Plan. Any nonconformance to these plans will be documented as such in this SysITR.

This document is based on the [Project Abbreviation] System Integration Test Report template with tailoring appropriate to the processes associated with the creation of the [Project Abbreviation]. The information contained in this SysITR has been created for [Customer Name] and is to be considered "for official use only."

Test Configuration. This section contains the following three subsections.

Test Environment. This subsection should summarize the deviations from the System Integration Test Plan–System Integration Test Environment. The following are examples:

> The system integration tests were performed at [Company Name], [Site Location]. All integration testing were conducted in the development center [Room Number]. The deviations from the System Integration Test Plan for the system and software used in the testing of [Project Abbreviation] are asterisked below:
>
> [System and software List]
>
> Several errors in Test Suite Alpha, Test Driver A1 of Test Harness A were discovered and corrected, resulting in retesting at both unit and subsystem levels and one day longer test interval, but no additional risk.

Test Identification. This subsection should summarize the deviations from the System Integration Test Plan—Test Identification. The following is an example:

> Subsystem A integration was delayed due to late arrival of Module A12345. This resulted in resequencing Substem B integration testing. A one week longer test interval resulted, but no additional risk.

Results/Outputs/Summary of Faults. This section should describe the results of system integration testing. The following are examples:

> For the 35 failures analyzed and localized to specific components or the test environment, the fault information was:
>
> > Case 1: Test Specification or Test Data
> > Case 2: Test Procedure Execution
> > Case 3: Test Environment
> > Case 4: Component Implementation
> > Case 5: Component Design
>
> The seven above Components with defects and corrected version #s were:
>
> > Subprogram078 v19   Config12343
> > Subprogramxxx vxx   Configxxxxx
> > . . .
>
> According to the System Integration Test Plan measurement objectives, 85% of all test scripts passed on the first pass.

Test Schedules. This section should summarize the deviations from the System Integration Test Plan—Test Schedule.

Risk Management/Requirements Traceability. This section should identify all high-level risks encountered and the actions taken. The following provides an example:

> While all components tested and integrated, with only a two-week extended test interval, several changes are needed in the System Test Plan to fully complete the revised Requirements Traceability Matrix.

Summary of Activities. This section should provide actual measurements for calculation of both test efficiency and test productivity ratios:

- Hours to complete testing, not counting test suspension
- Hours for setup
- Hours test environment used
- Size of testing scope

Notes. This section should provide any additional information not previously covered in the system integration test report.

Approvals. This section should include the approvals specified in the project plan.

OPERATION

Product Packaging Information

The following example of a product description based upon the recommendations of IEEE Std 1465, IEEE Standard Adoption of ISO/IEC 12119(E)—Information Technology—Software Packages—Quality Requirements and Testing [33]. Table A-17 provides an example of a product description and describes the type of information that should be present in every product description.

Table A-17. Example product description

Product description sheet, SecurTracker Version 1.0

SecurTracker—An Automated Security Tool

What is SecurTracker?

SecurTracker is a *comprehensive* security management tool that incorporates the efforts of the security professionals, administrative staff, IS team, and management team to integrate a complete security solution.

Architectural Goals

SecurTracker will support the five architectural goals identified below:

1. Meet user requirements
2. Promote system longevity
3. Support incremental system implementation
4. Aid iterative system refinement
5. Provide mechanisms for system support and maintenance

Why Use SecurTracker?

SecurTracker is the *only* application available that supports collateral, SCI, and SAR environments.

SecurTracker is a multiuser application that maintains all security management data in one centralized location. Centralization offers numerous significant advantages, including the elimination of duplicate records storage, reduction of risk related to inconsistent data, and a decrease in the amount of storage space (hardware costs) required to maintain security records.

SecurTracker was written using security professional expertise. Software development teams who read security-operating manuals to determine software requirements write other competitive security management tools. This approach inevitably omits the dynamic, real-world situations that cannot be captured in an operating manual.

SecurTracker is an interactive web-based solution. Client setup and maintenance is not required. Setup, configurations, and updates are

Table A-17. (continued)

accomplished at the server level, thus eliminating costly and time-consuming system maintenance.

SecurTracker Implementation Approach

SecurTracker is available as a single-user stand-alone system or as a multiuser web-based product. The user can accomplish the stand-alone installation. The installation process for multiuser implementation offers a two-phase approach; the first phase would consist of the database and web server setup; the second phase would include data migration, validation, and testing.

The implementation team will begin by extracting the security data from the legacy system(s). Once this is accomplished, the team will then extract and incorporate the data into SecurTracker. Data extraction can be accomplished remotely, thus eliminating travel cost.

After data extraction is accomplished, the SecurTracker system will be tested to verify the accuracy of the data conversion. Any necessary adjustment to the data import functions will be made at this time. Once data has been successfully extracted from one legacy system, the extraction process can be automated for other like systems, creating an efficient, repeatable process.

During the implementation phase, the implementation team will work closely with the onsite team lead and functional experts to determine an approach that ensures data integrity during migration. Final deployment includes user setup and maintenance item customization.

Data conversion, user training, and phone support are available on all solutions!

Points of Contact

Susan K. Land
SecurTracker Project Manager
(256) xxx—xxxx
e-mail address

System Requirements

*Database Server**
1 GB RAM; 60 GB hard drive; 1 GHz processer; network card, Windows NT 4.0 (SP6a)/Windows 2000 Server (SP3); Microsoft Excel; Oracle 8.1.6 or Oracle 9i.

*Web Server**
256 MB RAM; 20 GB hard drive; 700 MHz processor; network card; Windows NT 4.0 (SP6a)/Windows 2000 Server (SP3 and security patches); Oracle Forms and Reports 6i; Outlook Express Version 5.0/Microsoft Outlook 4.0 (with latest SP).

Web Client Machines
A Web Client machine can be any Internet-connected system that utilizes a browser; Video card supporting 256 colors, 800 x 600 resolution (or greater); Network Card.

*Although it is highly recommended that the web and database servers run on separate machines, a web/database server can operate on one machine. Using this option, the hardware requirements listed for the database server must be in place. Additionally, *all* software listed for the database *and* Web Server specifications must be installed on this server.

Stand-alone Configuration
512 MB RAM; 20 GB hard drive; 700 MHz processor; Windows NT 4.0 (SP6a/Windows 2000 (SP3); Personal Oracle 8.1.6 or 9i; Outlook Express 5.0/Microsoft Outlook 4.0; IE 5.5.

Frequently Asked Questions

Q: What languages/technologies were used in development?
A: SecurTracker is currently written in Oracle Forms and Reports 6i using an Oracle 9i database.

Q: Can the database be migrated to other versions of Oracle?
A: Yes. SecurTracker supports Oracle Standard and Enterprise editions of Oracle 8.1.6 or 9i.

Q: How is reporting implemented?
A: Reporting to web, printer, and e-mail are implemented by Oracle Reports and leverages .PDF format.

Q: What is the hard drive footprint for the web server?
A: 400 MB, which includes the application code.

(continued)

Table A-17. *(continued)*

Q: What is the hard drive footprint for the database server?
A: This depends upon the amount of space allocated for data. Initial setup is estimated at 500 MB.

Q: How is the standalone version different from the web version?
A: The standalone version will use Personal Oracle, which supports a single user. The web

version uses Oracle 8i or 9i, which will support any number of licensed users.

Q: Is there a limit on the number of records the system can handle?
A: The only limits occur due to the size of the hard drive; therefore, by adding more physical space as needed, the number of records is limitless.

MAINTENANCE

Change Enhancement Requests

The identification and tracking of application requirements is accomplished through some type of change enhancement request (CER). A CER is used to control changes to all documents and software under SCM control and for documents and software that have been released to SCM. Ideally, the CER is an online form that the software developers use to submit changes to software and documents to the SCM lead for updating the product. Table A-18 provides an example of the types of data that may be captured and tracked in support of application development or modification.

Table A-18. Typical elements in a CER

Element	Values
CER#	Unique CER identifier assigned by CRC.
Type	One of the following:
	BCR—Baseline Change Request; CER indicating additional or changed requirement
	SPR—Software Problem Report; CER indicating software problem identified externally during or after BETA test
	ITR—Internal Test Report; CER indicating software problem identified internally through validation procedures
	DOC—Documentation Change; CER indicting change to software documentation
Status	One of the following:
	OPEN—CER in queue for work assignment
	TESTING—CER in validation test
	WORKING—CER assignment for implementation as SCR(s)
	VOIDED—CER deemed not appropriate for software release
	HOLD—CER status to be determined
	FIXED—CER incorporation into software complete
Category	One of the following:
	DATA—Problem resulting from inaccurate data processing
	DOC—Documentation inaccurate
	REQT—Problem caused by inaccurate requirement

Table A-18. *(continued)*

Element	Values
Priority	One of the following: 1. Highest priority; indicates software crash with no work-around 2. Cannot perform required functionality; available work-around 3. Lowest priority; not critical to software performance 0. CER has no established priority
Date Submitted	The date CER initiated; defaults with current date
Date Closed	The date CER passes module test
Originator	Identifies the source of CER
Problem Description	Complete and detailed description of the problem, enhancement, or requirement
Short Title	Short title summarizing CER
SRS Ref#	Cross-reference to associated software requirements specification (SRS) paragraph identifier
SDD Ref#	Cross-reference to associated software design document (SDD) paragraph identifier(s)
STP Ref #	Cross-reference to associated software test plan (STP) paragraph identifier(s)
Targeted Version	Software module descriptor and version number; determined by SCM lead
File	Files affected by CER
Functions Affected	List of functions affected by CER
Time Estimated	Estimated hours until implementation complete
Time Actual	Actual hours to complete CER implementation
Module	Name of module affected by CER
Prob. Description	Description of original CER; add comments regarding implementation
Release	Release(s) affected by CER
CER#	Cross-reference number used to associate item with other relevant CERs
Software Engineering Name	Last name of engineer implementing CER

Baseline Change Request

The tracking of changes to the development baseline may be accomplished using a baseline change request (BCR). A BCR is a variation of the basic CER that is used to control changes to all documents and software under SCM control and for documents and software that have been released to SCM. Ideally, the BCR is an on-line form that the software developers use to submit changes to software and documents to the SCM lead for updating the product. Figure A-2 provides an example of the types of data that may be captured and tracked in support of application development or modification. *Baseline Change Request.doc,* on the companion CD-ROM, provides an electronic version of this BCR.

Originator		Module:		Priority:	
Name:		Title:		BCR No.:	
Organization:		SRS Revision:		Date:	
BCR Title:					
Change Summary:					
Reason For Change:					
Impact if Change Not Made:					
Org	Name	Yes	No	Desired Applicability	
				Release Available:	
				Release Unsupported:	
				CCB Approval	
				Name:	
				Date:	
				Signature:	
				_____ Approved	
				_____Disapproved	
				Comments:	

Figure A-2. Baseline change request.

Work Breakdown Structure for Postdevelopment

The example provided in Table A-19 provides suggested content for a WBS in support of the postdevelopment stage of a software development life cycle. This WBS content is illustrative only and should be customized to meet specific project requirements.

Table A-19. WBS content for postdevelopment activity

Product Installation
    Distribute product (software)
        Package and distribute product
        Distribute installation information
        Conduct integration test for product
        Conduct regression test for product
        Conduct user acceptance test for software
        Conduct reviews for product
        Perform configuration control for product
        Implement documentation for product
    Install product (software)
        Install product (packaged software) and database data
        Install any related hardware for the product
        Document installation problems
    Accept Product (software) in operations environment
        Compare installed product (software) to acceptance criteria
        Conduct reviews for installed product (software)
        Perform configuration control for installed product (software)
Operations and Support
    Utilize installed software system
    Monitor performance
    Identify anomalies
    Produce operations log
    Conduct reviews for operations logs
    Perform configuration control for operations logs
Provide Technical Assistance and Consulting
    Provide response to technical questions or problems
    Log problems
Maintain Support Request Logs
    Record support requests
    Record anomalies
    Conduct reviews for support request logs
Maintenance
    Identify product (software) improvements needs
        Identify product improvements
        Develop corrective/perfective strategies
        Produce product (software) improvement recommendations
    Implement problem reporting method
        Analyze reported problems
        Produce report log
        Produce enhancement problem reported information
        Produce corrective problem reported information
        Perform configuration control for reported information
    Reapply software life cycle methodology

Configuration Management

Configuration Control Board (CCB) Letter of Authorization. Figure A-3 is provided as an example of a formal letter of CCB authorization. This letter may be used to help officially sanction CCB groups. An electronic version of this letter, entitled *CCB Letter of Authorization.doc,* is provided on the companion CD-ROM.

MEMORANDUM FOR [Approving Authority]

DATE: [Date]

FROM: [Company Name]
         [Program Name]
         [Name, Program Manager]
         [Name, Project Manager

SUBJECT: Establishment of [Project Name] Configuration Control Board (CCB)

1. This is to formally propose the charter of the [Project Name] Configuration Control Board (CCB).
2. It is our desire that the [Project Name] CCB be chartered (Attachment no. 1) with the authority to act as the management organization responsible for ensuring the documentation and control of [Project Name] software development by ensuring proper establishment, documentation, and tracking of system requirements. In addition, the CCB will serve as the forum to coordinate software changes between represented agencies and assess the impacts caused by requirements changes.
3. Those recommended to participate in CCB voting will be [Approving Authority], including representatives from the following:

   Management Board:
   [Organization Name]

   Beta Test Site [Organization Name] Representatives:
   [List of sites]

   Contractor Associates:
   [Company Name]
   [Subcontractor Name]

4. The CCB will be the necessary "single voice," providing clear direction during [Project Name] requirements definition, product design, and system development. It is our goal to produce the highest quality product for the [Project Name] user. It is our desire that the [Project Name] CCB be chartered to provide [Project Name] development with consolidated and coordinated program direction.

Figure A-3. CCB letter of authorization.

Configuration Control Board Charter. The following is provided as the suggested example content and format of a configuration control board (CCB) charter. This charter may be used to establish project oversight, control, and responsibility. An electronic version of this charter example, entitled *CCB Charter.doc,* is provided on the companion CD-ROM.

Scope. This configuration control board (CCB) charter establishes guidelines for those participating in the [Project Name] CCB process. Any questions regarding this document may be directed to [Chartering Organization/Customer].

CCB Overview. The [Project Name] CCB is the management organization responsible for ensuring the documentation and control of [Project Name] software development by ensuring proper establishment, documentation, and tracking of system requirements. In addition, the CCB serves as the forum to coordinate software changes between represented agencies and assess the impacts caused by these changes.

CCB Meetings. The [Project Name] CCB will be conducted quarterly and will include the following agenda items:

1. Recommended changes to the CCB Charter
2. (Project Name) schedule
3. Discussion of proposed Baseline Change Requests (BCRs) for approval/disapproval
4. Review of new BCRs
5. Status of open action items
6. Review of new action items
7. Program risk assessment

Meeting dates and times will be determined by CCB consensus and announced at least 60 days prior to any scheduled meeting. Reminders will be distributed via DoD Messages, electronic mail, memoranda, and/or bulletin board announcements.

Formal minutes will be available and distributed within 30 days of the meeting's conclusion.

A CCB report will be prepared and maintained by the CCB chairman, listing CCB membership; individual representatives; interfacing modules, subsystems, or segments; pending or approved BCRs; status of software development activities such as open action items and risk tracking; and any other identified pertinent [Project Name] program information. Figure A-4 (on next page) provides an example of a baseline change request (BCR).

Participation. CCB participation is open to all government and contractor organizations that develop or use [Project Name] associated systems or [Project Name] elements. Participants belong to one of five categories:

1. Management Board (MB)
2. Chairman
3. Board Members (Command Representatives)
4. Implementing Members
5. Associate Members
6. Advisors

Discussions concerning [Project Name] BCRs will be limited to the organizations responsible for the implementation of the BCRs. If other organizations are interested in the operational usage of the BCRs, these issues should be brought to the attention of the Board Members.

Participants. The MB is composed of the following participants: (this should be three or less people at a very high level with the ability to make command decisions for the CCB). The purpose of the MB is to address software development problems that cannot be resolved at the CCB Board level. The MB will be chaired by _____ and co-chaired by _____.

Chairman _____ is responsible for establishing and running the CCBs in accordance with this plan, which includes the following:

Originator		Module:		
Name:		Title:		BCR No.:
Organization:		SRS Revision:		Date:
BCR Title:				
Change Summary:				
Reason For Change:				
Impact if Change Not Made:				
Org	Name	Yes	No	Desired Applicability
				Release Available:
				Release Unsupported:
				CCB Approval
				Name:
				Date:
				Signature:
				_____ Approved
				_____Disapproved
				Comments:

Figure A-4. Example baseline change request (BCR).

1. Leading the effort to document and control all module items associated with HIMS
2. Ensuring that a proper forum exists for coordinating BCRs and selecting the solution that is in the best interest of the government
3. Ensuring minutes and action items are recorded and distributed

The Chairman has the final approval authority for all BCRs.

Board Members. [Government organizations] are responsible for the following:

1. Reviewing BCRs to ensure they support operational requirements
2. Voting to approve/disapprove BCRs

Implementing Members. [Government organizations on contract] are responsible for the following:

1. Implementing BCRs
2. Reviewing BCRs for adequacy and program impact (cost or schedule)

Associate Members. [Government organizations not on contract] are responsible for BCRs for adequacy.

Advisors. [Government or contractor agencies invited to a specific CCB to support discussion of a specific BCR] are responsible for presenting information relevant to a BCR when requested by the sponsoring Operational, Implementing, or Associate Member.

CCB Process

General. All CCB actions are coordinated and take place through the BCR process. Any CCB Participant may propose BCRs, but proprietary BCRs will not be accepted or discussed. It is important to understand that [Project Name] technical interchange meetings (TIMs) provide the forum to brainstorm and discuss preliminary BCR ideas prior to the formal process described in this document. The CCB acts only as the final review and adjudication for BCRs.

Software Requirements Specifications (SRSs). [Project Name] software requirements are brought under CCB control in the BCR process. A BCR describing the request for a software requirements baseline is generated and distributed at the CCB. The BCR is reviewed using the accepted practices and, when approved, the SRS is accepted for CCB management.

BCR Format. BCRs are prepared by generating "is" pages and attaching those to a BCR cover sheet in the document. "Is" pages are change pages indicating how the page should read. A BCR request form is provided at the end of this document for reference.

BCR Process. BCRs can be coordinated using any of the following methods:

1. BCRs will be provided to _____ at least thirty (30) days prior to a CCB meeting so they may be reproduced and included in the agenda. Please fax to: (the BCRs should be sent to the Chair and distributed to all participants).
2. The Board, Implementing, or Associate Member sponsoring the BCR will be provided the opportunity to present a five-minute overview of why the change is needed and what is being changed.
3. Any Board, Implementing, or Associate Member who is potentially impacted by the BCR will identify himself/herself as a reviewing member for the BCR.
4. Reviewing members will submit their concurrence or nonconcurrence to the sponsoring member and Chair. (Silence is considered concurrence.) If a reviewing

member nonconcurs, any issues associated with implementing the BCR, including alternate solutions, will be documented and sent to the sponsoring member.

5. If issues are identified, the sponsoring member will work with all reviewing members to try and resolve conflicts prior to discussion at the CCB.

6. At the CCB, each potential BCR is reviewed for concurrence/nonconcurrence. BCRs that all reviewing agencies concur upon will be signed by the Board members and approved by the Chairman. These BCRs will be prioritized and scheduled for implementation.

7. BCRs that have received nonconcurs go through a formal review. The formal review consists of each reviewing member presenting a ten-minute presentation explaining their concurrence or nonconcurrence with the BCR, including their preferred approach and contractual/cost/schedule impacts.

8. The Chairman will decide whether to approve, disapprove, or defer the BCR. If the BCR is deferred, the process is repeated again starting with method 4 above and the reviewing agencies will use this time to attempt to reach concurrence.

Accelerated Process. Emergency BCRs are handled as follows:

1. BCRs will be provided to _____ and all affected CCB members at least 45 days prior to the CCB. Reviewing members will submit their concurrence or nonconcurrence to the sponsoring member and Chair. (Silence is considered concurrence.)

2. At the CCB, each potential BCR is reviewed for concurrence/nonconcurrence. BCRs that all reviewing agencies concur upon will be signed by Implementing Members and approved by the Chairman.

3. BCRs that have received nonconcurs will go through a formal review. The formal review consists of each reviewing member presenting a ten-minute presentation explaining why they concur or nonconcur with the BCR, including their preferred approach and contractual/cost/schedule impacts.

4. The Chairman will decide whether to approve, disapprove, or defer the BCR. If the BCR is deferred, the process is repeated starting with method 4 in the BCR unaccelerated process above. The reviewing agencies use this time to attempt to reach concurrence.

Contractual Direction. After a decision has been made by the CCB, all Implementing Members affected by the change will take the contractual action needed to ensure that the change is implemented.

Software Change Request Procedures

A change to software may be requested by one of the following change enhancement requests (CERs): Internal Test Report (ITR), Software Problem Report (SPR), or Baseline Change Request (BCR) from external or internal sources, or from the product specification/requirements. A Document Change Request (DCR) may be used to request a change to documentation. All of the above requested change procedures are referred to as a change request in the text below. These procedures are outlined below (Figure A-5) for [Project Abbreviation] maintenance and production life cycles.

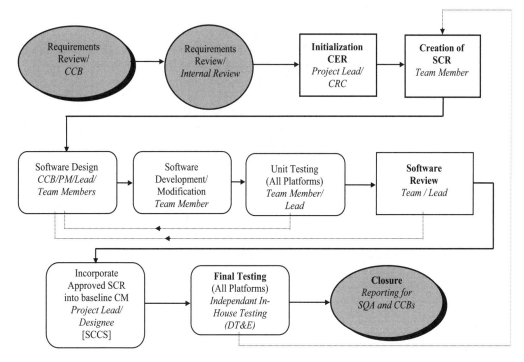

Figure A-5. Software baseline change process.

The Configuration Control Board (CCB) for each development team determines which of the change requests is required for each software release prior to the start of work for that release. Additional change requests are reviewed by the [Project Abbreviation] CCB to determine and assign the proper status to the change requests; these are held for CCB scheduling. Status is one of the following:

Open	Change request is to be implemented for current software version
Hold	Change request targeted for another software version
Voided	Change request is a duplicate of an existing CER or does not apply to existing software
Working	Change is currently being implemented
Testing	Implemented change request is under DT&E evaluation
Fixed	Change request is implemented, unit and integration test complete

The following steps define the procedure for each status. Each procedure starts with the change request being given to the Project Lead or designated Change Request Coordinator (CRC).

A. Open
   1. INITIALIZATION
     a) The CRC for the associated project creates a CER.
     b) The CRC notifies the Project Lead (PL) of the new CER.

2. CREATE SCR
   a) The PL reviews the CER and provides estimates of schedule impact to CCB.
   b) CCB prioritizes the CER and authorizes implementation.
   c) The PL assigns the CER to a Software Engineer (SE) to incorporate.
   d) The SE creates a Software Change Request (SCR).
3. SOFTWARE DESIGN
   The SE designs changes necessary to implement the CER:
   a) The SE identifies interface changes and consults with appropriate PLs.
   b) The SE determines changes needed.
   c) The SE identifies any changes needed for documentation.
   d) The SE estimates the work effort required for CER completion.
4. SOFTWARE DEVELOPMENT/MODIFICATION
   a) The SE makes needed changes and provides updated documentation for testing.
   b) The SE checks out the files from the configuration managed version (not the baselined version) of all the files needed to incorporate the CER.
   c) The SE fills in the CM sections of the SCR.
   d) The SE puts the changes in the configuration managed files.
5. UNIT TESTING
   The SE tests on all platforms; if a test fails go back to Step 3, Software Design.
6. SOFTWARE REVIEW
   a) The SE prepares a software review package:
      1) Hard copy of SCR.
      2) List of documentation changes.
   b) Software Review
      1) SE gives software review package to another team member for review.
      2) If SCR peer approved go to step 7.
      3) SCR not approved go to step 3, Software Design.
7. INCORPORATE APPROVED SCR INTO BASELINE CM
   a) The PL changes the SCR status to indicate that it is approved.
   b) The SE checks in the files.
   c) The SE fills in the CM sections of the SCR.
   d) The CRC reviews SCR for completeness.
   e) The CRC changes the status on the CER to indicate testing.
   f) The designated SCM gives a list of changed files to PLs for updating their working directories/executables.
   g) The designated SCM installs changes to baseline on all platforms.
8. FINAL TESTING
   a) The CRC identifies all CERs with testing status.
   b) Final testing by DT&E validation group. (If problems are found start over at Step 2a.)
9. CLOSURE
   The CRC changes final status on CER to indicate completion.

B. Hold

The CRC creates a CER marking status HOLD and indicates the targeted version for incorporating the change. The CRC files the original change request for the next version. The change will be considered for inclusion in the next version.

C. Voided

1. Duplicates:

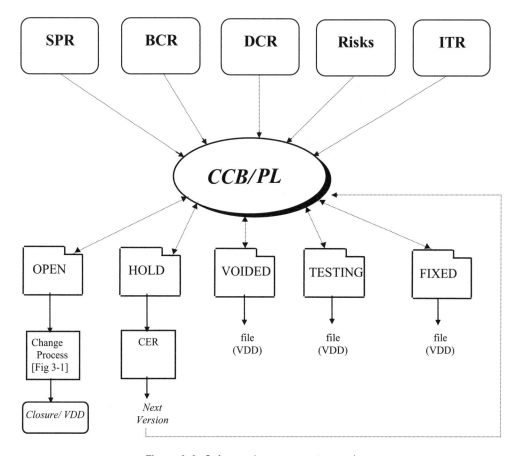

Figure A-6. Software change request procedures.

    a) The CRC marks the change requests as a duplicate, specifying which CER is duplicated.

    b) The CRC files the change request for the associated version.

  2. Other:

    a) The CRC marks the reason for cancellation.

    b) The CRC files the change request for associated version.

D. Testing

The CRC forwards the CER/SCR to PL for review. After review, the PL forwards the CER to DT&E test for validation. CER status is updated to indicate testing status.

E. Fixed

The CRC files the closed change request and updates the database. Figure A-6 provides an overview of the software change request procedures.

Version Description Document

The version description document (VDD) is the primary configuration control document used to track and control versions of software to be released to the operational environment. It is a summary of the features and contents for the software build. It identifies and

Table A-20. Version description document outline

INTRODUCTION
Roles and Responsibilities
Scope
Definitions and Acronyms
Key acronyms
Key terms
References
SYSTEM OVERVIEW
DOCUMENT OVERVIEW
ROLES AND RESPONSIBILITIES
VERSION DESCRIPTION
Inventory of Materials Released
Inventory of Software Contents
Changes Installed
Interface Compatibility
Installation Instructions
Possible Problems and Known Errors

describes the version of the software configuration item (CI) being delivered, including all changes to the software CI since the last VDD was issued. Every unique release of the software (including the initial release) should be described by a VDD. If multiple forms of the software CI are released at approximately the same time (i.e., to different sites) each must have a unique version number and a VDD.

The VDD is part of the software CI product baseline. The VDD, when distributed, should be sent with a cover memo that summarizes, on a single page, the significant changes that are included in the release. This will serve as an executive summary for the details found in the attached VDD.

The example provided in Table A-20 provides suggested content for a VDD in support of the postdevelopment stage of a software development life cycle. This VDD content is illustrative only and should be customized to meet specific project requirements. An electronic version of this VDD, entitled *Version Description Document.doc,* is provided on the companion CD-ROM.

QUALITY ASSURANCE

Example Life Cycle

A full development life cycle for a new software product and associated recommended software quality assurance activities is shown in Figure A-7. This information may be described in this SQA plan or in the associated SPMP. Steps 0 through 12 in Figure A-7 are described below and are summarized in Table A-21.

Step 0. Tasking is the process of receiving a statement of work (SOW) from the customer. It is the official requirement to which all software development efforts must be directed. The customer receives requirements from its associated Configuration Control Board (CCB).

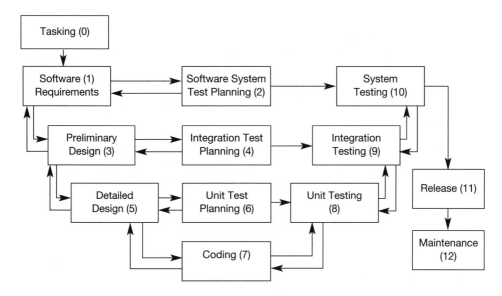

Figure A-7. Example life cycle and associated SQA activities.

Table A-21. Software development life cycle—responsibility

Step	Life Cycle Activity	Review by	Product
(0)	Tasking	CCB	Work Request
			ICWG Minutes
(1)	Software Requirements	[Customer Name], PM, PL, CCB	SRS
			SRS Review
(2)	Software System Test Planning	[Customer Name], PM, PL, CCB	Software System Test Plan
			Software System Test Plan Review
	Software Project Plan	[Customer Name], PM, PL, CCB	Software Project Plan
(3)	Preliminary Design	[Customer Name], PM, PL, CCB	DRAFT Design Document
			DRAFT Design Document Review
(4)	Integration Test Plannng	[Customer Name], PM, PL, CCB	Software Test Plan
			Software Test Plan Review
(5)	Detailed Design	[Customer Name], PM, PL, CCB	Software Project Plan
			SPP Review
(7)	Coding	PL, SQAM	Audit Documentation
		PL, SQAM	Walk-through Documentation
(8–10)	Unit, Integration, System Testing	[Customer Name], PM, PL, CCB	Software System Test Plan
			Software System Test Plan
			Review Documentation
(11)	Release	Project Lead	Baseline Release to Configuration Control
			Inspection Report
(12)	Maintenance	[Customer Name], PM	Maintenance Plan, Transition Plan

Step 1. Software requirements define the problem to be solved by analyzing the customer requirements of the solution in a software requirements specification (SRS). The Requirements portion of the life cycle is "the period of time in the software life cycle during which the requirements for a software product are defined and documented" [9].

Customer requirements should be in the form of a statement of work, tasking statement, or software system specification (SSS).

The associated Project Lead, Program Manager, and any associated CCB will review the SRS for technical content and clarity. The SRS will be signed by the [project abbreviation] Program Manager and a [customer name] representative after joint review. This constitutes their agreement on what is to be produced. The Project Lead, or designee, will review the SRS for adherence to the documentation template provided in Appendix [X] of the [project abbreviation] Software Requirements Management Plan. The SQA Lead will ensure that this review has been completed.

Software project/development planning follows software requirements in the software development life cycle. Software project/development planning requires the documentation of a proposed project's schedule, resources and manpower estimates. This Software project plan will be reviewed for technical content and clarity by the associated Project Lead (PL), Program Manager, Senior [company name] Management, and if appropriate, the associated CCB. The customer will also participate in plan review and authorize its implementation.

The PL, or a designee, will also review the software project plan for adherence to the documentation template provided for the [company name] software project planning plan and the checklist provided in section [X]. The [project abbreviation] Program Manger and a representative from [customer name] will review actual progress against existing plans to ensure tracking efficiency. These reviews will occur no less than bi-monthly. The PM will report progress tracking to [company name] senior management monthly.

Note: It is important to carefully consider document review coordination time and repair time when project planning.

Step 2. Software system test planning follows software requirements development in defining how the final product will be tested and what constitutes acceptable results. This step provides the basis for both acceptance and system testing. Issues identified during software test planning [also referred to as developmental test and evaluation (DT&E) test planning] may result in changes to the software requirements. A highly desirable by-product of system test planning is the further refinement of the software requirements. Errors found early in the software development process are less costly and easier to correct.

The software test plan will be reviewed for technical content and clarity by the associated Project Lead, Program Manager, and, if appropriate, the associated CCB. The customer will determine, and [company name] will agree to, test tolerances. The Project Lead, or a designee, will review the test plan to ensure that all requirements items identified in the SRS are also identified in the software test plan.

Step 3. Preliminary design breaks the specified problem into manageable components creating the architecture of the system to be built. The architecture in this step is refined until the lowest level of software components or CSUs are created. The CSUs will represent the algorithms to be used in fulfilling the customer's needs. The preliminary design will be documented as a draft software design document.

The PL will review any existing test plan to ensure that all requirements identified in the SRS are also identified in the software design document. The SQA Lead will ensure that this review has been completed.

Step 4. Integration test planning follows preliminary design by defining how the components of any related system architecture will be brought back together and successfully integrated. Issues identified during this step will help to define the ordering of component development. Other issues identified during this step include recognition of interfaces that may be difficult to implement. This saves on the final cost by allowing for early detection of a problem. Integration testing will be included as part of the software (DT&E) test plan.

This plan will be reviewed for technical content and clarity by the associated Project Lead, Program Manager, and, if appropriate, the CCB. The customer will determine and [company name] will agree to test tolerances. The Project Lead, or a designee will review the test plan to ensure that all requirements identified in the SDD are also identified in the software system test plan. The SQA Lead will ensure that this review has been completed.

Step 5. Detailed design defines the algorithms used within each of the CSUs. The design is "the period of time in the software life cycle during which the designs for architecture, software components, interfaces, and data are created, documented, and verified to satisfy requirements" [9].

The detailed design will be documented as the final version of the software design document (SDD). The software design document will be reviewed for technical content and clarity by the associated Project Lead, Program Manager, and, if appropriate, the [project abbreviation] CCB. The customer will also participate in the SDD review and authorize its implementation.

Step 6. Unit test planning follows the detailed design by creating the test plan for the individual CSUs. Of primary importance in this step is logical correctness of algorithms and correct handling of problems such as bad data. As in all other test planning steps, feedback to detailed design is essential. Unit testing will be included as part of the software (DT&E) test plan.

The associated Project Lead, Program Manager, and the CCB will review this plan for technical content and clarity. The customer will determine and [company name] will agree to test tolerances. The PL will review the test plan to ensure that all requirements items identified in the SRS are also identified in the software (DT&E) test plan. The SQA Lead will ensure that this review has been completed.

Step 7. Coding is the activity of implementing the system in a machine executable form. Implementation is "the period of time in the software life cycle during which a software product is created from documentation and debugged" [9].

The PL, or a designee, using existing coding standards to ensure compliance with design and requirements specifications, will review the implemented code. Periodic program SQA reviews will be conducted to ensure that code reviews are being properly implemented. The results of these reviews will comprise a "lessons learned" document. Results will also be reported to senior [company name] management.

Step 8. Unit testing executes the tests created for each CSU. This level of test will focus on the algorithms with equal emphasis on both "white" and "black" box test cases. White-box testing is testing that is based on an understanding of the internal workings of algorithms. Black-box testing focuses on final results and integrated software operation.

The PL or a designee will "walk through" software test procedures, ensuring that results are recorded, procedures are documented and based on the evaluation of documented requirements.

Step 9. Integration testing executes the integration strategy by combining CSUs into higher-level components in an orderly fashion until a full system is available. This level of test will focus on the interface.

The associated software/DT&E test plan will document system testing procedures and standards.

Step 10. System testing executes the entire system to ensure the requirements have been satisfied completely and accurately. This level of testing will focus on functionality and is primarily black-box testing. Black-box testing is focused on the functionality of the product (determined by the requirements); therefore, testing is based on input/output knowledge rather than processing and internal decisions.

The associated software/DT&E test plan will document integration testing procedures and standards.

Step 11. Release is the entry of the new or upgraded software into configuration management and the replacement of the previous executable code with the upgraded executable code as appropriate.

Step 12. Maintenance is the receipt of and response to problems encountered by the user and upgrades requested by the customer. Each problem or upgrade will spawn a new SQA process. (Refer to Figure A-8.)

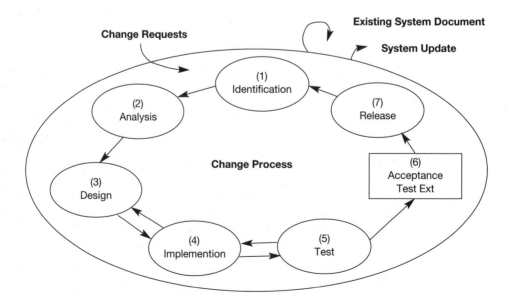

Figure A-8. Software maintenance life cycle.

Table A-22. Software maintenance life cycle products

Step	Life Cycle Activity	Review by	Product
(1)	Identification	PL, PM	CER categorization
			SCM Audit
(2)	Analysis	PL, PM, CCB, [Customer Name]	CER scheduling
			Functional Test Review
(3)	Design	PL, PM, CCB, [Customer Name]	Design Review
(4)	Implementation	PL, PM, SQAM	Walk-through Report
			Audit Documentation
(5)	Test	PL, PM, SQAM	Test Report
			Audit Documentation
(7)	Release	PL, PM	Staff Meeting Minutes

The maintenance life cycle of [project abbreviation] software consists of the steps shown in Figure A-8.

The software change process begins with the creation of a change request and is controlled by the software configuration management process for each program. For this discussion, change enhancement requests (CERs) will include software problem reports (SPRs), baseline change requests (BCRs), internal test reports (ITRs), and interface change notices (ICNs). These forms are shown in the [project abbreviation] software configuration management plan.

The steps in the software maintenance process are described below and are summarized in Table A-22.

Step 1. Identification and categorization of the CER. This step ensures that the change request is unique (not previously submitted), is valid (not the result of operator error or misunderstanding), and within the scope of the system. Categorization is concerned with the criticality of the change (urgent versus routine) and the type of change (corrective or enhancement).

The Project Lead assumes the initial responsibility for the determination of whether a CER is valid or invalid. Valid change requests require further analysis (Step 2). The SQAM will conduct periodic reviews of the software configuration management tracking procedures used for the identification, tracking, and prioritization of CERs.

Step 2. Analysis examines the source and reason for the change. The CERs will be reviewed for technical validity and project impact by the associated Project Lead, Program Manager, and, if appropriate, the associated CCB. The customer may also participate in CER review and authorize their implementation.

Corrective changes need to be analyzed to find the true source of the problem; enhancements need to be analyzed to determine the scope of the change.

Note: Enhancements are normally additions to the system that entail new or modified requirements. This is similar to starting at the top of Figure A-8. Part of this step is determining the change to functional testing that will be used upon completion of the change.

Those CERs identified as valid are prioritized and scheduled by the Project Lead, the Program Manager, the Customer, and associated CCBs.

Step 3. Design retrofit changes to the existing architecture and algorithms. As in development, the integration of changed CSUs and execution of algorithms must be planned early, to catch problems before testing occurs.

The SQAM will conduct periodic reviews to ensure that Project Lead reviews designs for technical appropriateness. Results of these reviews will be reported to the Program Manager.

Step 4. Implementation converts new solution into machine executable form. The implemented code will be reviewed using existing coding standards to determine compliance to design and requirements specifications. The code will be reviewed by the implementer and then by the PL, or a designee.

Periodic SQA reviews will be conducted to ensure that code reviews are being implemented. Results of these reviews will be reported to the Program Manager.

Step 5. Testing in this process is the combination of unit, integration, and system testing as described during development. This testing process is often called regression testing due to the need to modify and repeat testing previously performed. A critical issue at this point is the need for repeatability of testing. Changes are vital to the life of the system as is proper testing; therefore, the testing process should be developed for reuse.

The Project Leads are responsible for ensuring that all CERs scheduled for a production cycle are implemented in the software released for acceptance testing. The SQAM will conduct periodic reviews of the SCM procedures being implemented. Changes to the software baseline will be traced back to the initiating CERs reporting any anomalies.

Step 6. Acceptance testing is conducted to achieve the customer's agreement on completion of the change. Acceptance testing is controlled by the customer for [project abbreviation] software products, both maintenance and development.

Step 7. Release processes should be considered during identification of the change and follow an established pattern for the customer. Periodic releases of "builds" are easier to manage and provide more consistency; independent releases provide for quicker responses to changing demands. There are considerable trade-offs between the two extremes that should be negotiated with the customer to reach the best mix of speed and control.

The customer and/or associated determines both the "when" and "recipient" of a software release. The PM will, at a minimum, conduct biweekly staff reviews that will include review of release status and CER implementation progress. The SQA Lead will ensure that these reviews occur as described in the associated software project schedule.

Minimum Set of Software Reviews

Table A-23 presents a list of the recommended minimum set of software reviews as described in IEEE Std 730-2002, IEEE Standard for Software Quality Assurance Plans. The list of reviews provided here is not an absolute. Rather, these are provided as an example of what is contained in this standard. Each organization must determine its own recommended minimum set of software reviews.

Table A-23. Minimum set of software reviews

Software Specification Review
Architectural Design Review
Detailed Design Review
Verification and Validation Plan Review
Functional Audit
Physical Audit
In-Process Audit
Managerial Reviews
Software Configuration Management Plan Review
Postimplementation Review

SQA Inspection Log

An inspector uses the SQA inspection log form (Figure A-9) during his/her review of the inspection package materials. The recorder may also use it during an inspection. In preparing for an inspection, an inspector uses this form to identify all defects found. During an inspection, the recorder uses this form to record all official defects identified. An electronic version of this log form, entitled *SQA Inspection Log.doc,* is provided on the companion CD-ROM.

Log Number: _____		
Date:_____		
Software Item:_____		
Inspector: _____		
Moderator: _____		
Author: _____		
Time spent on review: _____		
Review Type: _____		
Location	Defect Description	Category

Figure A-9. SQA inspection log form.

Inspection Log Description

An explanation of the items in Figure A-9 follows:

Log Number. The log or tracking number assigned to this inspection:

Date. During an inspection, the recorder enters the date on which the inspection occurred. The Date block can be ignored by an inspector using this form during the preinspection review.

Software Item. The name of the software item being inspected.

Inspector. The Inspector block can be left blank when used by the recorder during an inspection. When using this form during a preinspection review, the inspector name is entered in this block.

Moderator. The moderator's name.

Author. The name of the author responsible for the inspection package is entered in this block.

Review Type. The type of inspection being conducted is entered in this block.

Category. A category is a two-character code that specifies the class and severity of a defect. A category has the format XY, where X is the severity (**Ma**jor/ m**I**nor) and Y is the class.

Severity:

Major = a defect that would cause a problem in program operation

m**I**nor = all other defects

Class:

M(issing) = required item is missing from software element

S(tandards Compliance) = nonconformance to existing standards

W(rong) = an error in a software element

E(xtra) = unneeded item is included in a software element

A(mbiguous) = an item of a software element is ambiguous

I(nconsistent) = an item of a software element is inconsistent

VERIFICATION

Inspection Log Defect Summary

The inspection log defect summary shown in Figure A-10 is provided as an example. This may used to record a summary of all defects identified during the inspection process. Once the inspection team has agreed to a disposition on the inspection, the leader should complete this form. An electronic copy of this log, *Inspection Log Defect Summary.doc,* is provided on the companion CD-ROM.

Log Number: _____   Date: _____
Software Item: _____
Moderator: _____   Review Type: _____

Defects found:

| | Major Defects | | | | | | | Minor Defects | | | | | | |
Insp.	M	S	W	E	A	I	TOTAL	M	S	W	E	A	I	TOTAL

Legend:

> Major = a defect that would cause a problem in program operation
> Minor = all other defects
> M(issing) = required item is missing from software element
> S(tandards Compliance) = nonconformance to existing standards
> W(rong) = an error in a software element
> E(xtra) = unneeded item is included in a software element
> A(mbiguous) = an item of a software element is ambiguous
> I(nconsistent) = an item of a software element is inconsistent

Figure A-10. Inspection log defect summary.

Inspection Log Defect Summary Description

An explanation of the items in Figure A-10 follows:

Log Number. The recorder will enter the log number assigned to this inspection.

Date. The recorder will enter the date of the inspection in this block.

Software Item. The recorder will enter the name of the software item that is being inspected.

Moderator. The name of the inspection moderator.

Inspection Type. The recorder will enter what type of inspection is being conducted in this block.

Defects Found. The recorder will record counts of all defects found during review of the inspection package. The defect counts will be recorded in the table as follows:

- The Insp. column will contain name of inspector
- The Major Defects columns (M, S, W, E, A, and I) will hold identifiers that correspond to the number of major defects found for a specific type in one of the six possible classes. As an example, if the "W" column held the number 3, then it would mean that there were three major defects found that were classed as Wrong.
- The first Total column will hold a total of all the major defects found for a type.
- The Minor Defects columns (M, S, W, E, A, and I) will hold numbers that correspond to the number of minor defects found for a specific type in one of the five possible classes. As an example, if the "A" column held the number 3, then it would mean that there were three minor defects found that were classed as Ambiguous.
- The second Total column will hold a total of all the minor defects found for a type.

Inspection Report

The inspection report form shown in Figure A-11 is provided as an example. This may

---

Log Number: _____   Date: _____
Software Item: _____
Inspection type: _____   Duration: _____
Number of inspectors: _____   Total prep time: _____
Size of materials inspected: _____
_____
_____

Disposition: Accept ____ Conditional ____ Reinspect ____

Estimated rework effort: _____ (Hours)
Rework to be completed by: _____
Reinspection scheduled for: _____

Inspectors:

_____ _____
_____ _____

Author(s):

_____ _____

Recorder: _____

Leader: _____

Leader certification: _____ Date: _____

Additional comments: _____

Figure A-11. Inspection report form.

used by the inspection leader to officially close the inspection. Once the inspection team has agreed to a disposition on an inspection, the inspection leader should complete this form. An electronic version of this report, *Inspection Report.doc,* is provided on the companion CD-ROM.

Inspection Report Description

An explanation of the items in Figure A-11 follows.

Log Number. The log or tracking number assigned to this inspection.

Date. The date of the inspection.

Software Item. The name of the software item being inspected.

Inspection Type. The type of inspection being conducted.

Duration. The duration of the inspection.

Number of Inspectors. The number of inspectors present during the inspection.

Total Prep Time. The total preparation time spent in preparation for the inspection.

Size of Materials. The size of the materials inspected.

Disposition. The disposition of the inspection.

Estimated Rework Effort. The estimated rework effort time, in hours.

Rework to Be Completed By. The estimated date of rework completion. If no rework is needed, enter "N/A."

Reinspection Scheduled For. The date for a reinspection of the software item to ensure rework completeness. It is the moderator's responsibility to decide if a reinspection is needed. If it is not needed, this block contains "N/A."

Inspectors. The names of the inspectors present at the inspection.

Author(s). The name(s) of the author(s) responsible for the software item.

Recorder. The name of the recorder present at the inspection.

Leader. The leader's name is entered in this block.

Leader Certification. The leader will sign and date this block to signal that the inspection is concluded.

Additional Comments. The leader will enter additional comments as necessary for clarification.

Requirements Walk-through Form

The form shown in Figure A-12 is provided as an example walkthrough checklist used in support of the peer review of software or system requirements. This type of form may be used to document issues and resolutions when performing project walk-throughs.

```
Software Project:

Software Item(s): _____

Review Date: _____

Reviewer(s):
_____   _____
_____   _____
_____   _____
_____   _____

Moderator, respond to the following:
                                                          Yes    No*
1. Is there a clear understanding of the problem that this system is designed   ____   ____
   to solve?
2. Are the external and internal interfaces properly defined?                   ____   ____
3. Are all the requirements traceable to the system level?                      ____   ____
4. Is prototyping conducted for the customer?                                   ____   ____
5. Is the performance achievable with constraints imposed by other system       ____   ____
   elements?
6. Are schedule, resources, and budget consistent with requirements?            ____   ____
7. Are validation criteria complete?                                            ____   ____

*If you answered "No" to any of the above, please attach problems found and suggestions to
 improve.

Please attach additional comments and a copy of requirement items.
```

Figure A-12. Requirements walk-through form.

Software Project Plan Walk-through Checklist

Figure A-13 is provided as an example walk-through checklist used in support of the peer review of a software project plan. This type of form may be used to document issues and resolutions when performing project walk-throughs.

Software Project:

Software Item(s): _____

Review Date: _____

Reviewer(s):

_____   _____
_____   _____
_____   _____
_____   _____

Moderator, please respond to the following:

	Yes	No*
1. Is the software scope unambiguously defined and bounded?	____	____
2. Is the terminology clear?	____	____
3. Are the resources adequate for scope?	____	____
4. Are the resources readily available?	____	____
5. Are the tasks properly defined and sequenced?	____	____
6. Is task parallelism reasonable given the available resources?	____	____
7. Have both historical productivity and quality data been used?	____	____
8. Have any differences in estimates been reconciled?	____	____
9. Are the preestablished budgets and deadlines realistic?	____	____
10. Is the schedule consistent?	____	____

*If you answered "No" to any of the above, please attach problems found and suggestions to improve.

Please attach additional comments and a copy of requirement items.

Figure A-13. Project plan walk-through checklist.

Preliminary Design Walk-through Checklist

Figure A-14 is provided as an example walkthrough checklist used in support of the peer review of preliminary design. This type of form may be used to document issues and resolutions when performing project walkthroughs.

```
Software Project:

Software Item(s): _____

Review Date: _____

Reviewer(s):

_____   _____
_____   _____
_____   _____
_____   _____

Moderator, respond to the following:
                                                          Yes      No*
1.  Are software requirements reflected in the software architecture?    ____    ____
2.  Is effective modularity achieved? Are modules functionally independent?    ____    ____
3.  Is the program architecture factored in?              ____    ____
4.  Are interfaces defined for modules and external system elements?    ____    ____
5.  Is the data structure consistent with the information domain?    ____    ____
6.  Is the data structure consistent with software requirements?    ____    ____
7.  Has maintainability been considered?                  ____    ____

*If you answered "No" to any of the above, please attach problems found and suggestions to
 improve.

 Please attach additional comments and a copy of requirement items.
```

Figure A-14. Preliminary design walk-through checklist.

Detailed Design Walk-through Checklist

Figure A-15 is provided as an example walk-through checklist used in support of the peer review of detailed design. This type of form may be used to document issues and resolutions when performing project walk-throughs.

Software Project:

Software Item(s): _____

Review Date: _____

Reviewer(s):

_____   _____

_____   _____

_____   _____

_____   _____

Moderator, please respond to the following:

	Yes	No*
1. Does the algorithm accomplish the desired function?	___	___
2. Is the algorithm logically correct?	___	___
3. Is the interface consistent with the architectural design?	___	___
4. Is logical complexity reasonable?	___	___
5. Has error handling been specified and built in?	___	___
6. Is the local data structure properly defined?	___	___
7. Are structured programming constructs used throughout?	___	___
8. Is the design detail amenable to the implementation language?	___	___
9. Which are used: operating system or language-dependent features?	___	___
10. Has maintainability been considered?	___	___

*If you answered "No" to any of the above, please attach problems found and suggestions to improve.

Please attach additional comments and a copy of requirement items.

Figure A-15. Design walk-through checklist.

Program Code Walk-through Checklist

Figure A-16 is provided as an example walk-through checklist used in support of the peer review of developed code. This type of form may be used to document issues and resolutions when performing project walk-throughs. *Note:* This checklist assumes that a design walk-through has already been done.

Software Project:

Software Item(s): _____

Review Date: _____

Reviewer(s):

_____   _____
_____   _____
_____   _____
_____   _____

Moderator, respond to the following:

	Yes	No*
1. Is design properly translated into code?	____	____
2. Are there misspellings or typos?	____	____
3. Has proper use of language conventions been made?	____	____
4. Is there compliance with coding standards for language style, comments, and module prologue?	____	____
5. Are incorrect or ambiguous comments present?	____	____
6. Are typing and data declaration proper?	____	____
7. Are physical contents correct?	____	____
8. Have all items in the design walk-through checklist been reapplied (as required)?	____	____

*If you answered "No" to any of the above, please attach problems found and suggestions to improve.

Please attach additional comments and a copy of requirement items.

Figure A-16. Program code walk-through checklist.

Test Plan Walk-through Checklist

Figure A-17 is provided as an example walk-through checklist used in support of the peer review of a software test plan. This type of form may be used to document issues and resolutions when performing project walk-throughs.

Software Project:

Software Item(s): _____

Review Date: _____

Reviewer(s):

_____   _____
_____   _____
_____   _____
_____   _____

Moderator, respond to the following:

	Yes	No*
1. Have major test activities been properly identified and sequenced?	____	____
2. Has traceability to validation criteria/requirements been established as part of software requirements analysis?	____	____
3. Are major functions demonstrated early?	____	____
4. Is the test plan consistent with the overall project plan?	____	____
5. Has a test schedule been explicitly defined?	____	____
6. Are test resources and tools identified and available?	____	____
7. Has a test record-keeping mechanism been established?	____	____
8. Have test drivers and stubs been identified, and has work to develop them been scheduled?	____	____
9. Has stress testing for software been specified?	____	____

*If you answered "No" to any of the above, please attach problems found and suggestions to improve.

Please attach additional comments and a copy of requirement items.

Figure A-17. Test plan walk-through checklist.

Walk-through Summary Report

Figure A-18 is provided as an example walk-through summary report. This may be used to document issues and resolutions when performing peer reviews.

Project Name:	Date:
Element Reviewed:	
Review Team:	
Problems Found:	
Proposed Solutions:	
Comments:	
Action Taken:	

Figure A-18. Walk-through summary report.

Classic Anomaly Class Categories

Anomaly classes provide evidence of nonconformance and may be categorized, as follows:

Missing

Extra (superfluous)

Ambiguous

Inconsistent

Improvement desirable

Not conforming to standards

Risk-prone; that is, the review finds that although an item was not shown to be "wrong," the approach taken involves risks (and there are known safer alternative methods)

Factually incorrect

Not implementable (e.g., because of system constraints or time constraints)

Editorial

VALIDATION

Example Test Classes

Check for Correct Handling of Erroneous Inputs

Test Objective. Check for proper handling of erroneous inputs: characters that are not valid for this field, too many characters, not enough characters, value too large, value too small, all selections for a selection list, no selections, all mouse buttons clicked or double clicked all over the client area of the item with focus. Test class # xx.

Validation Methods Used. Test.

Recorded Data. User action or data entered, screen/view/dialog/control with focus, resulting action.

Data Analysis. Was resulting action within general fault handling defined capabilities in the [project abbreviation] SysRS and design in [project abbreviation] SDD?

Assumptions and Constraints. None.

Check for Maximum Capacity

Test Objective. Check software and database maximum capacities for data; enter data until maximum number of records specified in the design is reached for each table, operate program and add one more record. Test class # xx.

Validation Methods Used. Test.

Recorded Data. Record number of records in each table, with resulting actions.

Data Analysis. Was resulting action to the maximum plus one normal?

Assumptions and Constraints. This test requires someone to create through some method a populated database with records several times greater than what actually exists in the sample data set. This takes a good deal of time. Integration and qualification test only.

User Interaction Behavior Consistency

Test Objective. Check that the interaction behavior of the user interface is consistent across the application or module under test: tab through controls, using mouse click and double click on all controls and in null area, maximize, minimize, normalize, switch focus to another application and then back, update data on one view that is included on another and check to see if the other view is updated when data is saved on the first view, use function keys and movement keys and other standard key combinations (clipboard combos, control key windows and program-defined sets, Alt key defined sets), enter invalid control and Alt key sets to check for proper handling. Test class # xx.

Validation Methods Used. Test, inspection.

Recorded Data. Record any anomalies of action resulting from user action not conforming to the behavioral standards for Windows programs.

Data Analysis. Was resulting action within behavioral standards of windows programs as defined in the [project abbreviation] SysRS and design in [project abbreviation] SDD? Was behavior consistent across the application or module as defined in the [project abbreviation] SysRS and design in [project abbreviation] SDD?

Assumptions and Constraints. If testing at module level, the multiple view portion of the test may not apply due to having only a single view.

Retrieving Data

Test Objective. Check that the data retrieved is correct for each dialog, list box, combo box, and other controls that show lists, and check the data displayed for correctness. Test class # xx.

Validation Methods Used. Test, inspection.

Recorded Data. Record data displayed and data sources (records from tables, resource strings, and code sections).

Data Analysis. Was data displayed correctly? Compare data displayed with sources.

Assumptions and Constraints. Requires alternate commercial database software to get records from the database.

Saving Data

Test Objective. Check that the data entered is saved to the database correctly for each dialog, list box, combo box, and other controls that show lists, and check the data entered and saved for correctness in the database. Test class # xx.

Validation Methods Used. Test, inspection.

Recorded Data. Record data entered and data destinations (records from tables).

Data Analysis. Was data saved correctly? Compare data entered with destination.

Assumptions and Constraints. Requires alternate commercial database software to get records from the database.

Display Screen and Printing Format Consistency

Test Objective. Are user interface screens organized and labeled consistently? Check that printouts are formatted as specified: enter data to maximum length of field in a printout and then print, show all screens (views, dialogs, print previews, OLE devices), and dump their image to paper. Test class # xx.

Validation Methods Used. Inspection.

Recorded Data. Screen dumps and printouts.

Data Analysis. Was the printout format correct? Were the fields with maximum length data not clipped? Were the labels and organization of screens consistent across the application or module as defined in the [project abbreviation] SDD?

Assumptions and Constraints. The module that performs forms printing is required with all other modules during their testing.

Check Interactions Between Modules

Test Objective. Check the interactions between modules; enter data and save it in one module and switch to another module that uses that data to check for latest data entered, then switch back and forth between all of the modules, manipulate data, and check for adverse results or program faults. Test class # xx.

Validation Methods Used. Demonstration.

Recorded Data. Screen dumps.

Data Analysis. Were resulting actions within specifications as defined in the [project abbreviation] SysRS and design in [project abbreviation] SDD?

Assumptions and Constraints. Requires customer participation. Requires all modules and supporting software.

Measure Time of Reaction to User Input

Test Objective. Check average response time to user input action: clock time from saves, retrieves, dialogs open and closed, views open and closed; and clock time from any response to user action that takes longer than 2 seconds. Test class # xx.

Validation Methods Used. Test, analysis.

Recorded Data. Record action and response clock time. Organize into categories and average their values. Are all average values less than the minimum response time specified?

Data Analysis. Organize into categories and average their values. Are all average values less than the minimum response time specified as defined in the [project abbreviation] SysRS and design in [project abbreviation] SDD?

Assumptions and Constraints. None.

Functional Flow

Test Objective. Exercise all menus, buttons, hotspots, and so on that cause a new display (view, dialog, OLE link) to occur. Test class # xx.

Validation Methods Used. Demonstration.

Recorded Data. Screen dumps.

Data Analysis. Were resulting actions in accord with specifications as defined in the [project abbreviation] SysRS and design in [project abbreviation] SDD?

Assumptions and Constraints. Requires customer participation. Requires all modules and supporting software.

Types of System Testing

The following are provided as examples of system testing.

Functional Testing. Testing against operational requirements is functional testing. Reference to the SysRS should be made to show a complete list of functional requirements and resulting test strategy.

Performance Testing. Testing against performance requirements is performance testing. Reference to the SysRS should be made to show a complete list of performance requirements and resulting test strategy.

Reliability Testing. Testing against reliability requirements is reliability testing. Reference to the SysRS should be made to show a complete list of reliability requirements and resulting test strategy.

Configuration Testing. Testing under different hardware and software configurations to determine an optimal system configuration is configuration testing. Reference to the

SysRS should be made to show a complete list of configuration requirements and resulting test strategy.

Availability Testing. Testing against operational availability requirements is availability testing. Reference to the SysRS should be made to show a complete list of availability requirements and resulting test strategy.

Portability Testing. Testing against portability requirements is portability testing. Reference to the SysRS should be made to show a complete list of portability requirements and resulting test strategy.

Security and Safety Testing. Testing against security and safety requirements is security and safety testing. Reference to the SysRS should be made to show a complete list of security and safety requirements and resulting test strategy.

System Usability Testing. Testing against usability requirements is system usability testing. Reference to the SysRS should be made to show a complete list of system usability requirements and resulting test strategy.

Internationalization Testing. Testing against internationalization requirements is internationalization testing. Reference to the SysRS should be made to show a complete list of internationalization requirements and resulting test strategy.

Operations Manual Testing. Testing against the procedures in the operations manual to see if it can be operated by the system operator is operations manual testing. Reference to the operations manual should be made to show a complete list of operator requirements and resulting test strategy.

Load Testing. Testing that attempts to cause failures involving how the performance of a system varies under normal conditions of utilization is load testing.

Stress Testing. Testing that attempts to cause failures involving how the system behaves under extreme but valid conditions (e.g., extreme utilization, insufficient memory, inadequate hardware, and dependency on overutilized shared resources) is stress testing.

Robustness Testing. Testing that attempts to cause failures involving how the system behaves under invalid conditions (e.g., unavailability of dependent applications, hardware failure, and invalid input such as entry of more than the maximum amount of data in a field) is robustness testing.

Contention Testing. Testing that attempts to cause failures involving concurrency is contention testing.

Test Design Specification

This example test design specification is based upon IEEE-Std 829, IEEE Standard for Software Test Documentation [15].

Purpose. This section should provide an overview describing the test approach and identifying the features to be tested by the design.

Outline. A test design specification should have the following structure:

1. Test design specification identifier. Specify the unique identifier assigned to this test design specification as well as a reference to any associated test plan.
2. Features to be tested. Identify all test items describing the features that are the target of this design specification. Supply a reference for each feature to any specified requirements or design.
3. Approach refinements. Specify any refinements to the approach and any specific test techniques to be used. The method of analyzing test results should be identified (e.g., visual inspection). Describe the results of any analysis that provides a rationale for test case selection, for example, a test case used to determine how well erroneous user inputs are handled.
4. Test identification. Include the unique identifier and a brief description of each test case associated with this design. A particular test case may be associated with more than one test design specification.
5. Feature pass/fail criteria. Specify the criteria to be used to determine whether the feature has passed or failed.

The sections should be ordered in the specified sequence. Additional sections may be included at the end. If some or all of the content of a section is in another document, then a reference to that material may be listed in place of the corresponding content. The referenced material must be attached to the test design specification or readily available to users.

Test Case Specification

This example test case specification is based upon IEEE-Std 829, IEEE Standard for Software Test Documentation [15].

Purpose. To define a test case identified by a test design specification.

Outline. A test case specification should have the following structure:

1. Test case specification unique identifier. Specify the unique identifier assigned to this test case specification.
2. Test items. Identify and briefly describe the items and features to be exercised by this test case. For each item, consider supplying references to the following test item documentation: requirements specification, design specification, user's guide, operations guide, and installation guide.
3. Input specifications. Specify each input required to execute the test case. Some of the inputs will be specified by value (with tolerances where appropriate), whereas others, such as constant tables or transaction files, will be specified by name. Specify all required relationships between inputs (e.g., timing).
4. Output specifications. Specify all of the outputs and features (e.g., response time) required of the test items. Provide the exact value (with tolerances where appropriate) for each required output or feature.

5. Environmental needs. Specify all hardware and software requirements associated with this test case. Include facility or personnel requirements if appropriate.

6. Special procedural requirements. Describe any special constraints on the test procedures that execute this test case.

7. Intercase dependencies. List the identifiers of test cases that must be executed prior to this test case. Summarize the nature of any dependencies.

If some or all of the content of a section is in another document, then a reference to that material may be listed in place of the corresponding content. The referenced material must be attached to the test case specification or readily available to users of the case specification.

A test case may be referenced by several test design specifications and enough specific information must be included in the test case to permit reuse.

Test Procedure Specification

This example test procedure specification is based upon IEEE-Std 829, IEEE Standard for Software Test Documentation [15].

Purpose. The purpose of the specification is to document the steps in support of test set execution or, more generally, the steps used to evaluate a set of features associated with a software item.

Outline. A test procedure specification should have the following structure:

1. Test procedure specification identifier. The unique identifier assigned to the test procedure specification. A reference to the associated test design specification should also be provided.

2. Purpose. Describe the purpose of this procedure, providing a reference for each executed test case. In addition, provide references to relevant sections of the test item documentation (e.g., references to usage procedures) where appropriate.

3. Special requirements. Identify any special requirements that are necessary for the execution. These may include prerequisite procedures, required skills, and environmental requirements.

4. Procedure steps. Describe the procedural steps supporting the test in detail, the following list provides guidance regarding the level of detail required:

   Log. Describe the method or format for logging the results of test execution, the incidents observed, and any other test events.

   Set up. Describe the sequence of actions necessary to prepare for execution of the procedure.

   Start. Describe the actions necessary to begin execution of the procedure.

   Proceed. Describe any actions necessary during execution of the procedure.

   Measure. Describe how the test measurements will be made.

   Shut down. Describe the actions necessary to immediately suspend testing if required.

   Restart. Identify any restart points and describe the actions necessary to restart the procedure at each of these points.

Stop. Describe the actions necessary to bring execution to an orderly halt.

Wrap up. Describe the actions necessary to restore the environment.

Contingencies. Describe the actions necessary to deal with anomalous events that may occur during execution.

The sections should be ordered as described. Additional sections, if required, may be included at the end of the specification. If some or all of the content of a section is in another document, then a reference to that material may be listed in place of the corresponding content. The referenced material must be attached to the test procedure or readily available to users of the procedure specification.

Test Item Transmittal Report

This example test item transmittal report is based upon IEEE-Std 829, IEEE Standard for Software Test Documentation [15].

Purpose. The purpose of this report is to identify all items being transmitted for testing. This report should identify the person responsible for each item, the physical location of the item, and item status. Any variations from the current item requirements and designs should be noted in this report.

Outline. A test item transmittal report should have the following structure:

1. Transmittal reports identifier. The unique identifier assigned to this test item transmittal report.
2. Transmitted items. Describe all test items being transmitted, including their version/revision level. Supply references associated with the documentation of the item including, but not necessarily limited to, the test plan relating to the transmitted items. Indicate the individuals responsible for the transmitted items.
3. Location. Identify the location of all transmitted items. Identify the media that contain the items being transmitted. Indicate media identification and labeling.
4. Status. Describe the status of the test items being transmitted, listing all incident reports relative to these transmitted items.
5. Approvals. List all transmittal approval authorities, providing a space for associated signature and date.

The sections should be ordered as described. Additional sections, if required, may be included at the end of the specification. If some or all of the content of a section is in another document, then a reference to that material may be listed in place of the corresponding content. The referenced material must be attached to the test item transmittal report or readily available to users of the transmittal report.

Test Log

This example test log is based upon IEEE-Std 829, IEEE Standard for Software Test Documentation [15].

Purpose. The purpose of this test log is to document the chronological record of test execution.

Outline. A test log should have the following structure:

1. Test log identifier. The unique identifier assigned to this test log.
2. Description. This section should include the identification of the items being tested, environmental conditions, and type of hardware being used.
3. Activity and event entries. Describe event start and end activities, recording the occurrence date, time, and author. A description of the event execution, all results, associated environmental conditions, anomalous events, and test incident report number should all be included as supporting information.

The sections should be ordered in the specified sequence. Additional information may be included at the end. If some or all of the content of a section is in another document, then a reference to that material may be listed in place of the corresponding content. The referenced material must be attached to the test log or readily available to users of the log.

Test Incident Report

This example test incident report is based upon IEEE-Std 829, IEEE Standard for Software Test Documentation [15].

Purpose. The purpose of this report is to document all events occurring during the testing process requiring further investigation.

Outline. A test incident report should have the following structure:

1. Test incident report identifier. The unique identifier assigned to this test incident report.
2. Summary. Provide a summary description of the incident, identifying all test items involved and their version/revision level(s). References to the appropriate test procedure specification, test case specification, and test log should also be supplied.
3. Incident description. Provide a description of the incident. This description should include the following items: inputs and anticipated results, actual results, all anomalies, the date and time and procedure step, a description of the environment, any attempt to repeat, and the names of testers and observers.
4. Related activities and observations that may help to isolate and correct the cause of the incident should be included (e.g., describe any test case executions that might have a bearing on this particular incident and any variations from the published test procedure).
5. Impact. Describe the impact this incident will have on existing test plans, test design specifications, test procedure specifications, or test case specifications.

The sections should be ordered in the sequence described above. Additional sections may be included at the end if desired. If some or all of the content of a section is in another document, then a reference to that material may be listed in place of the corresponding content. The referenced material must be attached to the test incident report or made readily available to users of the incident report.

Test Summary Report

This example test summary report is based upon IEEE-Std 829, IEEE Standard for Software Test Documentation [15].

Purpose. The purpose of this report is to provide a documented summary of the results of testing activities and associated result evaluations.

Outline. A test summary report should have the following structure:

1. Test summary report identifier. Include the unique identifier assigned to this test summary report.
2. Summary. Provide a summary of items tested. This should include the identification of all items tested, their version/revision level, and the environment in which the testing activities took place. For each test item, also supply references to the following documents if they exist: test plan, test design specifications, test procedure specifications, test item transmittal reports, test logs, and test incident reports.
3. Variances. Report any variances of the test items from their specifications, associating the reason for each variance.
4. Comprehensiveness assessment. Identify any features not sufficiently tested, as described in the test plan, and explain the reasons for this.
5. Summary of results. Provide a summary of test results, including all resolved and unresolved incidents.
6. Evaluation. This overall evaluation should be based upon the test results and the item-level pass/fail criteria. An estimate of failure risk may also be included.
7. Summary of activities. Provide a summary of all major testing activities and events.
8. Approvals. List the names and titles of all individuals who have approval authority. Provide space for the signatures and dates.

The sections should be ordered in the specified sequence. Additional sections may be included just prior to approvals if required. If some or all of the content of a section is in another document, then a reference to that material may be listed in place of the corresponding content. The referenced material must be attached to the test summary report or readily available to users.

JOINT REVIEW

Open Issues List

Figure A-19 is presented as an example form that may be used in support of the documentation and tracking of issues associated with a software project. These issues are typically captured during reviews or team meetings and should be reviewed on a regular basis. An electronic version of this work product, *Open Issues List.doc,* is provided on the companion CD-ROM.

Open Issues List
As of: [Date]

Identification

This document contains open issues from the _____ meeting:

Package	
Author	
Review date	
List version	

Open issues

The matrix below lists any *open issues*, and the response to those issues. The disposition column indicates the resolution of the issue.

Reference	Severity	Comment	Source	Response	Disposition

Figure A-19. Open issues list.

AUDIT

Status Reviews

Reviews should occur throughout the life cycle, as identified in the software project management plan. These reviews can be conducted as internal management or customer management reviews. The purpose of these types of reviews is to determine the current status and risk of the software effort. The data presented during these reviews should be used as the basis for decision making regarding the future path and progress of the project.

Internal Management Reviews. The following are suggested agenda items for discussion during an internal management review:

An overview of current work to date

Status versus project plan comparison (cost and schedule)

Discussion of any suggested alternatives (optional)

Risk review and evaluation

Updates to the project management plan (if applicable)

Customer Management Reviews. The following are suggested agenda items for discussion during a customer management review:

An overview of current work to date

Status versus project plan comparison (cost and schedule)

Discussion of any suggested alternatives (optional)

Risk review and evaluation

Customer approval of any updates to the project plan

Critical Dependencies Tracking

IEEE Std 1490, IEEE Adoption of PMI Standard, A Guide to the Project Management Body of Knowledge, describes the identification, tracking, and control of all items critical to the successful management of a project. The PMI refers to the identification and tracking of critical dependencies as integrated change control. According to IEEE Std 1490:

> Integrated change control is concerned with a) influencing the factors that create changes to ensure that changes are agreed upon, b) determining that a change has occurred, and c) managing the actual changes when and as they occur. The original defined project scope and the integrated performance baseline must be maintained by continuously managing changes to the baseline, either by rejecting new changes or by approving changes and incorporating them into a revised project baseline [35].

As described by IEEE Std 1490, there are three key inputs that support integrated change control. The content of the software project management plan provides key input and becomes the change control baseline. Status reporting provides key input providing interim information on project performance. Lastly, change requests provide a record of the changes requested to items critical to the success of the project.

The work products that support the tracking of critical dependencies are directly related to the three key inputs described above. All changes in the status of critical dependencies affect the project plan and must be reflected in the plan as plan updates. Corrective action, and the documentation of lessons learned, may be triggered as a result of status reporting. Any corrective action, and lessons learned, may be recorded as the resolution of status report action items. Requested changes may be recorded any number of ways. These may be recorded as notes from a status review or, more formally, as part of a change request process.

For integrated project management with other impacting projects, several planning and management tasks must be enhanced for interproject communications. Table A-24 details tasks specifically for IPM SP 2.2—Manage Dependencies.

Table A-24. Tasks in support of integrated project management

Critical Dependencies Tracking—Detailed Project Planning

1. Detail project definition and work plan narrative and dependences on other projects
2. Define work breakdown structure (WBS) and dependence on other project plan's WBS
3. Develop list of relevant stakeholders, including other projects stakeholders
4. Develop communication plan (internal, with other projects, and external)
5. Layout task relationships (proper sequence: dependencies, predecessor and successor tasks)
6. Identify critical dependencies
7. Estimate timelines
8. Determine the planned schedule and identify milestones
9. Determine the planned schedule and identify common milestones with other projects
10. Obtain stakeholder (including other projects) buyoff and signoff
11. Obtain executive approval to continue

Critical Dependencies Tracking—Detailed Project Plan Execution

12. Mark milestones (significant event markers) and critical dependencies
13. Assign and level resources against shared resources pool
14. Set project baseline
15. Set up issues log
16. Set up change request log
17. Set up change of project scope log
18. Record actual resource usage and costs, including other projects
19. Track and manage critical path and critical dependencies
20. Manage and report status as determined by project plan
21. Communicate with stakeholders, including other projects, as determined by project plan
22. Report monthly project status, including other projects
23. Close project by formal acceptance of all stakeholders
24. Report on project actuals vs. baseline estimates
25. Complete postimplementation review report—lessons learned
26. Complete and file all project history documentation in document repository
27. Celebrate

List of Measures for Reliable Software [4]

Table A-25 provides a list of suggested measures in support of the production of reliable software. This list is a reproduction of content found in IEEE Std 982.1, IEEE Standard Dictionary of Measures to Produce Reliable Software.

Table A-25. List of suggested measures for reliable software

Paragraph	Description of Measure
4.1	Fault Density
4.2	Defect Density
4.3	Cumulative Failure Profile
4.4	Fault-Days Number
4.5	Functional or Modular Test Coverage
4.6	Cause and Effect Graphics
4.7	Requirements Traceability
4.8	Defect Indices
4.9	Error Distribution
4.10	Software Maturity Index
4.11	Man Hours per Major Defect Detected
4.12	Number of Conflicting Requirements
4.13	Number of Entries and Exits per Module
4.14	Software Science Measures
4.15	Graph-Theoretic Complexity for Architecture
4.16	Cyclomatic Complexity
4.17	Minimal Unit Test Case Determination
4.18	Run Reliability
4.19	Design Structure
4.20	Mean Time to Discover the Next K Faults
4.21	Software Purity Level
4.22	Estimated Number of Faults Remaining
4.23	Requirement Compliance
4.24	Test Coverage
4.25	Data or Information Flow Complexity
4.26	Reliability Growth Function
4.27	Residual Fault Count
4.28	Failure Analysis Using Elapsed Time
4.29	Testing Sufficiency
4.30	Mean Time to Failure
4.31	Failure Rate
4.32	Software Documentation and Source Listings
4.33	Required Software Reliability
4.34	Software Release Readiness
4.35	Completeness
4.36	Test Accuracy
4.37	System Performance Reliability
4.38	Independent Process Reliability
4.39	Combined Hardware and Software Operational Availability

Example Measures—Management Category

Manpower Measures. Manpower measures are used primarily for project management and do not necessarily have a direct relationship with other technical and maturity measures. These measures should be used in conjunction with the development progress measures. The value of these measures is somewhat tied to the accuracy of the project plan, as well as to the accuracy of the labor reporting.

These measures provide an indication of the application of human resources to the development program and the ability to maintain sufficient staffing to complete the project. They can also provide indications of possible problems with meeting schedule and budget. They are is used to examine the various elements involved in staffing a software project. These elements include the planned level of effort, the actual level of effort, and the losses in the software staff measured per labor category. Planned manpower profiles can be derived from the associated project planning documents. Reporting examples are shown in Figures A-20 and A-21.

The planned level of effort is the number of labor hours estimated to be worked on a software module during each tasking cycle. The planned levels are monitored to ensure that the project is meeting the necessary staffing criteria.

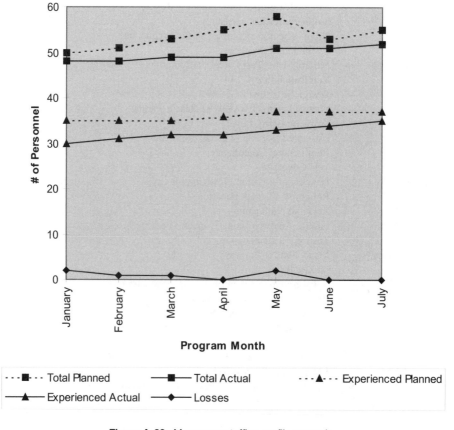

Figure A-20. Manpower staffing profile example.

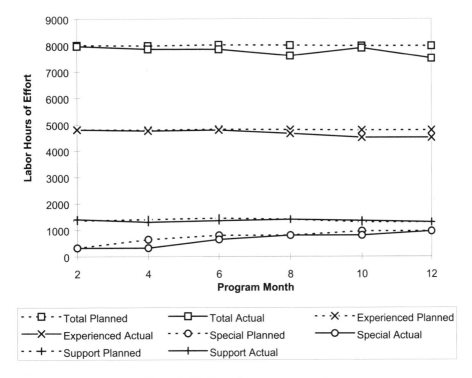

Figure A-21. Example manpower metric.

Life Cycle Application. The shape of the staff profile trend curve tends to start at a moderate level at the beginning of the contract, grow through design, peak at coding/testing, and diminish near the completion of integration testing. Individual labor categories, however, are likely to peak at different points in the life cycle. The mature result would show little deviation between the planned and actual level for the entire length of development and scheduled maintenance life cycles.

Algorithm/Graphical Display. The graphical display measure should reflect, by project deliverable software item, the actual versus planned labor hours of effort per month. This measure can be further refined by breaking labor hours into categories (i.e., experienced, novice, and special).

Significant deviations (those greater than 10%) of actual from planned levels indicate potential problems with staffing. Deviations between actual and planned levels can be detected, analyzed, and corrected before they negatively impact the development schedule. Losses in staff can be monitored to detect a growing trend or significant loss of experienced staff. This indicator assists in determining whether there have been a sufficient number of employees to produce the product in the tasking cycle.

Data Requirements. Software tasking documentation should reflect the expected staffing profiles and labor hour allocations. For each labor category tracked, indicate the labor category name:

Experienced

Programmer

Lead

Senior

Support

For each experience level per tasking cycle, provide the

Number of planned personnel staffing

Number of personnel actually on staff in the current reporting period

Number of unplanned labor hour losses

Number of labor hours that are planned to be expended in next reporting period (cumulative)

Number of labor hours that are actually expended in the current reporting period (cumulative)

Frequency and Type of Reporting. There should be no less than one report monthly. Reporting by [Project Abbreviation] Program Manager to [Company Name] corporate management should be in the form of a Program Manager's Review (PMR).

Use/Interpretation. Tracking by individual labor category may be done for projects in order to monitor aspects of a particular program that are deemed worthy of special attention. Total personnel is the sum of experienced and support personnel. Special skills personnel are counted within the broad categories of experienced and support, and may be tracked separately.

Special Skills Personnel. Special skills personnel are defined as those individuals who possess specialized software-related abilities defined as crucial to the success of the particular system. For example, Ada programmers are defined as having skills necessary for completing one type of software project but not others.

Experienced Personnel. Experienced personnel are defined as degreed individuals with a minimum of three years experience in software development for similar applications.

Support Personnel. Support personnel are degreed and nondegreed individuals with a minimum of three years experience in software development other than those categorized as experienced software engineers.

Development Progress Measures. Progress measures provide indications of the degree of completeness of the software development effort, and can be used to judge readiness to proceed to the next stage of software development. In certain instances, consideration must be given to a possible rebaselining of the software or the adding of modules due to changing requirements.

These measures are used to track the ability to keep computer software unit (CSU) design, code, test and integration activities on schedule. They should be used with the manpower measures to identify specific projects that may be having problems.

These measures should also be used with the breadth, depth, and fault profile testing measures to assess the readiness to proceed to a formal government test.

Life Cycle Application. Collection should begin at software requirements review (SRR) and continue for the entirety of the software development. The progress demonstrated in design, coding, unit testing, and integration of CSUs should occur at a reasonable rate. The actual progress shown in these areas versus the originally planned progress can indicate potential problems with a project's schedule.

Algorithm/Graphical Display. The progress measure graphical display should reflect, by project software deliverable item, the percent of planned and actual CSUs per tasking cycle milestone (for example, percent complete per month). Completed (actual) CSUs may be subcategorized into percent of CSUs 100% designed, percent of CSUs coded and successfully unit tested, and percent of CSUs 100% integrated. An example is provided in Figure A-22.

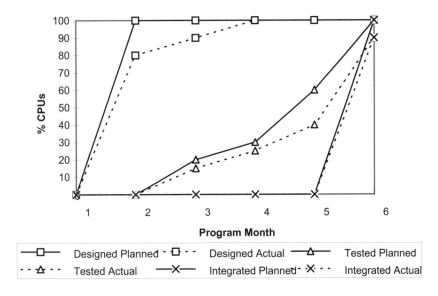

Figure A-22. Development progress measure.

Additionally, using the requirements traceability matrix, the developed and verified functionality versus time may be plotted as a measure of development progress.

Data Requirements. Project development, test, and integration schedules are required. For each project, provide:

1. The number of CSUs
2. Number of CSUs 100% designed
3. Number of CSUs 100% coded and successfully validated
4. Number of CSUs 100% integrated
5. Number of planned CSUs to be 100% designed and reviewed for the development cycle

Also include the number of planned CSUs to be 100% coded and successfully validated for the development cycle and the number of planned CSUs to be 100% integrated for the development cycle.

Frequency and Type of Reporting. The [Project Abbreviation] Project Manager should report progress measure information to the [Project Abbreviation] Program Manager no less than once a month. Also, there should be no less than one report monthly by the Program Manager to [Company Name] corporate management and the customer in the form of a Program Manager's Review (PMR).

Use/Interpretation. It is important to remember that these measures pass no judgment on whether the objectives in the development plan can be achieved. Special attention should be paid to the development progress of highly complex CSUs.

Schedule Measures. These measures indicate changes and adherence to the planned schedules for major milestones, activities, and key software deliverables. Software activities and delivery items may be tracked using the schedule measures. The schedule measures may be used with other measures to help judge program risk. For example, they could be used with the test coverage measures to determine if there is enough time remaining in the current schedule to allow for the completion of all testing.

Life Cycle Application. The Project Manager should begin data collection at the project's start and continue for the entire software development cycle.

Algorithm/Graphical Display. The Project Manager should plot planned and actual schedules for major milestones and key software deliverables as they change over the tasking cycle. Any milestone event of interest may be plotted.

The schedule measures may be plotted as they change over time (reflecting milestone movement), providing indications of problems in meeting key events or deliveries. The further to the right of the trend line for each event, the more problems are being encountered.

Data Requirements. For all projects, provide:

1. Software Project Plan
2. Major Milestone schedules
3. Delivery schedule for key software items

For the activity or event tracked during the tasking period, provide:

1. Event name
2. Date of report
3. Planned activity start date
4. Actual activity start date
5. Planned activity end date
6. Actual activity end date

Frequency and Type of Reporting. The [Project Abbreviation] Project Manager should report schedule information to the [Project Abbreviation] Program Manager and the customer no less than biweekly. This reporting is accomplished as part of scheduled staff meetings. Also, there should be no less than one report monthly by the Program Manager to [Company Name] corporate management and the customer, in the form of a Program Manager's Review (PMR).

Use/Interpretation. No formal evaluation criteria for the trend of the schedule measures are given. Large slippages are indicative of problems. Maintaining conformance with calendar-driven schedules should not be used as the basis for proceeding beyond milestones. The schedule measure passes no judgment on the achievability of the development plan. A reporting example is provided in Figure A-23.

STATUS

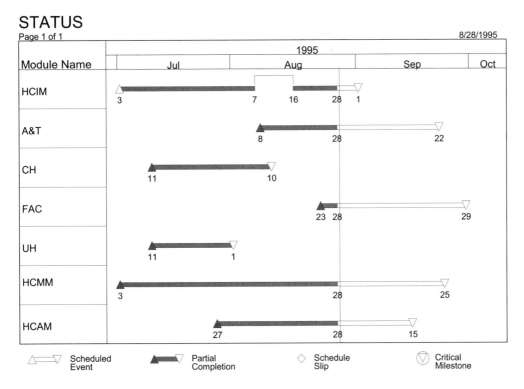

Figure A-23. Status reporting.

Example Measures—Requirements Category

Requirements Traceability Measures. The requirements traceability measures measure the adherence of the software products to their requirements throughout the development life cycle. The technique of performing this requires the development of a project software requirements traceability matrix (SRTM).

The SRTM is the product of a structured, top-down hierarchical analysis that traces the software requirements through the design to the code and test documentation. The requirements traceability measures should be used in conjunction with the test coverage measures (depth and breadth of testing) and the development progress measure (optional) to verify if sufficient functionality has been demonstrated to warrant proceeding to the next stage of development or testing. They should also be used in conjunction with the design stability and requirements stability measures.

Life Cycle Application. The Project Manager should begin tracing measures during user requirements definition, updating the trace in support of major milestones or at major software release points.

By the nature of the software development process, especially in conjunction with an evolutionary development strategy, the trace of requirements is an iterative process. That is, as new software releases add more functionality to the system, the requirements trace will have to be revisited and augmented.

Algorithm/Graphical Display. This measure is a series of percentages that can be calculated from the matrix described. These will include:

Percent baseline software requirements specification requirements in CSU design

Percent software requirements in code

Percent software requirements having test cases identified

Tracing from code to requirements should also be accomplished. Each CSU is examined for the requirements it satisfies and the percentage of open or additional requirements traced. Backward tracing of requirements at the early stages can identify requirements that have been added in the requirements decomposition process. Each requirement added to the specification should be attributable in some way to a higher level requirement from a predecessor requirement.

Data Requirements. For all projects, provide:

1. Documented/approved software requirements specification (SRS)
2. Documented/approved software design document (SDD)
3. Software test description in accordance with DOD-STD-498
4. Completed SRTM

For each project software deliverable, provide the number of SRS software requirements:

Total traceable to SRS

Traceable to SDD

Traceable to code

Traceable to system test

Frequency and Type of Reporting. The Program Manager should update periodically in support of milestones or major releases. This tracing should be a key tool used at all system requirement and design reviews. It can serve to indicate those areas of requirements or software design that have not been sufficiently thought out. The trend of the SRTM should be monitored over time for closure.

Use/Interpretation. A benefit of requirements traceability is that those modules that appear most often in the matrix (thus representing the ones that are most crucial in the respect that they are required for multiple functions or requirements) can be highlighted for earlier development and increased test scrutiny. An example follows.

Requirements Traceability Matrix
[Project Abbreviation] Requirements Traceability Matrix*

Requirement Name	Priority	Risk	SRS Paragraph	Formal Test Paragraph	Test Activity
Input Personnel Data	H	L	3.1.1	6.3.14	D
Store Personnel Data	M	H	3.1.2	6.3.17	T, A

*Verification Method(s): D = Demonstration, T = Test, A = Analysis, I = Inspection.

Field Descriptions:

1. Requirement Name   A short description of the requirement to be satisfied.
2. Priority   High (H), Medium (M), Low (L); to be negotiated with customer and remains fixed throughout the software development life cycle.
3. Risk   High (H), Medium (M), Low (L); determined by technical staff and will change thoughout the software development life cycle.
4. SRS Paragraph   Paragraph number from section 3 of the SRS.
5. Validation Method(s)   Method to be used to validate that the requirement has been satisfied.
6. Formal Test Paragraph   This column will provide a linkage between the SRS and the software test plan. This will indicate the test to be executed to satisfy the requirement.

Requirements Stability Measures. Requirements stability measures indicate the degree to which changes in the software requirements and/or the misunderstanding of requirements implementation affect the development effort. They also allow for determining the cause of requirements changes. The measures for requirements stability should be used in conjunction with those for requirements traceability, fault profiles, and the optional development progress.

Life Cycle Application. Changes in requirements tracking begin during requirements definition and continue to the conclusion of the development cycle. It is normal that when program development begins, the details of its operation and design are incomplete. It is normal to experience changes in the specifications as the requirements become better defined over time. When design reviews reveal inconsistencies, a discrepancy report is generated. Closure is accomplished by modifying the design or the requirements. When a change is required that increases the scope of the project a baseline change request (BCR) should be submitted.

*For additional information regarding requirements elicitation refer to [Project Acronym] Software Requirements Management Plan.

Algorithm/Graphical Display. This measure may be presented as a series of percentages representing:

1. The percent of requirements discrepancies, both cumulative and cumulative closed, over the life of the project, and cumulative requirements discrepancies over time versus closure of those discrepancies. Good requirements stability is indicated by a leveling off of the cumulative discrepancies curve with most discrepancies having reached closure.

2. The percent of requirements changed or added, noncumulative, spanning project development, and the effect of these changes in requirements in lines of code.

Several versions of this type of chart are possible. One version may show the number of change enhancement requests (CERs) and affected lines of code. Additionally, it is also possible to look at the number of modules affected by requirements changes.

The plot of open discrepancies can be expected to spike upward at each review and to diminish thereafter as the discrepancies are closed. For each engineering change, the amount of software affected should be reported in order to track the degree to which BCRs increase the difficulty of the development effort. Only those BCRs approved by the associated CCB should be counted.

Data Requirements. A "line of code" will differ from language to language, as well as from programming style to programming style. In order to consistently measure source lines of code, count all noncommented, nonblank, executable and data statements (standard noncommented lines of code or SNCLC).

For each project, provide:

1. Number of software requirements discrepancies as a result of each review (software requirements review, preliminary design review, etc.)

2. Number of baseline change requests (BCRs) generated from changes in requirements

3. Total number of SNCLCs

4. Total number of SRS requirements

5. Percentage of SRS requirements added due to approved BCRs

6. Percentage of SRS requirements modified due to approved BCRs

7. Percentage of SRS requirements deleted due to approved BCRs

Frequency and Type of Reporting. The Project Manager should update the measures periodically in support of milestones or major releases. This tracing should be a key tool used at all system requirement and design reviews.

Use/Interpretation. Causes of project turbulence can be investigated by looking at requirements stability and design stability together. If design stability is low and requirements stability is high, the designer/coder interface is suspect. If design stability is high and requirements stability is low, the interface between the user and the design activity is suspect. If both design stability and requirements stability are low, both the interfaces between the design activity and the code activity and between the user and the design activity are suspect (Figure A-24).

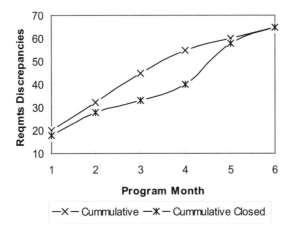

Figure A-24. Requirements stability.

Allowances should be made for higher instability in the case where rapid prototyping is utilized. At some point in the development effort, the requirements should be firm so that only design and implementation issues will cause further changes to the specification.

A high level of instability at the Critical Design Review (CDR) stage indicates serious problems that must be addressed prior to proceeding to coding.

Example Measures—Quality Category

Design Stability Measure. Design stability measures are used to indicate the quantity of change made to a design. The design progress ratio shows how the completeness of a design advances over time and helps give an indication of how to view the stability in relation to the total projected design. The design stability measures can be used in conjunction with the complexity measures to highlight changes to the most complex modules. They can also be used with the requirements measures to highlight changes to modules that support the most critical user requirements. Design stability does not assess the quality of design. Other measures (e.g., complexity) can contribute to such an evaluation.

Life Cycle Application. Begin tracking no later than preliminary design review (PDR) and continue for each version until completion.

Algorithm/Graphical Display. Items to be measured are:

M = Number of CSUs in current delivery/design

F_c = Number of CSUs in current delivery/design that include design

F_a = Number of CSUs in current delivery/design that are additions to previous delivery

F_d = Number of CSUs in previous delivery/design that have been deleted

T = Total modules projected for project

$$S \text{ (stability)} = [M - (F_a + F_c + F_d)]/M$$

Note: It is possible for stability to be a negative value. This may indicate that everything previously delivered has been changed and that more modules have been added or deleted.

$$DP \text{ (design progress ratio)} = M/T$$

Note: If some modules in the current delivery are to be deleted from the final delivery, it is possible for design progress to be greater than one.

Data Requirements. For each project and version, provide:

1. Date of completion
2. M
3. F_c
4. F_a
5. F_d
6. T
7. S
8. DP

Frequency and Types of Reporting. The Project Manager should update periodically in support of milestones or at major releases. Design stability should be monitored to determine the number and potential impact of design changes, additions, and deletions on the software configuration. A reporting example is provided in Figure A-25.

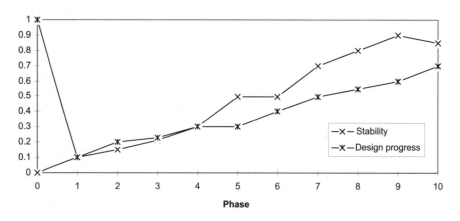

Figure A-25. Design stability/progress measure.

Use/Interpretation. The trend of design stability provides an indication of whether the software design is approaching a stable state, or leveling off of the curve at a value close to or equal to one. It is important to remember that during periods of inactivity, the magnitude of the change is relatively small or diminishing, and may be mistaken for stability.

In addition to a high value and level curve, the following other characteristics of the software should be exhibited:

1. The development progress measure is high.
2. Requirements stability is high.
3. Depth of testing is high.
4. The fault profile curve has leveled off and most software trouble reports (test problem reports (TPRs) or Internal Test Reports (ITRs)) have been fixed or voided.

This measure does not measure the extent or magnitude of the change within a module or assess the quality of the design. Other measures (e.g., complexity) can contribute to this evaluation.

Allowances should be made for lower stability in the case of rapid prototyping or other development techniques that do not follow the [Project Abbreviation] life cycle model for software development. (For additional information refer to the [Project Abbreviation] Software Quality Assurance Plan.) An upward trend with a high value for both stability and design progress is recommended before release.

Experiences with similar projects should be used as a basis for comparison. Over time, potential thresholds may be developed for similar types of projects.

Breadth of Testing Measures. Breadth of testing addresses the degree to which required functionality has been successfully demonstrated as well as the amount of testing that has been performed. This testing can be called "black box" testing, since one is only concerned with obtaining correct outputs as a result of prescribed inputs.

The customer should clearly specify what functionality should be in place at each milestone during the development life cycle. At each stage of testing (unit through system), emphasis should be placed on demonstrating that a high percentage of the functionality needed for that stage of testing is achieved.

Life Cycle Application. Begin collecting at the end of unit testing. Each priority level of the requirements test case to be used for evaluating the success of a requirement should be developed in coordination with project testing. This is to ensure that sufficient test cases are generated to adequately demonstrate the requirements.

Algorithm/Graphical Display. Breadth of testing consists of three different measures. One measure deals with coverage and two measures deal with success. These three subelements are portrayed in the following equation:

$$\underset{\text{Coverage}}{\frac{\text{\# CSUs tested}}{\text{total \# CSUs}}} \times \underset{\text{Test Success}}{\frac{\text{\# CSUs passed}}{\text{\#CSUs tested}}} = \underset{\text{Overall Success}}{\frac{\text{\# CSUs passed}}{\text{total \# CSUs}}}$$

Breadth of Testing Coverage. Breadth of testing coverage is computed by dividing the number of requirements that have been tested (with all applicable test cases under both representative and maximum stress loads) by the total number of requirements.

Breadth of Testing Success. Breadth of testing success is computed by dividing the number of requirements that have been successfully demonstrated through testing by the number of requirements that have been tested.

Breadth of Testing Overall Success. Breadth of testing overall success is computed by dividing the number of requirements that have been successfully demonstrated through testing by the total number of requirements.

Data Requirements. For each project, provide:

1. The number of SRS requirements and associated priorities
2. Number of SRS requirements tested with all planned test cases
3. Number of SRS requirements and associated priorities successfully demonstrated through testing
4. For each of the four priority levels for additional requirements (SPR, BCR), provide:
   a. Number of additional requirements tested with all planned test cases
   b. Number of additional requirements demonstrated through testing
5. For each of the four alpha/beta (ITR, TPR) test requirement priority levels, provide:
   a. Number of requirements
   b. Number of requirements validated with planned test cases
   c. Number of requirements successfully demonstrated through testing
6. Test Identification (e.g., unit test, alpha test, beta test)

Frequency and Types of Reporting. All three measures of breadth of testing should be tracked for all project requirements. The results of each measure should be reported at each project level throughout software functional and system level testing at each project milestone and to [Company Name] corporate management monthly.

It is suggested that coverage or success values be expressed as percentages by multiplying each value by 100 before display in order to facilitate understanding and commonality among measures presentations.

Use/Interpretation. One of the most important measures reflected by this measure is the categorization of requirements in terms of priority levels. With this approach, the most important requirements can be highlighted. Using this prioritization scheme, one can partition the breadth of testing measure to address each priority level. At various points along the development path, the pivotal requirements for that activity can be addressed in terms of tracing, test coverage, and test success.

Breadth of Testing Coverage. The breadth of testing coverage indicates the amount of testing performed without regard to success. By observing the trend of coverage over time it is possible to derive the full extent of testing that has been performed.

Breadth of Testing Success. The breadth of testing success provides indications about requirements that have been successfully demonstrated. By observing the trend of the overall success portion of breadth of testing over time, one gets an idea of the growth in successfully demonstrated functionality.

Failing one test case results in a requirement not being fully satisfied. If sufficient resources exist, breadth of testing may be addressed by examining each requirement in terms of the percent of test cases that have been performed and passed. In this way, partial credit for testing a requirement can be shown (assuming multiple test cases exist for a requirement), as opposed to an "all or nothing" approach. This method, which is not mandated, may be useful in providing additional granularity to the breadth of testing. An example is provided in Figure A-26.

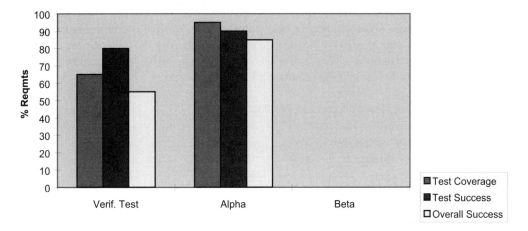

Figure A-26. Breadth of testing example.

Depth of Testing Measure. The depth of testing measure provides indications of the extent and success of testing from the point of view of coverage of possible paths/conditions within the software. Depth of testing consists of three separate measures, each of which is comprised of one coverage and two success subelements (similar to breadth of testing). The testing can be called "white-box" testing, since there is visibility into the paths/conditions within the software.

This measure should be used in conjunction with requirements traceability and fault profiles and the optional complexity and development progress measures. They must also be used with breadth of testing measures to ensure that all aspects of testing are consistent with beta test requirements.

Life Cycle Application. Begin collecting data at the critical design review (CDR) and continue through development as changes occur in design, implementation, or testing. Revisit as necessary.

Algorithm/Graphical Display. Graphically, this measure is represented as a composite of the module-level path measures. This should reflect test coverage and overall success of testing.

Module-Level Measures. There are four module-level measures:

1. Path measure. Defined as the number of paths* in the module that have been successfully executed at least once, divided by the total number of paths in the module.

2. Statement measure. The number of executable statements in each module that have been successfully exercised at least once, divided by the total number of executable statements in the module.

3. Domain measure. Number of input instances that have been successfully tested with at least one legal entry and one illegal entry in every field of every input parameter, divided by the total number of input instances in the module.

4. Decision point measure (optional). The number of decision points in the module that have been successfully exercised with all classes of legal conditions as well as one illegal condition (if any exist) at least once, divided by the total number of decision points in the module. Each decision point containing an "or" should be tested at least once for each of the condition's logical predicates.

Data Requirements. For each module, provide:

1. Software project name
2. Module identification
3. Number of paths
4. Number of statements
5. Number of input instances
6. Number of paths tested
7. Number of statements tested
8. Number of input instances tested
9. Number of decision points tested
10. Number of paths successfully executed (at least once)
11. Number of statements that have been successfully exercised
12. Number of inputs successfully tested with both legal and illegal entries
13. Number of decision points (optional)
14. Number of decision points that have been successfully exercised at least once with all legal classes of conditions and one illegal condition (optional)

Frequency and Type of Reporting. Report only those modules that have been modified or further tested after depth of testing values have been reported for the first time. In recognition of the effort required to collect and report this measure, the following rules are offered:

1. Always report the domain measure.
2. Compute the path and statement measure, if provided, with automated tools.

A reporting example is provided in Figure A-27.

---

*A path is defined as a logical traversal of a module, from an entry point to an exit point.

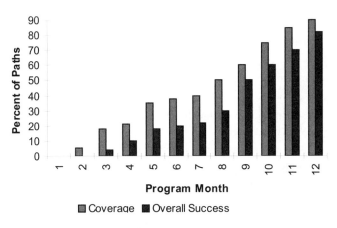

Figure A-27. Depth of testing example.

Use/Interpretation. The depth of testing measures addresses the issues of test coverage, test success, and overall success by considering the paths, statements, inputs, and decision points of a software module.

These measures are to be initially collected at the module level, but this may be extended to the system level if appropriate. Early in the testing process, it makes sense to assess depth of testing at the module level, but later it may make more sense to consider the system in its entirety. This determination is made by the [Project Abbreviation] Program Manager in conjunction with system test/validation.

Fault Profiles Measures. Fault profiles provide insight into the quality of the software, as well as the contractor's ability to fix known faults. These insights come from measuring the lack of quality (i.e., faults) in the software.

Life Cycle Application. Begin after completion of unit testing, when software has been brought under configuration control, and continue through software maintenance.

Algorithm/Graphical Display. For each software project plot, provide:

1. The cumulative number of detected and cumulative number of closed software faults (i.e., TPR/ITRs) will be tracked as a function of time. If manpower allows, one plot will be developed for each priority level.
2. The number of software faults detected and software faults closed tracked as a function of time.
3. Open faults, by module and priority.
4. Calculate the average age of closed faults as follows. For all closed TPR/ITRs, sum the days from the time the STR was opened and when it was closed. Divide this by the total number of closed problem reports. If manpower allows, this should be calculated for each problem priority or overall.

Data Requirements. The following information can be derived from all change enhancement requests (CERs) that are of the TPR, SPR, or ITR variety:

1. CER unique identifier
2. Date initiated
3. Descriptive title of problem
4. Detailed description of problem
5. Priority:
   a. 1 = causes mission-essential function to be disabled or jeopardizes personnel safety. No work-around.
   b. 2 = causes mission-essential function to be degraded. There is a reasonable work-around.
   c. 3 = causes operator inconvenience but does not affect a mission-essential function.
   d. 0 = all other errors.
6. Category:
   a. Requirements = baseline change request (BCR)
   b. Alpha test = internal test report (ITR)
   c. Beta test = test problem report (TPR)
   d. User-driven changes to data = software problem report (SPR)
   e. User-driven change to interfaces = interface change notice (ICN)
7. Status:
   a. Open
   b. Voided
   c. Hold
   d. Testing (alpha)
   e. Fixed
8. Date detected
9. Date closed
10. Software project, module and version
11. Estimated required effort and actual effort required
12. The following reports may be generated using the SCRTS reporting capabilities. These items are the building blocks for graphical representations and fault profiles (for each module by priority):
    a. Cumulative number of CERs
    b. Cumulative number of closed CERs
    c. Average age of closed CERs
    d. Average age of open CERs
    e. Average age of CERs (both open and closed)
    f. Totals for each CER category (described above)

Frequency and Type of Reporting. The Project Manager will update periodically in support of milestones or at major releases. Design stability should be monitored to determine the number and potential impact of design changes, additions, and deletions on the software configuration. Examples are provided in Figures A-28, A-29, and A-30. Types of items examined are:

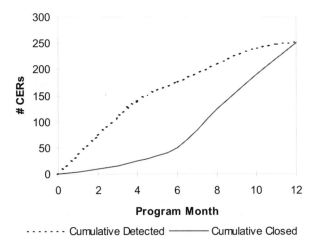

Figure A-28. Cumulative by month reporting.

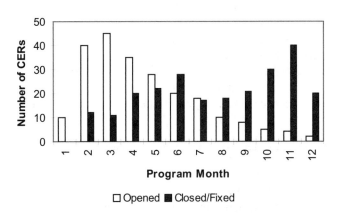

Figure A-29. Reporting by month.

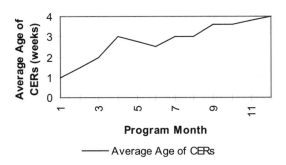

Figure A-30. Average age/open faults.

1. Displays of detected faults versus closed (verified) faults should be examined for each problem's priority level and for each module. Applied during the early stages of development, fault profiles measure the quality of the translation of the software requirements into the design. CERs opened during this time may suggest that requirements are either not being defined or interpreted incorrectly. Applied later in the development process, assuming adequate testing, fault profiles measure the implementation of requirements and design into code. CERs opened during this stage could be the result of having an inadequate design to implement those requirements, or a poor implementation of the design into code. An examination of the fault category should provide indications of these causal relationships.

2. If the cumulative number of closed ITR/TPRs remains constant over time and a number of them remain open, this may indicate a lack of problem resolution. The age of the open STRs should be checked to see if they have been open for an unreasonable period of time. If so, these areas need to be reported to the [Project Abbreviation] Program Manager as areas of increased risk.

3. The monthly noncumulative ITR totals for each project can be compared to the TPR totals to provide insights into the adequacy of the alpha validation program.

4. Open-age histograms can be used to indicate which modules are the most troublesome with respect to fixing faults. This may serve to indicate that the project team may need assistance.

5. Average-age graphs can track whether the time to close faults is increasing with time, which may be an indication that the developer is becoming saturated or that some faults are exceedingly difficult to fix. Special consideration should be paid to the gap between open and fixed CERs. If a constant gap or a continuing divergence is observed, especially as a major milestone is approached, the Program Manager should take appropriate action.

Once an average CER age has been established, large individual deviations should be investigated.

Use/Interpretation. Early in the development process, fault profiles can be used to measure the quality of the translation of the software requirements into the design. Later in the process, they can be used to measure the quality of the implementation of the software requirements and design into code.

If fault profile tracking starts early in software development, an average CER open age of less than three months may be experienced. After release, this value may be expected to rise.

Measurement Information Model in ISO/IEC 15939

Table A-26 provides three examples of instantiations of the Measurement Information Model as presented in ISO/IEC 15939, Information Technology—Systems and Software Engineering—Measurement Process [60].

Table A-26. Measurement information model in ISO/IEC 15939

	Productivity	Quality	Progress
Information Need	Estimate productivity of future project	Evaluate product quality during design	Assess status of coding activity
Measurable Concept	Project productivity	Product quality	Activity status
Relevant Entities	1. Code produced by past projects 2. Effort expended by past projects	1. Design packages 2. Design inspection reports	1. Plan/schedule 2. Code units completed or in progress
Attributes	1. C++ language statements (in code) 2. Time card entries (recording effort)	1. Text of inspection packages 2. Lists of defects found in inspections	1. Code units identified in plan 2. Code unit status
Base Measures	1. Project X Lines of code 2. Project X Hours of effort	1. Package X size 2. Total defects for package X	1. Code units planned to date 2. Code unit status
Measurement Method	1. Count semicolons in Project X code 2. Add time card entries together for Project X	1. Count number of lines of text for each package 2. Count number of defects listed in each report	1. Count number of code units scheduled to be completed by this date 2. Ask programmer for percent complete of each unit
Type of Measurement Method	1. Objective 2. Objective	1. Objective 2. Objective	1. Objective 2. Subjective
Scale	1. Integers from zero to infinity 2. Real numbers from zero to infinity	1. Integers from zero to infinity 2. Integers from zero to infinity	1. Integers from zero to infinity 2. Integers from zero to one hundred
Type of Scale	1. Ratio 2. Ratio	1. Ratio 2. Ratio	1. Ratio 2. Ordinal
Unit of Measurement	1. Line 2. Hour	1. Lines 2. Defects	1. Code unit 2. One-hundredth code unit
Derived Measure	Project X productivity	Inspection defect density	Progress to date
Measurement Function	Divide Project X lines of code by Project X hours of effort	Divide total defects by package size for each package	Add status for all code units planned to be complete to date
Indicator	Average productivity	Design defect density	Code status expressed as a ratio
Model	Compute mean and standard deviation of all project productivity values	Compute process center and control limits using values of defect density	Divide progress to date by (code units planned to date times 100)

(continued)

Table A-26. *(continued)*

	Productivity	Quality	Progress
Decision Criteria	Computed confidence limits based on the standard deviation indicate the likelihood that an actual result close to the average productivity will be achieved. Very wide confidence limits suggest a potentially large d eparture and the need for contingency planning to deal with this outcome.	Results outside the control limits require further investigations	Resulting ratio should fall between 0.9 and 1.1 to conclude that the project is on schedule

PROBLEM RESOLUTION

Risk Taxonomy

The taxonomy in Table A-27 was developed by the SEI [134] in support of risk management process development.

Table A-27. Risk taxonomy

A. Product Engineering
  1. Requirements
    a. Stability
    b. Completeness
    c. Clarity
    d. Validity
    e. Feasibility
    f. Precedent
    g. Scale
  2. Design
    a. Functionality
    b. Difficulty
    c. Interfaces
    d. Performance
    e. Testability
    f. Hardware Constraints
    g. Nondevelopmental Software
  3. Code and Unit Test
    a. Feasibility
    b. Testing
    c. Coding/ Implementation

  4. Integration and Test
    a. Environment
    b. Product
    c. System
  5. Engineering Specialties
    a. Maintainability
    b. Reliability
    c. Safety
    d. Security
    e. Human Factors
    f. Specifications
B. Development Environment
  1. Development Process
    a. Formality
    b. Suitability
    c. Process Control
    d. Familiarity
    e. Product Control
  2. Development System
    a. Capacity
    b. Suitability
    c. Usability
    d. Familiarity

    e. Reliability
    f. System Support
    g. Deliverability
  3. Management Process
    a. Planning
    b. Project Organization
    c. Management Experience
    d. Program Interfaces
  4. Management Methods
    a. Monitoring
    b. Personnel Management
    c. Quality Assurance
    d. Configuration Management
  5. Work Environment
    a. Quality Attitude
    b. Cooperation
    c. Communication
    d. Morale
C. Program Constraints
  1. Resources
    a. Schedule

Table A-27. (continued)

b. Staff	c. Dependencies	d. Prime Contractor
c. Budget	3. Program Interfaces	e. Corporate
d. Facilities	a. Customer	Management
2. Contract	b. Associate	f. Vendors
a. Type of contract	Contractors	g. Politics
b. Restrictions	c. Subcontractor	

Risk Taxonomy Questionnaire

The questionnaire in Table A-28 was developed by the SEI [134] in support of risk management process development. This questionnaire can be extremely helpful when trying to identify and categorize project risk.

Table A-28. Risk taxonomy questionnaire

A. Product Engineering
   Technical aspects of the work to be accomplished
   1. Requirements
      a. Stability
         Are requirements changing even as the product is being produced?
      b. Completeness
         Are requirements missing or incompletely specified?
      c. Clarity
         Are the requirements unclear or in need of interpretation?
      d. Validity
         Will the requirements lead to the product the customer has in mind?
      e. Feasibility
         Are there requirements that are technically difficult to implement?
      f. Precedent
         Do requirements specify something never done before or beyond the experience of program personnel?
      g. Scale
         Is the system size or complexity a concern?
   2. Design
      a. Functionality
         Are there any potential problems in designing to meet functional requirements?
      b. Difficulty
         Will the design and/or implementation be difficult to achieve?
      c. Interfaces
         Are internal interfaces (hardware and software) well defined and controlled?
      d. Performances
         Are there stringent response time or throughput requirements?
      e. Testability
         Is the product difficult or impossible to test?
      f. Hardware Constraints
         Does the hardware limit the ability to meet any requirements?
      g. Nondevelopmental Software
         Are there problems with software used in the program but not developed by the program?

(continued)

Table A-28. *(continued)*

3. Code and Unit Test
   a. Feasibility
      Is the implementation of the design difficult or impossible?
   b. Testing
      Is the specified level and time for unit testing adequate?
   c. Coding/Implementation
      Are the design specifications in sufficient detail to write code? Will the design be changing while coding is being done?
4. Integration and Test
   a. Environment
      Is the integration and test environment adequate? Are there problems in developing realistic scenarios and test data to demonstrate any requirements?
   b. Product
      Are the interface definition inadequate, facilities inadequate, time insufficient? Are there requirements that will be difficult to test?
   c. System
      Has adequate time been allocated for system integration and test? Is system integration uncoordinated? Are interface definitions or test facilities inadequate?
5. Engineering Specialties
   a. Maintainability
      Will the implementation be difficult to understand or maintain?
   b. Reliability
      Are reliability or availability requirements allocated to the software? Will they be difficult to meet?
   c. Safety
      Are the safety requirements infeasible and not demonstrable?
   d. Security
      Are there unprecedented security requirements?
   e. Human Factors
      Is there any difficulty in meeting the human factor requirements?
   f. Specifications
      Is the documentation adequate to design, implement, and test the system?

B. Development Environment
   Methods, procedures, and tools in the production of the software products.
   1. Development Process
      a. Formality
         Will the implementation be difficult to understand or maintain?
      b. Suitability
         Is the process suited to the development mode; e.g., spiral, prototyping? Is the development process supported by a compatible set of procedures, methods, and tools?
      c. Process Control
         Is the software development process enforced, monitored, and controlled using measures?
      d. Familiarity
         Are the project members experienced in use of the process? Is the process understood by all project members?
      e. Product Control
         Are there mechanisms for controlling changes in the product?
   2. Development System
      a. Capacity
         Are there enough workstations and processing capacity for all the staff?

Table A-28. *(continued)*

---

    b. Suitability

       Does the development system support all phases, activities, and functions of the program?

    c. Usability

       Do project personnel find the development system easy to use?

    d. Familiarity

       Have project personnel used the development system before?

    e. Reliability

       Is the system considered reliable?

    f. System Support

       Is there timely expert or vendor support for the system?

    g. Deliverability

       Are the definition and acceptance requirements defined for delivering the system to the customer?

  3. Management Process

    a. Planning

       Is the program managed according to a plan?

    b. Project Organization

       Is the program organized effectively? Are the roles and reporting relationships well defined?

    c. Management Experience

       Are the managers experienced in software development, software management, the application domain, and the development process?

    d. Program Interfaces

       Is there a good interface with the customer and is the customer involved in decisions regarding functionality and operation?

  4. Management Method

    a. Monitoring

       Are management measures defined and is development progress tracked?

    b. Personnel Management

       Are project personnel trained and used appropriately?

    c. Quality Assurance

       Are there adequate procedures and resources to assure product quality?

    d. Configuration Management

       Are the change procedures or version control, including installation site(s), adequate?

  5. Work Environment

    a. Quality Attitude

       Does the project lack orientation toward quality work?

    b. Cooperation

       Does the project lack team spirit? Does conflict resolution require management intervention?

    c. Communication

       Does the project lack awareness of mission or goals or communication of technical information among peers and managers?

    d. Morale

       Is there a nonproductive, noncreative atmosphere? Does the project lack rewards or recognition for superior work?

C. Program Constraints

  Methods, procedures, and tools in the production of the software products.

  1. Resources

    a. Schedule   *(continued)*

Table A-28. *(continued)*

Is the project schedule inadequate or unstable?
   b. Staff
       Is the staff inexperienced, lacking domain knowledge or skills, or not adequately sized?
   c. Budget
       Is the funding insufficient or unstable?
   d. Facilities
       Are the facilities inadequate for building and delivering the product?
2. Contract
   a. Type of contract
       Is the contract type a source of risk to the project?
   b. Restrictions
       Does the contract include any inappropriate restrictions?
   c. Dependencies
       Does the program have any critical dependencies on outside products or services?
3. Program Interfaces
   a. Customer
       Are there any customer problems such as a lengthy document-approval cycle, poor communication, or inadequate domain expertise?
   b. Associate Contractors
       Are there any problems with associate contractors such as inadequately defined or unstable interfaces, poor communication, or lack of cooperation?
   c. Subcontractor
       Is the program dependent on subcontractors for any critical areas?
   d. Prime Contractor
       Is the program facing difficulties with its prime contractor?
   e. Corporate Management
       Is there a lack of support or too much micromanagement from upper management?
   f. Vendors
       Are vendors unresponsive to program needs?
   g. Politics
       Are politics causing a problem for the program?

Risk Action Request

Table A-29. provides the suggested content for a risk action request that may be used to effectively support the assessment of project risk. The recommended content is based on the requirements found in IEEE Std 1540-2001, IEEE Standard for Software Lifecyle Processes—Risk Management annex B, Risk Action request.

Table A-29. Risk action request suggested content

Unique Identifier	Project Constraints
Date of Issue and Status	Risk Description
Approval Authority	Risk Liklihood and Timing
Scope	Risk Consequences
Request Originator	Risk Mitigation Alternatives
Risk Category	Descriptions
Risk Threshold	Recommendation
Project Objectives	Justificaton
Project Assumptions	Disposition

Risk Mitigation Plan

Table A-30 provides suggested content for a risk mitigation plan that may be used to document and control project risk. The recommended content is shown is based on the requirements found in IEEE Std 1540-2001, IEEE Standard for Software Lifecyle Processes—Risk Management, Annex C, Risk Treatment Plan.

Table A-30. Risk mitigation plan suggested content

Unique Identifier
Cross-reference to Risk Action Request Unique Identifier
Date of Issue and Status
Approval Authority
Scope
Request Originator
Planned Risk Mitigation Activities and Tasks
Performers
Risk Mitigation Schedule (including resource allocation)
Mitigation Measures of Effectiveness
Mitigation Cost and Impact
Mitigation Plan Management Procedures

Risk Matrix Sample

A risk matrix is a helpful tool when used to support risk analysis activities [139]. Table A-31 provides an example of a risk matrix. Tables A-32 and Table A-33 provide descriptions of the associated severity and probability levels.

Table A-31. Sample risk matrix

	Probability				
Severity	Frequent	Probable	Occasional	Remote	Improbable
Catastrophic	IN	IN	IN	H	M
Critical	IN	IN	H	M	L
Serious	H	H	M	L	T
Minor	M	M	L	T	T
Negligible	M	L	T	T	T

Legend: T = tolerable, L = low, M = medium, H = high, and IN = intolerable.

Table A-32. Risk severity levels description

Severity	Consequence
Catastrophic	Greater than 6-month slip in schedule; greater than 10% cost overrun; greater than 10% reduction in product functionality.
Critical	Less than 6-month slip in schedule; less than 1% cost overrun; less than 10% reduction in product functionality.
Serious	Less than 3-month slip in schedule; less than 5% cost overrun; less than 5% reduction in product functionality.
Minor	Less than 1-month slip in schedule; less than 2% cost overrun; less than 2% reduction in product functionality.

Table A-33. Risk probability levels description

Probability	Description
Frequent	Anticipate occurrence several times a year (>10 events)
Probable	Anticipate occurrence repeatedly during the year (2 to 10 events)
Occasional	Anticipate occurrence sometime during the year (1 event)
Remote	Occurrence is unlikely though conceivable (< 1 event per year)

MANAGEMENT

Work Breakdown Structure

A work breakdown structure (WBS) defines and breaks down the work associated with a project into manageable parts. It describes all activities that have to occur to accomplish the project. The WBS serves as the foundation for the development of project schedules, budget, and resource requirements.

A WBS may be structured by project activities or components, functional areas or types of work, or types of resources, and is organized by its smallest component—a work package. A work package is defined as a deliverable or product at the lowest level of the WBS. Work packages may also be further subdivided into activities or tasks. IEEE Std 1490-2003, IEEE Guide—Adoption of the PMI Standard, A Guide to the Project Management Body of Knowledge, recommends the use of nouns to represent the "things" in a WBS. Figure A-31 provides an example of a sample WBS organized by activity.

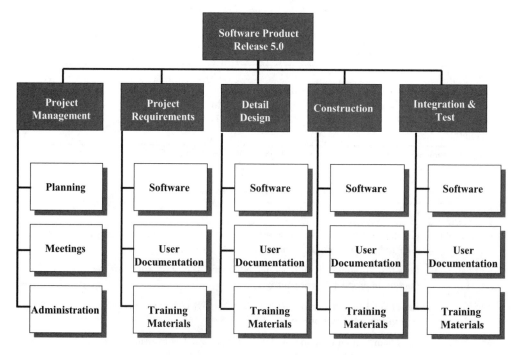

Figure A-31. WBS organized by activity [35].

Work Flow Diagram

Work flow may be defined as the set of all activities in a project from start to finish. A work flow diagram is a pictorial representation of the operational aspect of a work procedure: how tasks are structured, who performs them, what their relative order is, how they are synchronized, how information flows to support the tasks, and how tasks are being tracked. Work flow problems can be modeled and analyzed using Petri nets. Petri nets are abstract, formal models of information flow, showing static and dynamic properties of a system [9]. Figure A-32 provides an example of a workflow diagram.

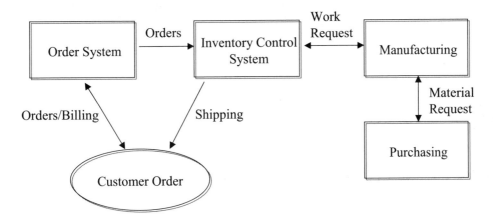

Figure A-32. Example work flow diagram.

Stakeholder Involvement Matrix. The matrix in Table A-34 presents a consolidated view of stakeholder involvement. This matrix can be used to define and document stakeholder involvement. An electronic version of this is provided on the companion CD-ROM as *Stakeholder Involvement Matrix.doc.*

Table A-34. Stakeholder involvement matrix [103]

Project Processes	Customer	Operations	Field Support	Program Management	System Engineering	Software Engineering	System Test	Integration and Test	Integrated Logistics Support	Quality Assurance	Configuration Management	Subcontracts	Help Desk	Suppliers
Proposal Generation	R			L	R	R	R	A	R			R		A
Project Planning	R		R	L	R	R	R	R	R			R		R
Project Monitoring and Control				L	R	R	A	R	R			R		A
Measures				R	L	R					O			
Risk Management	R			L	R	R	A	A	A					
Requirements Development	R	R	O		L	R	R		A	R				
Operations Telecons	R	R	O	O	L	R	O				O			A
Systems Engineering Review	R	R	R	R	L	R	R	R	R	R	R	R	R	R
Technical Exchange Meeting	R	A	R	R	L	R	R	A	A		O			A
Interface Control Working Group	R		R	R	L	R	R							A
Software Reviews					O	L	O			O	R			A
SW CCB				O	R	R	R	O		O	R	L		A
HW CCB				O	R		R	R			R	L		
Integration Testing					A	A	L			A				
Validation Testing		L	R		A	A	R			O				
Verification Testing			R				L			O				
Regression Testing			R				L			O				
Site Installation	R		R		A	A	A	L			R	A	A	
Leased Service Support				A	A	A	A	A	A		A	A	L	A
2nd Level Engineering Support					L	R	R		A		A		A	A
Subcontract Mgmt				A	A	A					A	L		A

Legend: R = required, O = optional, A = as needed, L = leader.

INFRASTRUCTURE

Organizational Policy Examples

The following information is provided for illustrative purposes. Each organization should focus on the development of policies and practices that best support its needs. Each organizational policy should include a top-level statement, any relevant responsibilities, and associated specific actions as desired. The following provides three examples of an organizational policy in support of configuration management activities.

Example 1. Individuals responsible for project leadership shall ensure that all software development, system integration, and system engineering projects conduct configuration management activities to maintain the integrity of their products and data. Each project's configuration management effort shall ensure the identification of configuration items

(CIs) and associated baselines, the use of a change management process for managing changes to baselines, the support of process audits, and configuration status tracking and reporting.

Example 2. Our goal is to deliver the highest quality products to our customers. To ensure that the integrity of the project's products is established and maintained throughout the project's life cycle, this configuration management policy has three primary requirements:

1. Each project must have a CM Lead to develop and carry out the CM process. This individual is responsible for defining, implementing, and maintaining the CM Plan.
2. Every project must have a CM Plan identifying the products to be managed by the CM process, the tools to be used, the responsibilities of the CM team, and the procedures to be followed. The CM Plan must be prepared in a timely manner.
3. Every project must have a Configuration Control Board (CCB). The CCB authorizes the establishment of product baselines and the identification of configuration items/units. It also represents the interests of the Project Manager and all groups who may be affected by changes to the configuration. It reviews and authorizes changes to controlled baselines and authorizes the creation of products from the baselines.

Example 3. In order to ensure that delivered products contain only high-quality content, engineering projects shall:

- Establish and maintain baselines of all identified work products
- Track and control changes to all identified work products
- Establish and maintain the integrity of all baselines

Software Life Cycle (SLC) Selection and Design

Software engineering (SE) organizations planning to use Lean Six Sigma (LSS) need a robust software life cycle for their projects to follow and to improve upon. Specifically, management needs to design, construct, and support the software project life cycle process (SPLCP) for the SE project staff to use and subsequently improve with the LSS tools. IEEE 1074, Standard for Developing a Project Software Life Cycle Process, provides the important steps to create the SPLCP.

The IEEE Std 1074 road map (Figure A-33) starts with the organization selecting their SPLCM on which to map the Activities of IEEE 1074 to produce the software project life cycle (SPLC). This standard requires selection of a user's SPLC based on the organization's mission, vision, goals, and resources. The various software life cycle models described are the waterfall, modified waterfall, V-shaped, incremental, spiral, synchronize and stabilize, rapid prototype, and code and fix.

With the SPLCM and IEEE 1074, the required activities are determined to create the SPLC. The SPLC is that portion of the entire software life cycle (SLC) that is applicable to a specific project. It is the sequence of activities created by mapping the activities of this standard onto a selected SPLC.

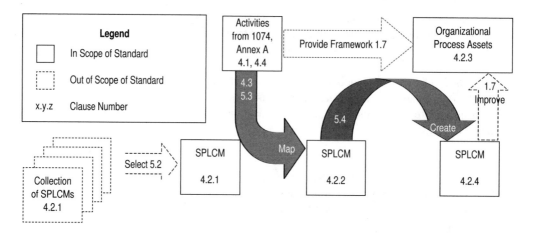

Figure A-33. Developing an SPLCP.

The next step, the available organizational process assets (OPA), should be applied to the SPLC activities, and known constraints should be reconciled. The output information generated by each activity should be assigned to the appropriate document(s). The SPLCP is used by the project members within their development environment to build and deliver the desired end system.

As previously stated, IEEE Std 1074 describes several types of software project life cycle models (SPLCM), such as waterfall, modified waterfall, V-shaped, incremental, spiral, synchronize and stabilize, rapid prototype, and code and fix.

Waterfall. The waterfall model uses a sequence of development stages in which the output of each stage becomes the input for the next. The focus is on control. This is most common with maintenance releases in which the control of changes is important to the customer.

Modified Waterfall. The modified waterfall or fountain model recognizes that although some activities cannot start before others; such as those that a design before you can start coding, there is a considerable overlap of activities throughout the development cycle.

V-Shaped. The V-shaped model is just like the waterfall model; the V-shaped life cycle is a sequential path of execution of processes. Each phase must be completed before the next phase begins. Testing is emphasized in this model moreso than in the waterfall model. The testing procedures are developed early in the life cycle before any coding is done, during each of the phases preceding implementation.

Incremental. The incremental model is an intuitive approach to the waterfall model. The incremental model divides the product into builds, so that sections of the project are created and tested separately. Multiple development cycles take place here, making the life cycle a "multiwaterfall" cycle. Cycles are divided up into smaller, more easily managed iterations. Each iteration passes through the requirements, design, implementation, and testing phases. This approach will likely find errors in user requirements quickly, since user feedback is solicited for each stage and because code is tested soon after it is written.

Spiral. The spiral life cycle model is the combination of the classic waterfall model and an element called risk analysis. This model is very appropriate for new projects and large software projects. The model consists of four main parts or blocks—planning, risk analysis, engineering, and customer evaluation—and the process is shown by a continuous spiral loop going from the outside towards the inside. The baseline spiral starts in the planning phase where requirements are gathered and risk is assessed. Each subsequent spiral builds on the baseline spiral. The spiral model emphasizes the need to go back and reiterate earlier stages a number of times as the project progresses. Each cycles producing an early prototype representing a part of the entire project. This approach helps demonstrate a proof of concept early in the cycle, and it more accurately reflects the disorderly, even chaotic, evolution of technology.

Synchronize and Stabilize. The synchronize and stabilize method combines the advantages of the spiral model with technology for overseeing and managing source code. This method allows many teams to work efficiently in parallel.

Rapid Prototype for New Projects. In the rapid prototyping (sometimes called rapid application development) model, initial emphasis is on creating a prototype that looks and acts like the desired product in order to test its usefulness. The prototype is an essential part of the requirements determination phase, and may be created using tools different from those used for the final product. Once the prototype is approved, it is discarded and the "real" software is written.

Code and Fix. Code and Fix is the crudest of the methods. If you do not use a methodology, it is likely that you are doing code and fix. Write some code, and then keep modifying it until the customer is happy. Without planning, this is very open-ended and can be risky. Code and fix rarely produces useful results. It is very dangerous as there is no way to assess progress, quality, or risk.

Process Definition

A template has been provided on the companion CD-ROM to assist project managers in identifying project-critical deliverables and assuring their completion as needed. This template can be used to assist in the project-specific mapping of information into the required project documentation. This template is named *SPLCP Mapping.doc.*

Process Definition Form

A defined, Lean Six Sigma activity should clearly state all inputs, entry criteria, associate activities, roles, measures, required verification, outputs, and exit criteria. Figure A-34 provides an example of a form that may be used in support of process definition and development.

OVERVIEW

Entry Criteria	Exit Criteria

Inputs	Outputs

Required Activities

Stakeholders   L = lead; S = support; R = review; O = optional (as necessary)

Measures

Organizational Process Improvement Information

Verification

Tailoring

Implementation Guidance

Supporting Documentation and Assets

Figure A-34. Process definition form [142].

Asset Library Catalog

IEEE Std 610.12 (R2002), IEEE Standard Glossary of Software Engineering Terminology, defines process as, "A set of activities. A sequence of steps performed for a given purpose; for example, the software development process." [9] Similarly, Lean Six Sigma defines process as set of interrelated or interacting activities that transforms inputs into outputs. These inputs and outputs can be considered organizational assets.

Process Assets are defined as

... artifacts that relate to describing, implementing, and improving processes (e.g., policies, measurements, process descriptions, and process implementation support tools). The term "process assets" is used to indicate that these artifacts are developed or acquired to meet the business objectives of the organization, and they represent investments by the organization that are expected to provide current and future business value. [168]

Organizational assets include the following [168]:

- An organization's set of standard processes, including the process architectures and process elements
- Descriptions of life cycle models approved for use
- Guidelines and criteria for tailoring the organization's set of standard processes
- An organization's measurement repository
- An organization's asset library

An organization's asset library may, therefore, be described as a library of information used to store process assets that may be used in the support of the definition, implementation, and management of processes within an organization. Some examples of this type of documentation include policies, processes, procedures, checklists, lessons learned, templates and plans, referenced standards, and training materials.

IMPROVEMENT

Organizational Improvement Checklist

The IDEAL[SM] (initiating, diagnosing, establishing, acting, and learning) model is an organizational improvement model that was originally developed to support CMM®-based software process improvement [159]. It serves as a road map for initiating, planning, and implementing improvement actions. This model can serve to lay the groundwork for a successful improvement effort (initiate), determine where an organization is in reference to where it may want to be (diagnose), plan the specifics of how to reach goals (establish), define a work plan (act), and apply the lessons learned from past experience to improve future efforts (learn).

This model serves as the basis for CMMI® process improvement. The information presented in Table A-35 provides a matrix of IDEALSM phases, activities, details, and deliverables in support of process improvement and may be used as a reference checklist in support of process improvement activities.

Table A-35. Organizational process improvement checklist using IDEAL^{SM} [171]

IDEAL^{SM} Phases (5)	IDEAL^{SM} Activities (14)	IDEAL^{SM} Details	Organization Milestone Details/Deliverables
Initiating	Stimulus for Change	Know what prompted this particular change: • Reaction to unanticipated events/ circumstances • Edict from "on high" • Enactment of change as part of proactive continuous improvement approach	Business Results Problem Reports
	Set Context	Identify where this change fits within the larger organizational context: • Within the organization's business strategy Specify business goals and objectives that will be realized or supported: • The organization's core mission • Business goals and objectives • A coherent vision for the future • A strategy for achieving the vision • Reference models Identify impact on other initiatives and ongoing work	Business goals and objectives
	Build Sponsorship	Secure the support required to give the change a reasonable chance of succeeding Effective Sponsors: • Give personal attention to the effort • Stick with the change through difficult times • Commit scarce resources to the effort • Change their own behavior to be consistent with the change being implemented • Modify the reward system	Sponsor(s) signed up
	Charter Infrastructure	Adjust relevant organizational systems to support the effort Establish an infrastructure for managing specifics of implementation: • An oversight group (senior-level people) • A change agency group (staff to the oversight body) • One or more technical working groups (TWGs)	Engineering Process Group (EPG) Charter
Diagnosing	Characterize Current and Desired States	Identify the organization's current state with respect to a related standard or reference model: • A reference model (e.g., capability maturity model)	Process Appraisal Plan, Assessment Plan

(continued)

Table A-35. *(continued)*

IDEAL^{SM} Phases (5)	IDEAL^{SM} Activities (14)	IDEAL^{SM} Details	Organization Milestone Details/Deliverables
Diagnosing (*cont.*)		• Standardized appraisal instruments and methodologies associated with the model • Business objectives aligned with the change effort Identify the organization's desired state against the same standard or model • Analyze the gap between the current and desired states • Determine aspects of current state to retain • Determine the work to reach the desired state • Align recommendations with business objectives • Identify potential projects where you may pilot or test	
	Develop Recommendations	Recommend actions that will move the organization from where it is to where you want it to be Identify potential barriers to the effort	Recommendation Report
Establishing	Set Priorities	Develop priorities among the recommended actions on the basis of such things as: • Availability of key resources • Dependencies between the actions recommended • Funding Organizational realities: • Resources needed for change are limited • Dependencies exist between recommended activities • External factors may intervene • Organization's strategic priorities must be honored	
	Develop Approach	Develop an approach that reflects priorities and current realities: • Strategy to achieve vision (set during initiating phase) • Specifics of installing the new technology • New skills and knowledge required of the people who will be using the technology • Organizational culture • Potential pilot projects • Sponsorship levels • Market forces	

(*continued*)

Table A-35. *(continued)*

IDEAL^{SM} Phases (5)	IDEAL^{SM} Activities (14)	IDEAL^{SM} Details	Organization Milestone Details/Deliverables
Establishing (*cont.*)	Plan Actions	Develop plans to implement the chosen approach: • Deliverables, activities, and resources • Decision points • Risks and mitigation strategies • Milestones and schedule • Measures, tracking, and oversight	Organizational Process Improvement Plan Process Action Plan (PAP)
Acting	Create Solution	Bring together everything available to create a "best guess" solution specific to organizational needs: • Existing tools, processes, knowledge, skills • New knowledge, information • Outside help Solution should: • Identify performance objectives • Finalize the pilot/test group • Construct the solution material and training with pilot/test group • Develop draft plans for pilot/test and implementation Technical working groups develop the solution with pilot/test group	Revised process description and/or process asset
	Pilot/Test Solution	Put the solution in place on a trial basis to learn what works and what does not: • Train the pilot or test group • Perform the pilot or test • Gather feedback	Process lessons learned
	Refine Solution	Revisit and modify the solution to incorporate new knowledge and understanding Iterations of pilot/test and refine may be necessary before arriving at a solution that is deemed satisfactory	Revised process description and/or process asset
	Implement Solution	When solution is deemed workable and sufficient, implement it throughout the organization: • Adjust the plan as necessary • Execute the plan • Conduct postimplementation analysis • Various approaches may be used: top-down, lateral, or staged	Process Lessons Learned
Learning	Analyze and Validate	Determine what was actually accomplished by the effort: • In what ways did it/did it not accomplish its intended purpose? • What worked well?	Change in Measurements

continued)

Table A-35. *(continued)*

IDEAL^{SM} Phases (5)	IDEAL^{SM} Activities (14)	IDEAL^{SM} Details	Organization Milestone Details/Deliverables
		• What could be done more effectively or efficiently?	
		Compare what was accomplished with the intended purpose for undertaking the change	
		Summarize/roll up lessons learned regarding processes used to implement IDEAL	
	Propose Future Actions	Develop recommendations concerning management of future change efforts using IDEAL:	
		• Improving an organization's ability to use IDEAL	
		• Address a different aspect of the organization's business	

Organization Process Appraisal Checklist

The information provided in Table A-36 may be used in support of the organization process appraisals. An electronic version of this checklist is provided on the companion CD-ROM, identified as *Process Appraisal Checklist.doc.*

Table A-36. Organization process appraisal checklist

Checklist	Dates + Resources
1. Identify Appraisal Scope	
a. Appraisal input subject to sponsor approval	
2. Develop Appraisal Plan [using this checklist]	
a. Select appraisal class	
b. Identify appraisal scope	
c. Select appraisal team members	
d. Develop and publish appraisal plan	
3. Prepare Organization for Appraisal	
a. Prepare and train appraisal team	
b. Brief appraisal participants	
c. Data collection and analysis	
d. Administer questionnaire	
e. Conduct initial document review	
f. Remediate, consolidate, triage responses	
g. Conduct opening meeting	
h. Interview project leaders	
i. Consolidate information	
j. Interview middle managers	
k. Consolidate information	
l. Interview FAR groups	
m. Consolidate information	
n. Verify and validate objective evidence	
o. Prepare draft findings	
p. Present draft findings	
q. Consolidate information	
4. Rate and prep final findings	
5. Present final findings	
6. Conduct executive session	
7. Present final findings to appraisal participants	
8. Wrap-up appraisal	

Lessons Learned

The documentation of lessons learned facilitates individual project knowledge sharing so as to benefit the entire organization. A successful lessons-learned program can support the repeat of desirable outcomes and avoid undesirable outcomes. Lessons learned should not only be used as a close-out activity; the continual recording of lessons learned throughout the project is the best way to ensure that information is accurately recorded. Figure A-35 is provided as a lessons-learned template. Please refer to the companion CD-ROM for an electronic version of this template, entitled *Lessons Learned.doc.*

[Project Name] Lessons Learned
Revision [xx], date

Team Meeting

During each project team meeting discuss what strategies contributed to success as well as areas of potential improvement. Enter your conclusions in the table below (insert rows as needed).

Strategies and Processes that Led to Success

Date	Description

Areas of Potential Improvement

Date	Description

Project Close-Out Discussion

All stakeholders should participate in the lessons learned close-out discussion. Use the questions below to trigger discussion and summarize findings.

1. List the project's three biggest successes

Description	Contributing Factors

2. Other relevant successes

Description	Contributing Factors

3. List improvement items and proposed strategies

Description	Improvement Strategy

4. Additional Comments

Document Approval

The following stakeholders have reviewed the information contained in this Project lessons learned document and agree with its content.

Name	Title	Signature	Date

Figure A-35. Lessons-learned template. [Adapted from a template written by Covansys for the City of Raleigh, NC, Enterprise PMO. Generic format by CVR-IT (www.cvr-it.com). May be used freely if referenced.]

Measures Definition for Organizational Processes

IEEE Std. 982.1-1998, IEEE Standard Dictionary of Measures to Produce Reliable Software; IEEE 1061-1998, IEEE Standard for a Software Quality Metrics Methodology; IEEE Std. 14143.1-2000, Implementation Note for IEEE Adoption of ISO/IEC 14143-1:1998 Information Technology—Software Measurement—Functional Size Measurement—Part 1: Definition of Concepts; ISO/IEC 15939:2007, Information Technology—Systems and Software Engineering—Measurement Process; and Lean Six Sigma materials were used as to develop the following suggestions for measures in support of organizational processes.

Measures are often classified as either process or product measures. Process measures are based upon characteristics associated with the development process or environment. Product measures are measures associated with the software product being developed. Examples of these types of measures are provided in Table A-37.

Table A-37. Process versus product measures

Process Measures

    Estimated and actual duration (calendar time): track for individual tasks, project milestones, and process support functions.

    Estimated and actual effort (labor hours): track for individual tasks, project milestones, and process support functions.

Product Measures

    Count lines of code, function points, object classes, numbers of requirements.

    Time spent in development and maintenance activities.

    Count the number of defects, identifying their type, severity, and status.

The following basic information in support of process measurement should be collected [149]:

- Released defect levels
- Product development cycle time
- Schedule and effort estimating accuracy
- Reuse effectiveness
- Planned and actual cost

TRAINING

Figure A-36 describes a form that may be used to request training. Project managers should note training needs within their project. Use as many forms as required to document all training needs. Training managers may use this form to analyze, evaluate, validate and prioritize training requests. An electronic version, *Training Request Form.doc,* is provided on the companion CD-ROM.

Project Name: _____ Project Manager: _____

Project Location: _____ Date: _____

Will project contract pay for requested training?   Yes   No   Labor only

Describe the Training Requirement:

Purpose and date of required training:

Number of employees to be trained:

Can employees be released for training during work hours?   Yes   No
Are subject matter experts available?   Yes   No
If so, Names:

Project training requirement priority (5 is the highest, 1 is the lowest)
5   4   3   2   1

Recommended provider (if known)

Recommended course title (if known)

Recommended action:

Program Manager: _____ Date: _____

Division Manager: _____ Date: _____

Status: (To be completed by Training Manager)

Training Manager: _____ Date: _____

SBU Training Requirement Priority (5 is the highest, 1 is the lowest)
5   4   3   2   1

ACTION TAKEN:

Invalid requirement* _____ Valid requirement_____

Placed on training schedule_____ No action taken*_____

Funding identified: _____

Training course identified: _____

Placed on development schedule: _____

Provider: _____

Course title: _____ Date: _____

*Remarks:

Figure A-36. Training request form.

Training Log

Figure A-37 describes a form may be used to document training. Project managers, or individuals delegated with this responsibility, should record all completed training. Training managers may use this form to analyze, evaluate, validate, and prioritize training requests. An electronic version, *Training Log.doc,* is included on the companion CD-ROM.

Trainee Name	Date	Location	Description of Training	No. Hours	ID #	Comments

Figure A-37. Training log.

PLANNING FOR SMALL PROJECTS

PROCESS DEFINITION AND SMALL PROJECTS

Many organizations are intimidated by the thought of large-scale organizational software process improvement efforts. Lean Six Sigma provides an ideal model for improving the processes and practices associated with smaller software development efforts. The lean methodology stresses that improvements should be found through the application of simple methods and that these gains should be team based. Proponents of Six Sigma emphasize that their techniques are best applied with common logic [80]. Lean Six Sigma can be summarized as an attempt to understand what part of the process or product to be improved is affected by variation and what part is affected by waste.

The likelihood that small software development efforts will experience implementation problems is potentially greater due to

- Minimal available resources
- Difficulty in understanding and applying the standards
- Costs involved in setting up and maintaining a quality-management system

One thing that distinguishes small settings from medium-to-large settings is that the resources an organization would expect to be used to support process improvement efforts (appraisals, consulting, infrastructure building) would be a large percentage of its operating budget.

With only a few people involved, communications in a small business can often be simple and more direct. Individuals are expected to undertake a wide variety of tasks within the business. Decision making is confined to a few people (or even one). The key to implementing Lean Six Sigma on a smaller scale is having defined process procedures and practices. Also, instead of having all the documentation relate to a single project, the supporting plans (i.e., software requirements management, configuration management, quality assurance, etc.) describe the processes adopted by the managing organization. In a typical organization, you will find three layers of management, from bottom to top: project, program, and organization.

To provide support for the small software project, the policies might be developed at the organizational level and the supporting plans at the program level. The project would then be required to follow these program-level plans and only then develop project-specific processes and practices. This approach provides several immediate benefits. It allows the technical manager to get down to the business of managing development, encourages standardization across the organization, and encourages participation in process definition activities at the organizational level.

PROJECT MANAGEMENT PLAN—SMALL PROJECTS

The purpose of the small project plan (SPP) is to help project managers develop the processes and practices associated with the management of a small software project. This template is useful for projects with time estimates ranging from one person-month to one person-year. It is designed for projects that strive to follow repeatable and defined processes. This project plan is used to refer to existing organizational policies, plans, and procedures. Exceptions to existing policies, plans, and procedures must be documented with additional information appropriate to each project's SPP.

The SPP provides a convenient way to gather and report the critical information necessary for a small project without incurring the overhead of documentation necessary to support the additional communication channels of a larger project. Software quality assurance (SQA) procedures will be captured in Appendix A of the software development plan (SDP). Appendix B will be used to document software configuration management (SCM) activities. Project management and oversight activities (i.e., risk tracking, problem reporting) will be documented in Appendix C of the SDP. Requirements for the project will be captured in Appendix D of the SPP rather than a separate software requirements specification (SRS). Table B-1 provides a suggested outline for a SDP.

Table B-1. Small software project management plan document outline

Title Page
Revision Page
Table of Contents
1. Introduction
    1.1 Identification
    1.2 Scope
    1.3 Document Overview
    1.4 Relationship to Other Plans
2. Acronyms and Definitions
3. References
4. Overview
    4.1 Relationships
    4.2 Source Code
    4.3 Documentation

(continued)

Table B-1. *(continued)*

This small software project management plan template is designed to facilitate the definition of processes and procedures relating to software project management activities.

Small Software Project Management Plan Document Guidance

This template was developed using IEEE Std 12207.0 and 12207.1, Standards for Information Technology—Software Life Cycle Processes [40]; IEEE Std 1058, IEEE Standard for Software Project Management Plans [13]; IEEE Std 730-2002, IEEE Standard for Software Quality Assurance Plans0; IEEE Std 828, IEEE Standard for Software Configuration Management Plans [10]; IEEE Std 830, IEEE Recommended Practice for Software Requirements Specifications [6]; IEEE Std 1044, Standard Classification for Software Anomalies [7]; IEEE Std 982.1, Standard Dictionary of Measures to Produce Reliable Software [8]; IEEE Std 1045, Standard for Software Productivity Metrics [12]; and IEEE Std 1061, Software Quality Metrics Methodology [17], which have been adapted to provide the foundational support necessary for Lean Six Sigma practitioners.

This plan must be supplemented with separate supporting plans. It may be used to support the development of a SDP for a project operating under the same organizational configuration management, quality assurance, and risk management policies and procedures. This eliminates redundant documentation, encourages conformity among development efforts, facilitates process improvement through the use of a common approach, and more readily allows for the cross-project transition of personnel.

Introduction. This section should provide a brief summary description of the project and the purpose of this document.

Identification. This subsection should briefly identify the system and software to which this SPP applies. It should include such items as identification number(s), title(s), abbreviations(s), version number(s), and release number(s). It should also outline deliverables, special conditions of delivery, or other restrictions/requirements (e.g., delivery media). An example follows:

> This document applies to the software development effort in support of the development of [Software Project] Version [xx]. A detailed development time line in support of this effort is provided in the supporting project Master Schedule.

Scope. This subsection should briefly state the purpose of the system and software to which this SPP applies. It should describe the general nature of the existing system/software and summarize the history of system development, operation, and maintenance. It should also identify the project sponsor, acquirer, user, developer, support agencies, current and planned operating sites, and relationship of this project to other projects. This overview should not be construed as an official statement of product requirements. It will only provide a brief description of the system/software and will reference detailed product requirements outlined in the SRS, Appendix D. An example follows:

> [Project Name] was developed for the [Customer Information]. [Brief summary explanation of product.]
>
> This overview shall not be construed as an official statement of product requirements. It will only provide a brief description of the system/software and will reference detailed product requirements outlined in the [Project Name] SRS.

Document Overview. This subsection should summarize the purpose, contents, and any security or privacy issues that should be considered during the development and implementation of the SPP. It should specify the plans for producing scheduled and unscheduled updates to the SPP and methods for disseminating those updates. This subsection should also explain the mechanisms used to place the initial version of the SPP under change control and to control subsequent changes to the SPP. An example is provided:

> The purpose of the [Project Name] Software Development Plan (SDP) is to guide [Company Name] project management during the development of [Product Identification]. Software Quality Assurance (SQA) procedures are captured in Appendix A of the SDP. Appendix B is used to document Software Configuration Management (SCM) activities when these activities deviate from [Organization Name] SCM Plan. Project management and oversight activities (i.e., risk tracking, problem reporting) are documented in Appendix B of the SDP when these activities deviate from the [Organization Name] Risk Management Plan. Requirements for the project will be captured in the [Project Name] Software Requirements Specification (SRS). This plan will be placed under configuration management controls according the [Organization Name] Software Configuration Management Plan. Updates to this plan will be handled according to relevant SCM procedures and reviews as described in the [Organization Name] Software Quality Assurance Plan.

Relationship to Other Plans. This subsection should describe the relationship of the SPP to other project management plans. If organizational plans are followed, this should be noted here. An example is provided:

> There are several other [Organization Name] documents that support the information contained within this plan. These documents include: [Organization Name] Software Configuration Management Plan, [Organization Name] Software Quality Assurance Plan,

and [Organization Name] Measurement and Measurements Plan. [Project Name] project-specific documentation relating to this plan include: [Project Name] Software Requirements Specifications and [Project Name] Master Schedule. This plan has been developed in accordance with all [Organization Name] software development processes, policies, and procedures.

Acronyms and Definitions. This section should identify acronyms and definitions used within the SPP for each project.

References. This section should identify the specific references used within the SPP. Reference to all associated software project documentation, including the statement of work and any amendments should be included.

Overview. This section of the SPP should list the items to be delivered to the customer, delivery dates, delivery locations, and quantities required to satisfy the terms of the contract. This list should not be construed as an official statement of product requirements. It will only provide an outline of the system/software requirements and will reference detailed product requirements outlined in the SRS, Appendix D. An example follows:

> The project schedule in support of [Project Name] was initiated [Date] with a target completion date of [Date]. Incremental product deliveries may be requested, but none are identified at this time. The delivery requirements are to install [Project Name] on the current [Project Name] external website. [Project Name] will be delivered once deployment has occurred and [Customer Name] accepts the product. [Customer Name] may request the installation of [Product Name] at one additional location. A [Project Name] user's manual shall also be considered to be required as part of this delivery. Detailed schedule guidance is provided by [document name], located in the [document location]. Detailed requirements information is provided by [document name], located in the [document location].

Relationships. This subsection should identify interface requirements and concurrent or critical development efforts that may directly or indirectly affect product development efforts.

Source Code. This subsection should identify source code deliverables. An example is provided:

> There are no requirements for source code deliverables. If [customer name] requests the source code, it will be delivered on CD-ROM.

Documentation. This subsection should itemize documentation deliverables and their required format. It should also contain, either directly or by reference, the documentation plan for the software project. This documentation plan may either be a separate document, stating how documentation is going to be developed and delivered, or may be included in this plan as references to existing standards with documentation deliverables and schedule detailed herein. An example follows:

> [Project Name] will have on-line help; a user's manual will be created from this online help system. The user's manual will be posted on the [Project Name] website and will be available for download by requesting customers. Please refer to the [Project Name] master schedule for a development timeline of associated user's on line help and manual.

Project Resources. This subsection should describe the project's approach by describing the tasks (e.g., req. → design → implementation → test) and efforts (update documentation, etc.) required to successfully complete the project. It should state the nature of each major project function or activity and identify the individuals who are responsible for those functions or activities.

This subsection should also describe the makeup of the team, project roles, and internal management structure of the project. Diagrams may be used to depict the lines of authority, responsibility, and communication within the project. Figure B-1 is an example of an organizational chart.

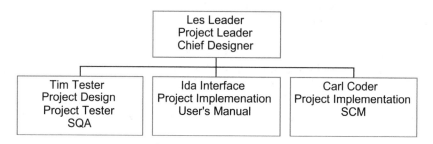

Figure B-1. Example project organization.

The relationship and interfaces between the project team and all nonproject organizations should also be defined by the SPP. Minimum interfaces include those with the customer, elements (management, SQA, SCM, etc.), and other support teams. Relationships with other contracting agencies or organizations outside the scope of the project that will impact the project require special attention within this section.

Project Constraints. This subsection should describe the limits of the project, including interfaces with other projects, the application of the program's SCM and SQA (including any divergence from those plans), and the relationship with the project's customer. This section should also describe the administrative and managerial boundaries between the project and each of the following entities: parent organization, customer organization, subcontracted organizations, or any organizational entities that interact with the project.

This subsection should capture the anticipated volume of the project through quantifiable measurements such as lines of code, function points, number of units modified, and number of pages of documentation generated or changed. It may be useful, if the project is well defined in advance, to break down project activities and perform size estimates on each individual activity. This information can then be tied to the project schedule as defined in Section 5.

Software Process. This section contains the following subsections.

Software Development Process. This subsection of the SPP should define the relationships among major project functions and activities by specifying the timing of major milestones, baselines, reviews, work products, project deliverables, and signature approvals. The process model may be described using a combination of graphical and textual notations that include project initiation and project termination activities.

Life Cycle Model. This subsection of the SPP should identify the life cycle models used for the software development process. It should describe the project organizational structure, organizational boundaries and interfaces, and individual responsibilities for the various software development elements.

Software Engineering Activities. This subsection contains the following five subsections:

Handling of Critical Requirements. This subsection will identify the overall software engineering methodologies to be used during requirements elicitation and system design. An example follows:

> Please refer to the [Organization Name] Software Requirements Management Plan for information regarding the elicitation and management of customer requirements.

Recording Rationale. This subsection should define the methodologies used to capture design and implementation decisions. An example is provided:

> All initial design and implementation decisions will be recorded and posted in the [Project Name] portal project workspace. All formal design documentation will follow the format as described in the [Organization Name] Design Documentation Template.

Computer Hardware Resource Utilization. This subsection should describe the approach to be followed for allocating computer hardware resources and monitoring their utilization. Example text is provided:

> All computer hardware resource utilization is identified in the [Project Name] Spend Plan. Resource utilization is billed hourly as an associated site contract charge. Twenty thousand dollars has been allocated to support the procurement of a development environment. Charges in support of project equipment purchases will be recorded by the Project Manager, [Name], and reported to senior management weekly.

Reusable Software. This subsection should describe the approach for identifying, evaluating, and reporting opportunities to develop reusable software products. The following example is provided:

> It is the responsibility of each developer to identify reusable code during the development process. Any item identified will be placed in the [Organization Name] Reuse library that is hosted on the [location]. It is the responsibility of each programmer to determine whether items in the reuse library may be employed during [Project Name] development.

Software Testing. This subsection should be further divided to describe the approach for software implementation and unit testing. An example follows:

> This project will use existing templates for software test plans and software unit testing when planning for testing. These separate documents will be hosted [location].

Schedule. This section of the SPP should be used to capture the project's schedule, including milestones and critical paths. Options include Gantt charts (*Milestones Etc.*™, *Microsoft Project*™), Pert charts, or simple time lines. Figure B-2 is an example schedule created with *Milestones, Etc.*™ for a short (two month) project.

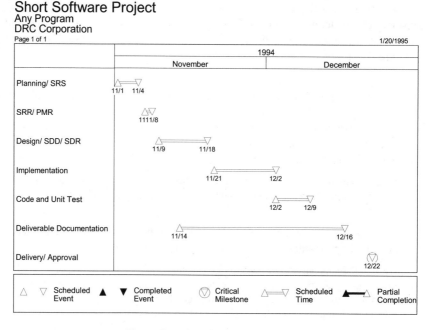

Figure B-2. Sample project schedule

Appendix A. Software Quality Assurance. This appendix should be divided into the following sections to describe the approach to software quality assurance (SQA): software quality assurance evaluations, software quality assurance records, independence in software quality assurance, and corrective action. This appendix should reference the organizational-, or program-, level quality assurance policies, plans, and procedures and should detail any deviations from the organizational standard. If a separate SQA plan has been created in support of the development work, simply provide a reference to that document as follows:

> Refer to the [Organization Name] Software Quality Assurance Plan for all applicable processes and procedures.

Appendix B. Software Configuration Management. This appendix should be divided into the following sections to describe the approach for software configuration management (SCM): configuration identification, configuration control, configuration status accounting, configuration audits, packaging, storage, handling, and delivery. This appendix should reference the organizational-, or program-, level configuration management policies, plans, and procedures and should detail any deviations from the organizational standard. If a separate SCM plan has been created in support of the development work, simply provide a reference to that document:

> Refer to the [Organization Name] Software Configuration Management Plan for all applicable processes and procedures.

Appendix C. Risk Tracking/Project Oversight. This appendix should be divided into the following sections to describe the approach for project risk identification and tracking: risk management strategies, measurement activities, and identified project risks. This appendix should reference the organizational-, or program-, level risk and project management policies, plans, and procedures and should detail any deviations from the organizational standard. If separate measures and measurement plan have been created in support of the development work, simply provide a reference to that document:

> The following measures will be taken as part of project oversight activities. Please refer to the [Organization Name] Software Project Measurement Plan for descriptions:
>
> [List of measures]

Appendix D. Software Requirements Specification. This appendix is the equivalent of the software requirements specification (SRS) required for large projects. It is produced early in the project life cycle and contains descriptions of all project requirements. This appendix should reference the organizational- or program-level requirements management policies, plans, and procedures and should detail any deviations from the organizational standard. If a separate SRS has been created in support of the development work, simply provide a reference to that document.

CD-ROM REFERENCE SUMMARY

The following table is provided as a summary reference of the template items found on the companion CD-ROM.

Process	Template	File Name
Software Engineering Fundamentals		
Acquisition	Acquisition Strategy Checklist	Acquisition Strategy Checklist.doc
Acquisition	Software Acquisition Plan	Software Acquisition Plan.doc
Acquisition	Supplier Checklist	Supplier Checklist.doc
Acquisition	Supplier Performance Standards	Supplier Performance Standards.doc
Acquisition	Decision Matrix	Decision Matrix.xls
Audit	Software Measurement and Measures Plan	Software Measurement and Measures Plan.doc
Configuration Management	Software Configuration Management Plan	Software Configuration Plan.doc
Configuration Management	CCB Charter	CCB Charter.doc
Configuration Management	CCB Letter of Authorization	CCB Letter of Authorization.doc
Development	Software Project Management Plan	Software Project Management Plan.doc
Development	Interface Control Document	Interface Control Document.doc
Development	Software Transition Plan	Software Transition Plan.doc
Development	Software Users Manual	Software Users Manual.doc
Development	System Integration Test Report	System Integration Test Report.doc
Development	Unit Test Report	Unit Test Report.doc
Development	Software Design Document	Software Design Document.doc
Development	Software Requirements Specification	Software Requirements Specification.doc

Process	Template	File Name
Development	System Requirements Specification	System Requirements Specification.doc
Development, Management	Software Project Management Plan	Software Project Management Plan.doc
Development, Management	Software Requirements Management Plan	Software Requirements Management Plan.doc
Development, Validation	System Integration Test Plan	System Integration Test Plan.doc
Improvement	Lessons Learned	Lessons Learned.doc
Improvement	SPLCP Mapping	SPLCP Mapping.doc
Joint Review	Open Issues List	Open Issues List.doc
Maintenance and Configuration Management	Baseline Change Request	Baseline Change Request.doc
Maintenance and Configuration Management	Version Description Document	Version Description Document.doc
Management	Risk Management Plan	Risk Management Plan.doc
Management	Stakeholder Involvement Matrix	Stakeholder Involvement Matrix.doc
Quality Assurance	Process Appraisal Checklist	Process Appraisal Checklist.doc
Quality Assurance	Software Quality Assurance Plan	Software Quality Assurance Plan.doc
Quality Assurance	SQA Inspection Log	SQA Inspection Log.doc
Supplier	Concept of Operations Document	ConOps Document.doc
Training	Training Log	Training Log.doc
Training	Training Plan	Training Plan.doc
Training	Training Request Form	Training Request Form.doc
Validation	Software Test Plan	Software Test Plan.doc
Validation	System Test Plan	System Test Plan.doc
Verification	Inspection Log Defect Summary	Inspection Log Defect Summary.doc
Verification	Inspection Report	Inspection Report.doc

Lean Six Sigma

Process	Template	File Name
Analyze	Multivote	Multivote.xls
Measure	Prioritization Matrix	PrioritizationMatrix.xls
Create Solution	Project Charter	
Analyze	Pugh Concept Selection Matrix	Pugh Matrix.xls
Create Solution	Quality Function Deployment (House of Quality)	QFD HouseofQuality.xls
	QFD House of Quality2	QFD HouseofQuality2.xls
	SIPOC	SIPOC.xls
	Lean Six Sigma Project Charter	LSS Project Charter.doc

Appendix D

IEEE STANDARDS ABSTRACTS

IEEE Standard No.	ISO/IEC No.	IEEE Standard Topic	Standard's Description
IEEE 610.12	24765	Glossary of Software Engineering Terminology	Identifies and establishes definitions for terms currently in use in the field of software engineering.
IEEE 730		Software Quality Assurance Plans	Provides software quality assurance plans (SQAPs) preparation and content for the development and maintenance of critical software.
IEEE 828		Software Configuration Management Plans	Provides software configuration management plan (SCMP) preparation and contents. The specific activities to be addressed and their requirements for any portion of a software product's life cycle are defined.
IEEE 829		Software Test and System Documentation	Describes a set of basic software and system test documents in terms of their form and content.
IEEE 830		Software Requirements Specifications (Recommended Practice)	Describes software requirements specification (SRS) content and qualities, and presents several sample SRS outlines.
IEEE 982.1		Dictionary of Measures to Produce Reliable Software	Provides a set of measures indicative of software reliability that can be applied to the software product as well as to the development and support processes. Provides a common, consistent definition of a set of measures that may meet those needs for measures that can be applied early in the development process, and that may be indicators of the reliability of the delivered product.
IEEE 1008		Software Unit Testing	Describes the software unit testing approach, the software engineering concepts and testing assumptions, and provides guidance and resource information to assist with the implementation and usage of the software unit testing approach.
IEEE 1012		Software Verification and Validation	Establishes a common framework for verification and validation (V&V) processes, activities, and tasks in support of all software life cycle processes, including acquisition, supply, development, operation, and maintenance processes. Defines the V&V tasks, required inputs, and required outputs. Identifies the minimum V&V tasks corresponding to software integrity levels using a four-level scheme. Defines the content of a software V&V plan (SVVP).
IEEE 1016		Software Design Descriptions (Recommended Practice)	Describes software design descriptions (SDDs) information content and recommendations for an organization. SDD is a representation of a software system that is used as a medium for communicating software design information and is applicable to paper documents, automated databases, design description languages, or other means of description.

Standard	ISO No.	Title	Description
IEEE 1028		Software Reviews	Defines five types of software reviews, together with procedures required for the execution of each review type. Review types include management reviews, technical reviews, inspections, walk-throughs, and audits.
IEEE 1044		Classification for Software Anomalies	Provides classification of anomalies found in software and its documentation. Describes the processing of anomalies discovered during any software life cycle phase, and the comprehensive lists of software anomaly classifications and related data items that are helpful to identify and track anomalies.
IEEE 1045		Software Productivity Metrics	Provides consistent ways to measure the elements that go into defining computing software productivity. Software productivity metrics terminology is given to ensure an understanding of measurement data for both source code and document production.
IEEE 1058 (P16326)	16326	Software Project Management Plans	Describes software project management plans format and contents. Identifies elements that should appear in all software project management plans.
IEEE 1061		Software Quality Metrics Methodology	Defines methodology that spans the entire software life cycle, for establishing quality requirements and identifying, implementing, analyzing, and validating the process and product software quality metrics.
IEEE 1062		Software Acquisition (Recommended Practice)	Describes a set of useful quality practices that can be selected and applied during one or more steps in a software acquisition process.
IEEE 1063	26514	Software User Documentation	Describes the structure, information content, and format of electronic or paper documentation. Addresses the interests of software acquirers, producers, and users in standards for consistent, complete, accurate, and usable documentation.
IEEE 1074		Developing a Software Project Life Cycle Process	Provides a process for creating a software project life cycle process (SPLCP) along with the selection of an appropriate software project life cycle model (SPLCM) for use on a specific project.
IEEE 1175.1		CASE Tool Interconnections—Classification and Description (Guide)	Introduces and characterizes the problem of interconnecting computer-aided software engineering (CASE) tools with their environment. Distinguishes four interrelated contexts for interconnection. Partitions interconnection concerns into issues of protocol, syntax, and semantics.

(*continued*)

IEEE Standard No.	ISO/IEC No.	IEEE Standard Topic	Standard's Description
IEEE 1175.2		CASE Tool Interconnections—Characterization of Interconnections (Recommended Practice)	Identifies a standard set of attributes that characterize the contexts in which a CASE tool operates. These contexts are organizations, users, platforms, and other tools. The attributes in each context summarize the major factors affecting interconnection of the tool with that context.
IEEE 1175.3		CASE Tool Interconnections—Reference Model for Specifying Software Behavior (Guide)	Identifies a common set of modeling concepts found in CASE tools for describing the operational behavior of a software product. Establish a uniform, integrated model and a textual syntax for expressing the common properties (attributes and relationships) of those concepts as they have been used to model software behavior.
P1175.4		CASE Tool Interconnections—Reference Model for Specifying System Behavior	Provides an explicitly defined meta-model for specifying system and software behavior. Defines a semantic basis of observables that allows each tool, whatever its own internal ontology, to communicate facts about the behavior of a subject system as precisely as the tools meta-model allows. Conventional tool model elements are reduced into simpler, directly observable fact statements about system behavior.
IEEE 1220	26702	Application and Management of the Systems Engineering Process	Defines the interdisciplinary tasks required throughout a system's life cycle to transform customer needs, requirements, and constraints into a system solution. Specifies the systems engineering process and its application throughout the product life cycle.
IEEE 1228	15026	Software Safety Plans	Establishes software safety plan content for the development, procurement, maintenance, and retirement of safety-critical software.
IEEE 1233		Developing System Requirements Specifications (Guide)	Provides for the development of the set of requirements, system requirements specification (SyRS) that satisfy expressed needs. Developing a SyRS includes the identification, organization, presentation, and modification of the requirements. Also addressed are the conditions for incorporating operational concepts, design constraints, and design configuration requirements into the specification.
IEEE 1320.1		Functional Modeling Language—Syntax and Semantics for IDEF0	IDEF0 function modeling is designed to represent the decisions, actions, and activities of an existing or prospective organization or system. IDEF0 graphics and accompanying texts are presented in an organized and systematic way to gain understanding, support analysis,

Standard	Title	Description
		provide logic for potential changes, specify requirements, and support system-level design and integration activities.
IEEE 1320.2	Conceptual Modeling Language Syntax and Semantics for IDEF1X97 (IDEF object)	IDEF1X97 consists of two conceptual modeling languages. The key-style language supports data/information modeling and the identity-style language is based on the object model with declarative rules and constraints.
IEEE 1362	System Definition—Concept of Operations (ConOps) Document (Guide)	Describes concept of operations (ConOps) document format and contents. A ConOps document is a user-oriented document that describes system characteristics for a proposed system from the users' viewpoint and communicates overall quantitative and qualitative system characteristics to the user, buyer, developer, and other organizational elements such as the user organization(s), mission(s), and organizational objectives.
IEEE 1420.1	Software Reuse—Data Model for Reuse Library Interoperability: Basic Interoperability Data Model (BIDM)	Provides the minimal set of information about assets that reuse libraries should be able to exchange to support interoperability.
IEEE 1420.1a	Software Reuse—Data Model for Reuse Library Interoperability: Asset Certification Framework	Defines a consistent structure for describing a reuse library's asset certification policy in terms of an asset certification framework, along with a standard interoperability data model for interchange of asset certification information.
IEEE 1420.1b	Software Reuse—Data Model for Reuse Library Interoperability: Intellectual Property Rights Framework	Incorporates intellectual property rights issues into software asset descriptions for reuse library interoperability as defined in IEEE Std 1420.1, Basic Interoperability Data Model.
IEEE 1462 14102	Evaluation and selection of CASE Tools (Guideline)	Defines both a sequence of processes and a structured set of CASE tools characteristics for use in the technical evaluation and selection of CASE tools.
IEEE 1465 12119	Software packages—Quality Requirements and Testing	Establishes quality requirements for software packages and instructions on how to test a software package against these requirements.
IEEE 1471 42010	Architectural Description of Software Intensive Systems (Recommended Practice)	Establishes a conceptual framework for architectural description. Defines the content of an architectural description. Addresses the activities of the creation, analysis, and sustainment of architectural descriptions.

(continued)

IEEE Standard No.	ISO/IEC No.	IEEE Standard Topic	Standard's Description
IEEE 1490	PMI Std	A Guide to the Project Management Body of Knowledge (Guide)	Identifies and describes the knowledge and project management practices applicable to most projects, most of the time, about which there is widespread consensus as to their value and usefulness.
IEEE 1517		Software Life Cycle Processes—Reuse Processes	Provides a common framework for extending the software life cycle processes of IEEE 12207.0 to include the systematic practice of software reuse. This standard specifies the processes, activities, and tasks to be applied during each phase of the software life cycle to enable a software product to be constructed from reusable assets. It also specifies the processes, activities, and tasks to enable the identification, construction, maintenance, and management of assets supplied.
P1633		Software Reliability (Recommended Practice)	Provides both practitioners and researchers with a common baseline for discussion and to define a procedure for assessing the reliability of software. Software reliability (SR) models have been evaluated and ranked for their applicability to a systems approach to prediction and to various situations.
P1644		Software Nomenclature—Software Naming Conventions for Application Software (Recommended Practice)	Describes a software naming convention (SNC) to be used for developing nomenclature for software applications. SNC uses a unique reference structure that links the software name to its related support system, functionality, and software release date. SNC improves the tracking, controlling, and managing of software releases
P1648		Establishing and Managing Software Development Efforts Using Agile Methods (Recommended Practice)	Describes a process that a software development client should adopt and use in contracting with and working with an "agile" software developer, to control the three aspects of software development: development status of specific features, progress through the development cycle, and expenditure of contract funds.
P9126.1	25010	Product quality—Part 1: Quality Model	Describes a two-part model for software product quality: (a) internal quality and external quality characteristics, and (b) quality in use.
IEEE 12207.0	12207	System and Software Engineering — Software Life Cycle Processes	Describes a common software life cycle processes framework containing processes, activities, and tasks, and well-defined terminology for developing and managing software.

IEEE 12207.1	15289	Content of Systems and Software Life Cycle Process Information Products (Documentation)	Provides guidance for recording life cycle data resulting from the execution of the activities and tasks of the software life cycle processes of IEEE 12207.0.
IEEE 12207.2	TR 24748	Industry Implementation of ISO/IEC 12207—Software Life Cycle Processes—Implementation Considerations	Describes implementation consideration guidance for the software life cycle processes of IEEE 12207.0.
IEEE 14143.1	14143-1	Software Measurement—functional size measurement—Part 1: Definition of Concepts	Defines the fundamental concepts of Functional Size Measurement (FSM) and describes the general principles for applying an FSM method.
P14471	14471	Guidelines for the Adoption of CASE tools	Identifies major factors that are critical to success in CASE adoption, including a comprehensive set of technical, managerial, organizational, and cultural factors for successfully introducing CASE technology into an organization.
IEEE 14764 (IEEE 1219)	14764	Software Life Cycle Processes—Maintenance	Describes the process for managing and executing software maintenance activities.
P15026	15026	System and Software Assurance	Provides requirements for the life cycle, including development, operation, maintenance, and disposal of systems and software products that are critically required to exhibit and be shown to possess properties related to safety, security, dependability, or other characteristics. It defines an assurance case as the central artifact for planning, monitoring, achieving, and showing the achievement and sustainment of the properties and for related support of other decision making.
IEEE 15288	15288	System Life Cycle Processes	Establishes a common framework of processes and associated terminology for describing the life cycle of systems.
P15939	15939	Measurement Process	Identifies the activities and tasks that are necessary to successfully identify, define, select, apply, and improve software measurement within an overall project or organizational measurement structure. Provides definitions for measurement terms.
IEEE 16085 (IEEE 1540)	16085	System and Software Life Cycle Processes—Risk Management	Defines a process for the management of risk in the life cycle of software as defined by the IEEE 12207, or it can be used independently.

(continued)

IEEE Standard No.	ISO/IEC No.	IEEE Standard Topic	Standard's Description
P20000.1	20000-1	Service Management—Part 1: Specification	Defines the requirements for a service provider to deliver managed services of an acceptable quality for its customers. Provides the high-level specification of requirements for delivery of managed services. Supports managed service acquisition, proposal, consistency, benchmarking or appraisal, and process improvement.
P20000.2	20000-2	Service Management—Part 2: Code of Practice	Describes quality standards and best practices for service management processes in conjunction with ISO/IEC 20000-1.
IEEE 23026 (IEEE 2001)	23026	Internet Practices—Web Page Engineering—Intranet/Extranet Applications (Recommended Practice)	Defines practices for web page engineering. Addresses the needs of webmasters and managers to effectively develop and manage World Wide Web projects (internally via an intranet or in relation to specific communities via an extranet). Discusses life cycle planning: identifying the audience, the client environment, objectives, and metrics; and continues with recommendations on server considerations and specific web page content.
P25051 (IEEE 1465)	25051	Requirements for Quality of Commercial Off-the-Shelf (COTS) Software Product and Instructions for Testing	Establishes quality requirements for COTS software products; requirements for test documentation for the testing of COTS software products, including test requirements, test cases, and test reporting; and instructions for conformity evaluation of COTS software products.
P90003	90003	Guidelines for the Application of ISO 9001:2000 to Computer Software	Provides guidance for organizations in the application of ISO 9001 to the acquisition, supply, development, operation, and maintenance of computer software.
SWEBOK	19759	Software Engineering Body of Knowledge	Provides a consensually validated characterization of the bounds of the software engineering discipline and a topical access to the body of knowledge supporting that discipline.

IEEE 12207 AND IEEE STANDARDS
MAPPED TO ISO/IEC/IEEE 12207:2008

IEEE has jointly worked with ISO/IEC to develop and adopt the 2008 version of 12207, which reorders the software life cycle processes to fit into the system engineering life cycle processes.

ISO/IEC/IEEE 12207:2008 Processes	IEEE 12207 Clause	Relevant IEEE Standards
6 System Life Cycle Processes		
6.1 Agreement Processes		
6.1.1 Acquisition Process	5.1	1062
6.1.2 Supply Process	5.2	
6.2 Organizational Project-Enabling Processes		
6.2.1 Life Cycle Model Management Process	7.1	1074
6.2.2 Infrastructure Management Process	7.2	1175, 1462
6.2.3 Project Portfolio Management Process	NA	
6.2.4 Human Resource Management Process	7.4	
6.2.5 Quality Management Process	7.3	90003
6.3 Project Processes		
6.3.1 Project Planning Process	NA	1058 (16326), 1228
6.3.2 Project Assessment and Control Process	NA	
6.3.3 Decision Management Process	NA	
6.3.4 Risk Management Process	NA	1540 (16085)
6.3.5 Configuration Management Process	NA	
6.3.6 Information Management Process	NA	
6.3.7 Measurement Process	NA	982.1, 1045, 1061, 14143.1
6.4 Technical Processes		
6.4.1 Stakeholder Requirements Definition Process	NA	1362
6.4.2 System Requirements Analysis Process	5.3.2	1233, 1320.1, 1320.2
6.4.3 System Architectural Design Process	5.3.3	1471 (42010)

ISO/IEC/IEEE 12207:2008 Processes	IEEE 12207 Clause	Relevant IEEE Standards
6.4.4 Implementation Process	NA	
6.4.5 System Integration Process	5.3.10	
6.4.6 System Qualification Testing Process	5.3.11	
6.4.7 Software Installation Process	5.3.12	
6.4.8 Software Acceptance Support Process	5.3.13	
6.4.9 Software Operation Process	5.4	
6.4.10 Software Maintenance Process	5.5	14764
6.4.11 Software Disposal Process	NA	
7 Software Life Cycle Processes		
7.1 Software Implementation Processes		
7.1.1 Software Implementation Process	5.3.1	
7.1.2 Software Requirements Analysis Process	5.3.4	830
7.1.3 Software Architectural Design Process	5.3.5	1471 (42010)
7.1.4 Software Detailed Design Process	5.3.6	1016
7.1.5 Software Construction Process	5.3.7	1008
7.1.6 Software Integration Process	5.3.8	829
7.1.7 Software Qualification Testing Process	5.3.9	829
7.2 Software Support Processes		
7.2.1 Software Documentation Management Process	6.1	1063, 12207.1 (15289)
7.2.2 Software Configuration Management Process	6.2	828
7.2.3 Software Quality Assurance Process	6.3	730, 1061, 1465 (25051)
7.2.4 Software Verification Process	6.4	1012
7.2.5 Software Validation Process	6.5	1012
7.2.6 Software Review Process	6.6	1028
7.2.7 Software Audit Process	6.7	1028
7.2.8 Software Problem Resolution Process	6.8	1044
7.3 Software Reuse Processes	1420.1, 1517	
7.3.1 Domain Engineering Process	NA	
7.3.2 Reuse Asset Management Process	NA	
7.3.3 Reuse Program Management Process	NA	

Source : ISO/IEC/IEEE 12207:2008 Annex G (informative)

ACRONYMS

Acronym	Meaning
ACR	Average Completion Rate
AFSO21	Air Force's Smart Operations for the 21st Century
ANOVA	Analysis of Variance
BCR	Baseline Change Request
C&E	Cause and Effect
CASE	Computer-Aided Software Engineering
CBT	Computer-Based Training
CCB	Configuration Control Board; Change Control Board
CDR	Critical Design Review
CER	Change Enhancement Request
CI	Configuration Item
CRC	Change Request Coordinator
CTQ	Critical to Quality
CMM®	Capability Maturity Model
CMMI®	Capability Maturity Model Integration
ConOps	Concept of Operations (document)
COTS	Commercial Off-the-Shelf (software)
CR	Change Request
CUSUM	Cumulative Sum
DCR	Interface Change Notice
DMADV	Define, Measure, Analyze, Design, Verify
DMAIC	Define, Measure, Analyze, Improve, Control
DT&E	Developmental Test and Evaluation
EMEA	Error Modes and Effects Analysis
EPG	Engineering Process Group
EVOP	Evolutionary Operation
EWMA	Exponentially Weighted Moving Average

Acronym	Meaning
FFRDC	Federally Funded Research and Development Center
FMEA	Failure Modes and Effect Analysis
GEG	Government Electronics Group
HR	Human Resources
ICD	Interface Control Document
IDEALSM	Initiating, Diagnosing, Establishing, Acting, and Learning
IEC	International Electrotechnical Commission
ISO	International Organization for Standardardization
IT	Information Technology
ITR	Internal Test Report
JIT	Just In Time
JTC1	Joint Technical Committee 1
LCL	Lower Control Limit
LESAT	Lean Enterprise Self Assessment Tool
LSS	Lean Six Sigma
NIST	National Institute of Standards and Technology
NPV	Net Present Value
MMP	Measurement and Measures Plan
OSD	Office of the Secretary of Defense
PAP	Process Action Plan
PCE	Process Cycle Effeciency
PDCA	Plan–Do–Check–Act
PDR	Preliminary Design Review
PL	Project Lead
PM	Program Manager
PMR	Program Manager's Review
PMBOK	Project Management Body of Knowledge
PMR	Program Management Review
QA	Quality Assurance
QC	Quality Control
QFD	Qualify Function Deployment
ROI	Return on Investment
ROIC	Return on Invested Capital
RFP	Request for Proposal (document)
RTM	Requirements Traceability Matrix
S2ESC	Software and Systems Engineering Standards Committee
SAB	Standards Activities Board
SAP	Software Acquisition Plan
SC7	Subcommittee 7
SCM	Software Configuration Management or Manager (use dependent)
SCMP	Software Configuration Management Plan
SCR	Software Change Request
SDD	Software Design Document
SDL	Software Developmental Library
SDLC	Software Development Life Cycle
SDP	Software Development Plan
SEI	Software Engineering Institute
SIPOC	Supplier, Inputs, Process, Outputs, and Customers
SITR	System Integration Test Report
SLC	Software Life Cycle

Acronym	Meaning
SPLCP	Software Project Life Cycle Process
SPLCM	Software Project Life Cycle Model
SMM	Software Measurement and Measures (plan)
SPC	Statistical Process Control
SPMP	Software Project Management Plan
SPP	Small Project Plan
SRMP	Software Requirements Management Plan
SRS	Software Requirements Specification
SRTM	Software Requirements Traceability Matrix
SSS	Software System Specification
SysITP	System Integration Test Plan
SysRS	System Requirements Specification
SQA	Software Quality Assurance
SQAM	Software Quality Assurance Manager
SQAP	Software Quality Assurance Plan
SQC	Statistical Quality Control
STP	Software Test Plan
STR	Software Test Report
SWEBOK®	Software Engineering Body of Knowledge
SysTP	System Test Plan
TAG	Technical Advisory Group
TBD	To Be Determined
TCSE	Technical Council on Software Engineering
TIP	Things–in–Process
TPM	Total Productive Maintenance
TPR	Test Problem Report
TPS	Toyota Production System
TQM	Total Quality Management
TWG	Technical Working Group
UCL	Upper Control Limit
UTR	Unit Test Report
V&V	Verification and Validation
VDD	Version Description Document
VOC	Voice of the Customer
WBS	Work Breakdown Structure
WIP	Work in Progress; Work in Process

REFERENCES

IEEE Publications

[1]   An American National Standard—IEEE Standard for Software Unit Testing, ANSI/IEEE Std 1008-1987(R1993), Reaffirmed Dec. 2002, IEEE Press, New York, 2002.

[2]   IEEE Guide for CA Software Engineering Tool Interconnections—Classification and Description, IEEE Std 1175.1-2002 (Nov. 11), IEEE Press, New York, 2002.

[3]   IEEE/ANSI. IEEE Guide to software configuration management. ANSI/IEEE Std 1042-1987, IEEE Press, New York, 1987. 92 pages.

[4]   IEEE Recommended Practice for Software Acquisition, IEEE Std 1062-1998 Edition (Dec. 2), Reaffirmed Sept 2002, IEEE Press, New York, 2002.

[5]   IEEE Recommended Practice for Software Design Descriptions, IEEE Std 1016-1998 (Sep. 23), IEEE Press, New York, 1998.

[6]   IEEE Recommended Practice for Software Requirements Specifications, IEEE Std 830-1998 (June 25), IEEE Press, New York, 1998.

[7]   IEEE Standard Classification for Software Anomalies, IEEE Std 1044-1993 (Dec. 2), Reaffirmed Sept. 2002, IEEE Press, New York, 2002.

[8]   IEEE Standard Dictionary of Measures to Produce Reliable Software, IEEE Std 982.1-1988 (June 9), IEEE Press, New York, 1988.

[9]   IEEE Standard Glossary of Software Engineering Terminology, IEEE Std 610.12-1990 (Sep. 28), Reaffirmed Sep. 2002, IEEE Press, New York, 2002.

[10]   IEEE Standard for Software Configuration Management Plans, IEEE Std 828-1998 (June 25), IEEE Press, New York, 1998.

[11]   IEEE Standard for Developing Software Life Cycle Processes, IEEE Std 1074-1997 (Dec. 9), IEEE Press, New York, 1997.

[12]   IEEE Standard for Software Productivity Metrics, IEEE Std 1045-1992 (Sep. 17) Reaffirmed Dec. 2002, IEEE Press, New York, 2002.

[13]   IEEE Standard for Software Project Management Plans, IEEE Std 1058-1998 (Dec. 8), IEEE Press, New York, 1998.

[14]   IEEE Standard for Software Reviews, IEEE Std 1028-1997 (Mar. 4), Reaffirmed Sep. 2002, IEEE Press, New York, 2002.

[15]   IEEE Standard for Software Test Documentation, IEEE Std 829-1998 (Sep. 16), IEEE Press, New York, 1998.

[16]   IEEE Standard for Software Quality Assurance Plans, IEEE Std 730-2002 (Sep.), IEEE Press, New York, 2002.

[17]   IEEE Standard for a Software Quality Metrics Methodology, IEEE Std 1061-1998 (Dec. 8), IEEE Press, New York, 1998.

[18]   IEEE Standard for Software User Documentation, IEEE Std 1063-2001 (Dec. 5), IEEE Press, New York, 2001.

[19]   IEEE Standard for Software Verification and Validation, IEEE Std 1012-1998 (Mar. 9), IEEE Press, New York, 1998.

[20]   Supplement to IEEE Standard for Software Verification and Validation: Content Map to IEEE/EIA 12207.1-1996, IEEE Std 1012a-1998 (Sep. 16), IEEE Press, New York, 1998.

[21]   IEEE Standard Reference Model for Computing System Tool Interconnections, IEEE Std 1175-1991 (Dec. 5), IEEE Press, New York, 1991.

[22]   IEEE Standard for Software Maintenance, IEEE Std 1219-1998 (June 25), IEEE Press, New York, 1998.

[23]   IEEE Standard for the Application and Management of the Systems Engineering Process, IEEE Std 1220-1998 (Dec. 8), IEEE Press, New York, 1998.

[24]   IEEE Standard for Software Safety Plans, IEEE Std 1228-1994 (Mar. 17), Reaffirmed Dec. 2002, IEEE Press, New York, 2002.

[25]   IEEE Guide for Developing System Requirements Specifications, IEEE Std 1233, 1998 Edition (Apr. 17), Reaffirmed Sep. 2002, IEEE Press, New York, 2002.

[26]   IEEE Standard for Functional Modeling Language—Syntax and Semantics for IDEF0, IEEE Std 1320.1-1998 (June 25), IEEE Press, New York, 1998.

[27]   IEEE Standard for Conceptual Modeling Language Syntax and Semantics for IDEF1X 97 (IDEF object), IEEE Std 1320.2-1998 (June 25), IEEE Press, New York, 1998.

[28]   IEEE Guide for Information Technology—System Definition—Concept of Operations (ConOps) Document, IEEE Std 1362-1998 (Mar. 19), IEEE Press, New York, 1998.

[29]   IEEE Standard for Information Technology—Software Reuse—Data Model for Reuse Library Interoperability: Basic Interoperability Data Model (BIDM), IEEE Std 1420.1-1995 (Dec. 12), Reaffirmed June 2002, IEEE Press, New York, 2002.

[30]   Supplement to IEEE Standard for Information Technology—Software Reuse—Data Model for Reuse Library Interoperability: Asset Certification Framework, IEEE Std 1420.1a-1996 (Dec. 10), Reaffirmed June 2002, IEEE Press, New York, 2002.

[31]   IEEE Trial-Use Supplement to IEEE Standard for Information Technology—Software Reuse—Data Model for Reuse Library Interoperability: Intellectual Property Rights Framework, IEEE Std 1420.1b-1999 (June 26), Reaffirmed June 2002, IEEE Press, New York, 2002.

[32]   IEEE Standard—Adoption of International Standard ISO/IEC 14102: 1995—Information Technology—Guideline for the Evaluation and Selection of CASoftware Engineering Tools, IEEE Std 1462-1998 (Mar. 19), IEEE Press, New York, 1998.

[33]   IEEE Standard—Adoption of International Standard ISO/IEC 12119: 1994(E)—Information Technology—Software Packages—Quality Requirements and Testing, IEEE Std 1465-1998 (June 25), IEEE Press, New York, 1998.

[34] IEEE Recommended Practice for Architectural Description of Software-Intensive Systems, IEEE Std 1471-2000 (Sep. 21), IEEE Press, New York, 2000.

[35] IEEE Guide—Adoption of PMI Standard—A Guide to the Project Management Body of Knowledge, IEEE Std 1490-2003 (Dec. 10), Replaces 1490-1998 (June 25), IEEE Press, New York, 2003.

[36] EIA/IEEE Interim Standard for Information Technology—Software Life Cycle Processes—Software Development: Acquirer-Supplier Agreement, IEEE Std 1498-1995 (Sep. 21), IEEE Press, New York, 1995.

[37] IEEE Standard for Information Technology—Software Life Cycle Processes—Reuse Processes, IEEE Std 1517-1999 (June 26), IEEE Press, New York, 1999.

[38] IEEE Standard for Software Life Cycle Processes—Risk Management, IEEE Std 1540-2001 (Mar. 17), IEEE Press, New York, 2001.

[39] IEEE Recommended Practice for Internet Practices—Web Page Engineering—Intranet/Extranet Applications, IEEE Std 2001-2002 (Jan. 21, 2003), IEEE Press, New York, 2003.

[40] Industry Implementation of International Standard ISO/IEC 12207:1995, Standard for Information Technology—Software Life Cycle Processes—Life Cycle Data, IEEE/EIA 12207.0/.1/.2-1996 (Mar.), IEEE Press, New York, 1996.

[41] IEEE, IEEE Standards Collection, Software Engineering, 1994 Edition, IEEE Press, Piscataway, NJ, 1994.

[42] IEEE, IEEE Software Engineering Standards Collection, IEEE Press, Piscataway, NJ, 2003.

[43] IEEE, Software and Systems Engineering Standards Committee Charter Statement, http://standards.computer.org/S2ESC/S2ESC_pols/S2ESC_Charter.htm, 2003.

[44] S2ESC Guide for Working Groups, http://standards.computer.org/S2ESC/S2ESC_wgre-sources/S2ESC-WG-Guide-2003-07-14.doc, 2003.

[45] IEEE, *Guide to the Software Engineering Body of Knowledge (SWEBOK),* Trial Version, IEEE Press, Piscataway, NJ, 2001.

[46] McConnell, S., "The Art, Science, and Engineering of Software Development," *IEEE Software Best Practices,* Vol. 15 No.1, 1998.

ISO Publications

[47] International Standard 9000, *Quality Management Systems—Fundamentals and Vocabulary,* ISO 9000:2000(E), Geneva, Switzerland, 2000.

[48] International Standard 9001, Quality Management Systems—Requirements, ISO 9001:2000(E), Geneva, Switzerland, 2000.

[49] International Standard 90003, *Software and System Engineering—Guidelines for the Application of ISO 9001:2000 to Computer Software,* ISO/IECF 90003:2003 (E), Geneva, Switzerland, 2003.

[50] International Standard 9004, *Quality Management Systems—Guidelines for Performance Improvements,* ISO 9004:2000(E), Geneva, Switzerland, 2000.

[51] *ISO 9000 Introduction and Support Package module: Guidance on ISO 9001:2000 Subclause 1.2 "Application,"* 524R4; ISO/TC 176/SC 2, London, 2004.

[52] *ISO 9000 Introduction and Support Package Module: Guidance on ISO 9001:2000 Guidance on the Documentation Requirements of ISO 9001:2000,* 525R, ISO/TC 176/SC 2, London, 2001.

[53] *ISO 9000 Introduction and Support Package Module: Guidance on ISO 9001:2000 Guidance on the Terminology Used in ISO 9001:2000 and ISO 9004:2000,* 526R, ISO/TC 176/SC 2, 2001.

[54] *ISO 9000 Introduction and Support Package Module: Guidance on ISO 9001:2000 Guidance on the Concept and Use of the Process Approach for Management systems,* 544R2, ISO/TC 176/SC 2, London, 2004.

[55] *ISO 9000 Introduction and Support Package Module: Guidance on ISO 9001:2000 Guidance on Outsourced Processes,* 630R2, ISO/TC 176/SC2, London, 2003.

[56] *ISO/IEC 15504 International Standard for Software Process Assessment,* ISO/IEC TR 15504, Geneva, Switzerland, 2003/2005.

[57] *ISO 19011 Guidelines for Quality and Environmental Management Systems Monitoring,* BS EN ISO 19011:2002, London, 2002.

[58] *Information Technology—Software Measurement—Functional Size Measurement—Definition of Concepts,* ISO/IEC JTC1/SC7, ISO/IEC 14143.1:1999, Canada, 1998.

[59] *Systems Engineering—System Life Cycle Processes,* ISO/IEC JTC1/SC7, ISO/IEC 15288:2002, Canada, 2002.

[60] *Information Technology—Systems and Software Engineering —Measurement Process,* ISO/IEC JTC1/SC7, ISO/IEC 15939:2007, Canada, 2007.

Lean Six Sigma References

[61] *Kaizen Rapid Process,* U.S. Environmental Protection Agency, http://www.epa.gov/lean/thinking/kaizen.htm, Dec. 2006.

[62] *NIST/SEMATECH e-Handbook of Statistical Methods,* http://www.itl.nist.gov/div898/handbook.

[63] *QI Story: Tools and Techniques,* Third Edition, Six Sigma Qualtec, 1999.

[64] *Six Sigma Quality Resources for Achieving Six Sigma Results,* http://www.isixsigma.com, iSixSigmaLLC, Dec. 2006.

[65] *Why the Lean in Lean Six Sigma?* Poppendieck, LLC, Sep., 2006, http://www.poppendieck.com/lean-six-sigma.htm

[66] Air Academy Associates, *Key Inputs to Maximize LSS/DFLSS Business Impact,* http://www.airacad.com/LeanSixSigmaImplementations.aspx, Dec. 2006.

[67] Adams, S.D., *One IT Size Does Not Fit All, Lean Six Sigma Can Help,* iSix Sigma Software/It, http://software.isixsigma.com/library/content/c061122b.asp, Dec. 2006.

[68] Adams, M., Kiemele, M., Pollock, L., and Quan, T., *Lean Six Sigma: A Tools Guide,* Second Edition, Air Academy Assoc, LLC., 2004.

[69] Akao, Y., *Quality Function Deployment: Integrating Customer Requirements into Product Design,* 2004.

[70] Bertels, T., *Integrating Lean and Six Sigma,* http://www.isixsigma.com, iSixSigmaLLC, 2006.

[71] Brassard, M., and Ritter, D., *The Memory Jogger II,* GOAL/QPC, 1994.

[72] Burton, T., *Is This a Six Sigma, Lean, or Kaizen Project?* http://www.isixsigma.com, iSixSigma LLC, 2006.

[73] Carpenter, E.D., *Prioritization Matrix is Made Easier with a Template,* http://www.isixsigma.com, iSixSigma LLC, 2007.

[74] Cockburn, A., *What Engineering Has in Common with Manufacturing and Why It Matters, Humans and Technology,* Technical Report, HaT TR 2006.04, Sep. 6, 2006.

[75] Dalal, S.R., Horgan, J.R., and Kettenring, J.R., "Reliable Software and Communication: Software Quality, Reliability, and Safety," In *Proceedings of the 15th International Conference on Software Engineering,* 17–21 May 1993

[76] Dasari, R.K., Lean Software Development, The Project Perfect White Paper Collection, http://www.projectperfect.com.au/downloads/Info/info_lean_development.pdf, Dec. 2006.

[77] Davis, G., Zannier, C., and Geras, A., "QFD for Software Requirements Management," University of Southern California–Davis, Irvine. August, 2007.

[78] Gack, G. A., *Building a Business Case for Software Defect Reduction,* iSix Sigma Software/IT, http://software.isixsigma.com/library/content/c051012b.asp, Dec. 2006.

[79] Gack, G.A., *Mad Belt Disease: Over-Emphasis on Certification,* iSix Sigma Software/IT, http://software.isixsigma.com/library/content/c040218b.asp, Dec. 2006.

[80] George, M., Maxey, J., Rowlands, D., and Price, M., *The Lean Six Sigma Pocket Toolbook: A Quick Reference Guide to 100 Tools for Improving Quality and Speed,* George Group, New York, 2005.

[81] George, M., Rowlands, D., and Kastle, B., *What is Lean Six Sigma?* McGraw-Hill, New York, 2004.

[82] Goel, D., Software Infrastructure Bottlenecks in J2EE, O'Reilly on Java.com, Jan. 2005, http://www.onjava.com/pub/a/onjava/2005/01/19/j2ee-bottlenecks.html.

[83] Goldratt, E.M., *Critical Chain: A Business Novel,* North River Press, Great Barrington, MA, 1997.

[84] Goldratt, E.M., and Cox, Jeff, *The Goal, A process of Ongoing Improvement,* Third Revised Edition, North River Press, Great Barrington, MA, 1984.

[85] Hayes, B.J., *Six Sigma for Software . . . More Than a New Tool,* iSix Sigma Software/IT, http://software.isixsigma.com/library/content/c030328a.asp, Dec. 2006.

[86] Hayes B.J., *What CIOs and CTOs Need to Know about Lean Six Sigma,* http://www.isixsigma.com, iSixSigma LLC, 2006.

[87] Hinckley, C.M., and Barkan, P., "The Role of Variation, Mistakes, and Complexity in Producing Nonconformities," *Journal of Quality Technology,* Volume 27, No. 3, pp. 242–249, 1995.

[88] Hallowell, D. L., "Causal Loop Diagrams: An Orientation in Software Context," iSix Sigma Software/IT, http://software.isixsigma.com/library/content/c050330a.asp, July 2007.

[89] Hallowell, D.L., "QFD, When and How Does it Fit in Software Development?" iSix Sigma Software/IT, http://software.isixsigma.com/library/content/c040707b.asp, August 2007.

[90] Hallowell, D. L., *Software Development Convergence: Six Sigma-Lean-Agile,* iSix Sigma Software/IT, http://software.isixsigma.com/library/content/c050302b.asp, Dec. 2006.

[91] Jacowski, J., *5 Laws of Lean Six Sigma,* Ezine Articles, ezinearticles.com, Dec. 2006.

[92] Jones, D., and Womak, J., *Seeing the Whole: Mapping the Extended Value Stream,* Lean Enterprise Institute, New York, 2002.

[93] Jones, D. T., "Heijunka—Leveling by Volume and Mix," in Kalpakjian, S., and Schmid, S., *Manufacturing, Engineering, and Technology,* Fifth Edition, 2006.

[94] Kano, N., *Guide to TQM in Service Industries,* Productivity Press, New York, 1996 (English).

[95] Kano, N., Seraku, N., Takahashi, F., and Tsjui, S., "Attractive Quality and Must-be Quality," Hinshitsu, Vol. 14, No. 2, 147–156.

[96] Liker, J., *The Toyota Way,* McGraw-Hill, New York, 2004.

[97] Miller, J., "Know Your Takt Time," Gemba Research LLC, 2003, http://www.gemba.com.

[98] Morrison, M., *Six-Sigma Pitfalls: Are Six-Sigma Programs Delivering on their Promises? Industry Week,* September 7, 2005.

[99] Neave, H.R., and Wheeler, D.J., "Shewhart's Charts and the Probability Approach," In *Ninth Annual Conference of the British Deming Association,* London, May 15, 1996.

[100] Nelson, C., and Morris, M., "Integrating Rational Unified Process and Six Sigma," *The Ra-

tional Edge, e-zine for the rational community, Nov. 2003, www.128.ibm.com/developer-works/rational/library/content/Rationaledge/Nov03.

[101] Ohno, T., *Toyota Production System: Beyond Large Scale Production,* Productivity Press, New York, 1988.

[102] Pande, P. S., Neuman, R. P., and Cavanagh, R. R., *The Six Sigma Way Team Fieldbook,* McGraw-Hill, New York, 2002.

[103] Pyzdek, Thomas, *The Six Sigma Handbook,* McGraw-Hill, New York, 2003.

[104] Pascal, D., *Andy & Me: Crisis and Transformation on the Lean Journey,* Productivity Press, New York, 2005.

[105] Paul, A., *Harvesting Benefits of Both Lean and Six Sigma in IT,* http://www.isixsigma.com, iSixSigma LLC, 2006.

[106] Poppendiech, M., and Poppendiech, T., *Implementing Lean Software Development: From Concept to Cash,* Addison-Wesley, Reading, MA, 2007.

[107] Poppendieck, M., "Lean Software Development," *C++ Magazine,* Methodology Issue, Fall 2003.

[108] Radice, R., "Statistical Process Control in Level 4 and 5 Organizations Worldwide," in *Proceedings of the 12th Annual Software Technology Conference,* Salt Lake City, UT, 2000. (Also at http:// www.stt.com/.)

[109] Rains, P., "Lean HR: Six Sigma Philosophy Process Improvement," Transactional HR Processing, http://www.transacthr.com/assets/userfiles/transact_hr/000211.ppt#256,1,'Lean HR' Six Sigma Philosophy Process Improvement, Dec. 2006.

[110] Raulerson, A.B., and Sparks, P., "Lean Six Sigma at Anniston Army Depot," *Army Logistician:Professional Bulletin of United States Army Logistics,* PB 700-06-06, Vol. 38, Issue 6, Nov–Dec 2006.

[111] Rizzardo, D., and Brooks, R., "Understanding Lean Manufacturing," Maryland Technology Extension Service, MTES Tech Tip, http://www.mtech.umd.edu/MTES/understand_lean.html, 2006.

[112] Roberts, J., "Total Productive Maintenance," *theTechnologyInterface,* Texas A&M University-Commerce, Fall, 1997.

[113] Robinson, C.J., and Ginder, A.P., *Implementing TPM,* Productivity Press, New York, 1995.

[114] Schutta, J.T., *Business Performance through Lean Six Sigma: Linking the Knowledge Worker, the Twelve Pillars, and Baldrige,* ASQ Quality Press, Milwaukee, WI, 2006.

[115] Senge, P., *The Fifth Discipline: The Art and Practice of the Learning Organization,* Doubleday Currency, New York, 1990.

[116] Shingo, S., *Zero Quality Control: Source Inspection and the Poka-yoke System,* trans. A.P. Dillion, Productivity Press, New York, 1985.

[117] Simon, K., "Why Control Chart Your Processes?" iSix Sigma Software/IT, http://www.isixsigma.com/library/content/c020708a.asp, July 2007.

[118] Sugiyama, T., *The Improvement Book,* Productivity Press, New York, 1989.

[119] Nikkan Kogyo Shimbun, Ltd., *Poka-Yoke: Improving Product Quality By Preventing Defects,* Productivity Press, New York, 1987 (Japanese), 1988 (English).

[120] Tague, N. R., *The Quality Toolbox,* Second Edition, ASQ Quality Press, Milwaukee, WI, 2004.

[121] Thomas, D., Turning Customer Data into Critical-to-Satisfaction Data, iSix Sigma Software/IT, http://www.isixsigma.com/library/content/c040913a.asp, Feb. 2007.

[122] Ungvari, S., "TRIZ Within the Context of the Kano Model or Adding the Third Dimension to Quality," *The TRIZ Journal,* http://www.triz-journal.com/archives/1999/10/e/, April 2007.

[123] Wait, P., "General Dynamics Wins Air Force Smart Ops 21 Deal," *Washington Technology,* http://www.washingtontechnology.com/online/1_1/30557-1.html, May 2007.

[124] Welsh, W., "BearingPoint to Help AF Improve Work Processes," Washington Technology, http://www.washingtontechnology.com/online/1_1/30531-1.html, April 2007.

[125] Yavuz, M., and Akcali, E., "Production Smoothing in Just-In-Time Production Systems: A Review of the Models and Solution Approaches." *International Journal of Production Research,* Vol. 45, No. 16, 15 Aug., pp. 3579–3597, 2007.

[126] Zultner, R., "QFD Schedule Deployment: Doing Development Faster with QFD," Zultner & Company, New York, www.zultner.com, 1998.

Other References

[127] Arthur, L., *Software Evolution: The Software Maintenance Challenge,* Wiley, New York, 1988.

[128] Babich, W., *Software Configuration Management,* Addison-Wesley, Reading, MA, 1986.

[129] Balanced Scorecard Institute, "Module 4: Affinity Diagram," Accesssed August 2007, http://www.balancedscorecard.org/files/affinity.pdf.

[130] Basili, V.R. et al., "A Reference Architecture for the Component Factory," *ACM Trans. Software Eng. and Methodology,* Vol 1, No. 1, Jan., pp. 53–80, 1992.

[131] Bersoff, E., Henderson, V., and Siegel, S., *Software Configuration Management: A Tutorial,* IEEE Computer Society Press, New York, 1980, pp. 24–32.

[132] Bounds, N.M., and Dart S.A., *Configuration Management (CM) Plans: The Beginning to Your CM Solution,* Software Engineering Institute, Carnegie Mellon University, Pittsburgh, PA, 1993.

[133] Bredemeyer Consulting, *The Architecture Discipline—Software Architecting Success Factors and Pitfalls,* http://www.bredemeyer.com/CSFs_pitfalls.htm, 2004.

[134] Carr, M. et al., *Taxonomy-Based Risk Identification,* Software Engineering Institute, Carnegie Mellon University, Pittsburgh, PA, Technical Report, CMU/SEI-93-TR-006, 1993.

[135] Croll, P., "Eight Steps to Success in CMMI-Compliant Process Engineering, Strategies and Supporting Technology," in *Third Annual CMMI®* Technology Conference and Users Group, Salt Lake City, UT, 2003.

[136] Croll, P., "How to Use Standards as Best Practice Information Aids for CMMI-Compliant Process Engineering," in *14th Annual DoD Software Technology Conference,* Salt Lake City, UT, 2002.

[137] Croll, P., and Land, S. K., S2ESC: "Setting Standards for Three Decades," *IEEE Computer Magazine,* January 2005.

[138] Davis, A., *Software Requirements: Analysis and Specification,* Prentice Hall, Upper Saddle River, NJ, 1990.

[139] Department of Defense, Software Transition Plan, Data Item Description DI-IPSC-81429.

[140] Department of Energy Quality Managers Software Quality Assurance Subcommittee, "Software Risk Management, A Practical Guide," SQAS21.01.00-1999, 2000.

[141] Department of Justice Systems Development Life Cycle Guidance, Interface Control Document Template, Appendix B-17; http://www.usdoj.gov/jmd/irm/life cycle/table.htm.

[142] Draper, G., and Hefner, R., *Applying CMMI®* Generic Practices with Good Judgment, SEPG Conference Tutorial, http://www.sei.cmu.edu/cmmi/presentations/sepg04.presentations/apply-gps.pdf, 2004.

[143] Dunn, R.H., and Ullman, R.S., *TQM for Computer Software,* Second Edition, McGraw Hill, New York, 1994.

[144] Ford, G., and Gibbs, N., "A Mature Profession of Software Engineering," Software Engineering Institute, Carnegie Mellon University, Technical Report, CMU/SEI-96-TR-004, 1996.

[145] Freedman, D.P., and Weinberg, G.M., *Handbook of Walkthroughs, Inspections, and Technical Reviews, Evaluating Programs, Projects, and Products,* 3Third Edition, Dorset House, New York, 1990.

[146] Gremba, J. and Myers, C., "The IDEAL Model: A Practical Guide for Improvement," *Bridge,* Issue 3, Software Engineering Institute, Carnegie Mellon University, Pittsburgh, PA, 1997.

[147] Hass, A.M.J., *Configuration Management Principles and Practice,* Addison-Wesley, Reading, MA, 2003.

[148] Jalote P., *An Integrated Approach to Software Engineering,* Second Edition, Springer, New York, 1997.

[149] Kasunic, Mark, "An Integrated View of Process and Measurement," Software Engineering Institute, Carnegie Mellon University, Pittsburgh, PA, Presentation, http://www.sei.cmu.edu/sema/pdf/integrated-view-process.pdf, 2004.

[150] Land, S.K., "First User's of Software Engineering Standards Survey," IEEE Software and Systems Engineering Standards Committee (S2ESC), 1997.

[151] Land S.K., "Second User's of Software Engineering Standards Survey," IEEE Software and Systems Engineering Standards Committee (S2ESC), 1999.

[152] Land, S.K., "IEEE Standards User's Survey Results," in *ISESS '97 Conference Proceedings,* IEEE Press, Piscataway, NJ, 1997.

[153] Land, S.K., "Second IEEE Standards User's Survey Results," in *ISESS '99 Conference Proceedings,* IEEE Press, Piscataway, NJ, 1999.

[154] Land, S.K., *Jumpstart CMM® /CMMI® Software Process Improvement Using IEEE Software Engineering Standards,* Wiley, Hoboken, NJ, 2004.

[155] Land, S.K., and Walz, J., *Practical ISO 9001 Software Process Documentation Using IEEE Software Engineering Standards,* Wiley/IEEE Press, Hoboken, NJ, 2006.

[156] Land, S.K., and Walz, J., *Practical Support for CMMI Software Process Documentation Using IEEE Software Engineering Standards,* Wiley/IEEE Press, Hoboken, NJ, 2006.

[157] McConnell, S., *Professional Software Development,* Addison-Wesley, Reading, MA, 2004.

[158] McConnell, S., "Software Engineering Principles," *IEEE Software,* March/April 1999.

[159] McFeeley, B., *IDEAL: A User's Guide to Software Process Improvement,* Software Engineering Institute, Carnegie Mellon University, Handbook, Pittsburgh, PA, CMU/SEI-96-HB-001, USA, 1996.

[160] Moore, J., "Increasing the Functionality of Metrics through Standardization," in *Conference on Developing Strategic I/T Metrics,* 1998.

[161] Moore, J., *Road Map to Software Engineering—A Standards Based Guide,* Wiley, Hoboken, NJ, 2005.

[162] Phifer, E., "DAR Basics: Applying Decision Analysis and Resolution in the Real World," Software Engineering Institute, Carnegie Mellon University, Pittsburgh, PA, SEPG Presentation, http://www.sei.cmu.edu/cmmi/presentations/sepg04.presentations/dar.pdf, 2004.

[163] Pressman, R., *Software Engineering,* McGraw-Hill, New York, 1987.

[164] Royce, W., "CMM® vs. CMMI, From Conventional to Modern Software Management," *The Rational Edge,* 2002.

[165] Schach, S.R., *Classical and Object-Oriented Software Engineering,* Third Edition, Irwin Press, Homewood, IL, 1993.

[166] SEI, *The Capability Maturity Model: Guidelines for Improving the Software Process,* v.1.1, Software Engineering Institute, Carnegie Mellon University, Pittsburgh, PA, 1997.

[167] SEI, *A Framework for Software Product Line Practice, V 4.2,* Software Engineering Institute, Carnegie Mellon University, Pittsburgh, PA, Web Report, http://www.sei.cmu.edu/plp/framework.html#outline, 2004.

[168] SEI, *Capability Maturity Model Integration (CMMI) for Software Engineering,* v.1.1 Staged Representation, Software Engineering Institute, Carnegie Mellon University, Pittsburgh, PA, Technical Report, CMU/SEI-2002-TR-029, 2002.

[169] SEI, Chrissis, M.B., Konrad, M., and Shrum, S., *CMMI®* Second Edition, Guidelines for Process Integration and Product Improvement, Addison-Wesley, Reading, MA, 2007.

[170] SEI, *Integrated Product Development (IPD)-CMM®,* Software Engineering Institute, Carnegie Mellon University, Pittsburgh, PA, Model Draft, 1997.

[171] SEI, *Organizational Process Improvement Checklist Using IDEAL^{SM},* Software Engineering Insititute, Carnegie Mellon University, Pittsburgh, PA, Presentation, www.sei.cmu.edu/ideal/ideal.present.

[172] Sykes, A., "An Introduction to Regression Analysis," The Inaugural Coase Lecture. Accessed August 10, 2007, available from http://www.law.uchicago.edu/Lawecon/WkngPprs_01-25/20.Sykes.Regression.pdf.

[173] USAF Software Technology Support Center (STSC), *CMM-SE/SW V1.1 to SW-CMM® V1.1 Mapping,* 2002.

[174] Veenendall, E.V., Ammerlaan, R., Hendriks, R., van Gensewinkel, V., Swinkels, R., and van der Zwan, M., "Dutch Encouragement: Test Standards We Use in Our Projects," *Professional Tester,* No. 16, October 2003.

[175] Westfall, L.L., "Seven Steps to Designing a Software Metric," BenchmarkQA, Whitepaper, 2002.

[176] Whitgift, D., *Methods and Tools for Software Configuration Management,* Wiley, Chichester, 1991.

[177] Wiegers, K.E., *Creating A Software Engineering Culture,* Dorset House Publishing, New York, 1996.

[178] Williams and Wegerson, *Evolving the SEPG to a CMMI World,* http://www.sei.cmu.edu/cmmi/presentations/sepg03.presentations/williams-wegerson.pdf, 2002.

[179] Wikipedia, The Free Encyclopedia, Various definitions; http://en.eikipedia.org, Dec. 2006. (All Wikipedia references were cross-checked against other reference sources for validation purposes. The authors felt it was important to include Wikipedia references as this is becoming a much used reference site; however, users should be aware that Wikipedia information should be validated.)

[180] Zubrow, D., Hayes, W., Siegel, J., and Goldenson, D., *Maturity Questionnaire,* Special Report, CMU/SEI-94-SR-7, Software Engineering Institute, Carnegie Mellon University, Pittsburgh, PA, 1994.

[181] Zultner, R., "Software Quality Deployment—Adapting QFD to Software," in *Second Symposium on Quality Function Deployment,* QFD Institute, Ann Arbor, MI, 1990.

[182] Young, R., *Requirements Engineering Handbook,* Artech House, Norwell, MA, 2003.

[183] Osborn, A., *Applied Imagination: Principles and Procedures of Creative Problem Solving,* New York, New York: Charles Scribner's Sons. 1953.

[184] Monden, Y., *Toyota Production Systems: An Integrated Approach to Just-In-Time,* Third Edition. Engineering & Management Press, Norcross, GA, 1998.

[185] *Lean Lexicon: A Graphical Glossary for Lean Thinkers,* Third Edition, Lean Enterprise Institute, Cambridge, MA, 2004.

[186] National Institute of Standards and Technology (NIST), *Engineering Statistics Handbook: Online Glossary of Terms,* http://www.itl.nist.gov/div898/handbook/glossary.htm#E.

[187] Kaplan, R. S. and Norton, D. P., "The Balanced Scorecard: Measures that Drive Performance," *Harvard Business Review,* Jan–Feb, pp. 71–80. 1992.

[188] Capers Jones, Software Productivity Research Inc. http://www.stsc.hill.af.mil/Crosstalk/2001/02/jones.html.

INDEX

Abstraction tree diagram, Lean Six Sigma life cycle, 177–179

Acceptance criteria
software requirements specifications, 45
transition planning and, 64

Acquisition process, software life cycle
IEEE 12207 standard, 20–38
alternative solution screening criteria matrix, 245, 247
concept of operations, 23–27
cost–benefit ratio, 247–248
make/buy decision matrix, 245–247
SAP document guidance, 20–23
supply process, 248–249
software test plan, 97
system integration testing, 112
system test plan, 105

Acronyms, table of, 379–382

Action planning stage, Lean Six Sigma life cycle, 175–176

Affinity diagram, Lean Six Sigma life cycle, 177–178

Algorithms
development progress measurements, 315–317
management requirements, manpower measures, 312–313
requirements traceability matrix, 317–319
scheduling measures, 316–317
software stability measurements, 321–323

Alternative solution screening criteria matrix, 245, 247

Ambiguity resolution, software requirements management plan, 140

Analysis of variance (ANOVA), Lean Six Sigma life cycle, 199

Analyze stage, Lean Six Sigma life cycle variants, 198–213
analysis of variance, 199
brainstorming, 200
cause-and-effect diagrams, 201–202
5 Whys Analysis, 199
flowcharts, y-to-x flowdown diagrams, 202–204
multivoting, 204–205
Pugh Concept Selection Matrix, 205–206
regression analysis, 206–207
root cause analysis, 207–208
scatter plots, 208–209
stratification, 209–210
value-stream mapping, 210–211
waste identification and elimination, 211–212

Anomaly data
classic categories, 298
software verification
classification, 90, 93
inspections, 87–91
ranking, 90–91, 93

Artifacts
FMEA/EMEA techniques, 192
software engineering process, 15

Asset library catalog, software project life cycle process, 344–345

Audits and reviews
configuration management, 76–77
IEEE Standard 12207, 309
critical dependencies tracking, 309–310

ABOUT THE AUTHORS

Susan K. Land is employed by MITRE, a not-for-profit organization chartered to operate in the public interest that manages three federally funded research and development centers (FFRDCs) for the U.S. government. She has more than 20 years of industry experience in the practical application of software engineering methodologies, the management of information systems, and leadership of software development teams. Ms. Land is currently a Member of the IEEE Computer Society Board of Governors, serving as 2009 President. A former vice president of the Standards Activities Board (SAB) for the Computer Society, she serves now as a member of this board and the Software and Systems Engineering Standards Committee (S2ESC).

Douglas B. Smith is a Northrop Grumman Information Technology Technical Fellow and Program Manager of the Northrop Grumman IT Defense Group Lean Six Sigma Program. He is a Lean Six Sigma Black Belt and has been involved in process improvement since 1987, applying Quality Improvement, Lean Six Sigma, and the Software Engineering Institute's Capability Maturity Models (SW-CMM and CMMI) as a practitioner, assessor, team leader, and manager. He is a Member of the IEEE and IEEE Computer Society.

John W. Walz retired as senior manager of Supply Chain Management for Lucent Technologies. His 30-year career at Lucent and AT&T was highlighted by customer-focused and dynamic results-oriented management with an excellent technical background and more than 20 years of management/coaching experience. He held leadership positions in hardware and software development, engineering, quality planning, quality auditing, quality standards implementation, and strategic planning. He was also responsible for Lucent's Supply Chain Network quality and was a valuable team member for Lucent's quality strategic planning. He is the current vice president for the SAB and also serves as a member of the S2ESC.

Additional titles available by these authors:

Jumpstart CMM®/CMMI® Software Process Improvement, Using IEEE Software Engineering Standards

Practical Support for CMMI-SW® Software Project Documentation, Using IEEE Software Engineering Standards

Practical Support for ISO 9001 Software Project Documentation, Using IEEE Software Engineering Standards